"This book fills a critical void in the IPM literature and serves as a practical how-to-do-it manual for the novice, as well as a valuable resource for all those involved in various aspects of IPM from fundamental research through practical field application. The book should help to substantially advance the necessary research and testing of new economic injury level and economic threshold information." — Marcos Kogan, Oregon State University

Integrated pest management (IPM) is an ecologically based approach for modifying the impact of pests to tolerable levels. Thresholds are based on the concept of economic injury level (EIL), which includes economic, management effectiveness, pest biology, and host stress considerations.

Economic Thresholds for Integrated Pest Management draws on scientific advances in entomology, plant pathology, and weed science. The book discusses the history of decision making in IPM, EILS, and approaches to developing economic thresholds. The seventeen contributors stress the importance of understanding the pest-host relationship and of taking into account such factors as environmental risk, pesticide resistance, and delayed and cumulative effects. Along with other pressing challenges in pest science, including new pests, new governmental policies, and growing demands on agriculture, the need for better understanding of thresholds has never been greater.

Leon G. Higley is an associate professor of entomology at the University of Nebraska–Lincoln. Larry P. Pedigo is a professor of entomology at Iowa State University. Both are the authors, with Laura L. Karr, of the *Manual of Entomology and Pest Management*.

Volume 9 in the series *Our Sustainable Future*

Series Editors

Lorna M. Butler
Washington State University

Cornelia Flora
Iowa State University

Charles A. Francis
University of Nebraska–Lincoln

William Lockeretz
Tufts University

Paul Olson
University of Nebraska–Lincoln

Marty Strange
Center for Rural Affairs

Economic Thresholds for Integrated Pest Management

EDITED BY
LEON G. HIGLEY AND
LARRY P. PEDIGO

University of Nebraska Press
Lincoln and London

© 1996 by the University of Nebraska Press
All rights reserved
Manufactured in the United States of America
⊖ The paper in this book
meets the minimum requirements of American
National Standard for
Information Sciences – Permanence of Paper
for Printed Library Materials,
ANSI Z39.48-1984.

Library of Congress Cataloging
in Publication Data
Economic thresholds for integrated pest
management/
edited by Leon G. Higley and Larry P. Pedigo.
p. cm – (Our sustainable
future; 9) Includes bibliographical references
(p.) and index.
ISBN 0-8032-2363-3 (cloth: alkaline paper)
1. Agricultural pests – Integrated
control – United States – Economic aspects.
2. Agricultural pests – Integrated
control – United States. I. Higley, Leon G.
II. Pedigo, Larry P.
III. Series: Our sustainable future; v. 9.
SB950.2.AIE36 1997
338.1'62.–dc20 96-12173
CIP

We dedicate this volume to
Vernon Stern, Ray Smith, Robert van den Bosch,
and Kenneth Hagen, *whose classic work,*
The Integrated Control Concept,
*laid the foundations for decision making
in pest management.*

Contents

List of Tables and Figures	ix
Preface *Leon G. Higley and Larry P. Pedigo*	xi

PART 1: THEORY

CHAPTER 1
Introduction to Pest Management and Thresholds   3
Larry P. Pedigo and Leon G. Higley

CHAPTER 2
The EIL Concept   9
Leon G. Higley and Larry P. Pedigo

CHAPTER 3
The Biological Basis of the EIL   22
Leon G. Higley and Robert K. D. Peterson

CHAPTER 4
General Models of Economic Thresholds   41
Larry P. Pedigo

CHAPTER 5
Limitations to EILs and Thresholds   58
Ronald B. Hammond

CHAPTER 6
Alternative Approaches to Decision Making   74
Michael D. Duffy

PART 2: METHODS

CHAPTER 7
Economic Thresholds for Weed Management   89
David A. Mortensen and Harold D. Coble

CHAPTER 8
Thresholds for Plant-Disease Management   114
Paul A. Backman and James C. Jacobi

CHAPTER 9
Economic Thresholds for Insect Management   128
G. David Buntin

PART 3: NEW DEVELOPMENTS

CHAPTER 10
The Status of Economic-Decision-Level
Development   151
Robert K. D. Peterson

CHAPTER 11
Economic Thresholds for Veterinary Pests   179
John B. Campbell and Gustave D. Thomas

CHAPTER 12
Aesthetic Thresholds and Their Development   203
Clifford S. Sadof and Michael J. Raupp

CHAPTER 13
Thresholds for Interseasonal Management   227
Stephen C. Welter

CHAPTER 14
Thresholds and Environmental Quality   249
Leon G. Higley and Wendy K. Wintersteen

CHAPTER 15
Thresholds Involving Plant Quality and
Phenological Disruption   275
Scott H. Hutchins

Afterword: Pest Science at a Crossroads   291
Leon G. Higley and Larry P. Pedigo

Selected Bibliography:
Economic-Decision-Level Literature,
1959–1993   297
Compiled by Robert K. D. Peterson

The Contributors   313

Index   317

Tables and Figures

FIGURES

2.1	Appropriate inclusion of injury in a threshold	18
2.2	Incorporation of past injury in the determination of an EIL	19
3.1	The damage curve	31
4.1	ET/EIL relationship, allowing time for management action	43
4.2	Growth of a green cloverworm population on soybean	50
4.3	Economic and subeconomic populations of green cloverworm in central Iowa	53
7.1	Soybean yield loss as a function of total weed-competitive load	105
9.1	Three generalized crop-yield responses to insect injury	138
10.1	Cumulative EIL papers, 1959–1993	153
10.2	EIL and ET papers by pest type	155
10.3	EIL and ET papers by insect order	156
10.4	EIL and ET papers by arthropod-injury type	157
12.1	Leaf discoloration caused by *T. urticae* on *E. alatus*	216
12.2	Relationship between density of *T. urticae* and average discoloration ranking	217
12.3	Customer responses to leaf discoloration	218
12.4	Predicted customer dissatisfaction and cumulative *T. urticae* density	220
12.5	Predicted customer dissatisfaction with plants managed at Hybrid EIL and ratio of control costs to replacement value	222
13.1	Effects of herbivory on perennial crops	229
15.1	Photograph of soybean damage from green stink bug nymphs	280
15.2	Relationship of pest-induced injury to harvest-time market value	283

TABLES

4.1	Green cloverworm leaf consumption and equivalence coefficients on soybean leaves	45
4.2	Mortality-adjusted injury equivalence of green cloverworm instars	51
4.3	Time-sequential sampling form for green cloverworm larvae in Iowa soybean	54
9.1	Yield response of oats to infestation of cereal leaf beetle larvae	133
9.2	Application number, yield, control costs, and revenue of managing lepidopterous larvae in cauliflower	134
9.3	EILs for alfalfa snout beetle in alfalfa	139
10.1	Taxa, species, and commodities with published economic-decision levels	158
12.1	Investigations of AILs, Hybrid EILs, and their components	207
12.2	Hybrid EILs for *T. urticae* on *E. alatus* 'compacta'	219
14.1	Environmental risks and costs of formulated field-crop insecticides	261
14.2	Example environmental ETs for second-generation European corn borer on pretassel stage corn	265
14.3	Sample environmental EIL calculation worksheet	266
15.1	Relationship of potato leafhopper densities on digestible energy yields of alfalfa	285
15.2	Relationship of potato leafhopper densities on date of first bloom for alfalfa	288
15.3	Regression equation for potato leafhopper density versus days of phenological delay	288
15.4	Economic time delays and EILs for potato leafhopper	289

Leon G. Higley and Larry P. Pedigo

Preface

The development of sampling methods and the use of simple economic-decision rules have been among the most important components in the advancement of pest science. Indeed, these components set modern integrated pest management (IPM) apart from the "identify-and-spray" methods of the past and significantly characterize state-of-the-art pest management programs. Of the economic-decision rules thus far proposed, none has been as pervasive and influential as the economic injury level/threshold (EIL) concept of Stern, Smith, van den Bosch, and Hagen. Although many refinements have taken place in the thirty-some years since its inception, the concept still forms the basis of our present theory on decision making in IPM. Indeed, the concepts have assumed even greater importance recently as mechanisms to maintain environmental quality and to provide sustainable pest management. Particularly important for its widespread acceptance is the concept's simplicity and practicality for most agricultural needs.

Even though the EIL concept was developed for use in insect management, other pest disciplines, including weed science and plant pathology, have recently recognized its importance. In fact, the development of economic thresholds for weed species and complexes is presently a major thrust in weed science. There, the objective is to establish a more rational herbicide-use policy. An important outcome of an interdisciplinary involvement in threshold development will be improved integration of decision making for production systems, a long-sought goal of IPM.

New initiatives in developing more sustainable agricultural systems and in reducing pesticide use highlight the need for greater attention to how we manage pests. Calls for 50% reductions in pesticide use in Europe and North America, as well as governmental proposals for widespread adoption of IPM, provide the impetus for real change in agricultural practices. A key component in efforts to produce more sustainable, less intrusive agricultural systems is through the use of pest management programs. Because

EILs are the cornerstones of such programs, understanding and expanding EIL theory and development is essential.

Given the central importance of the EIL concept in IPM theory and practice, there have been surprisingly few publications exclusively dedicated to the idea. Some of the most useful resources are subject reviews; however, these reviews are scattered throughout the literature and are mostly discipline oriented. Previously, no book had been written on EILs to cover major aspects of theory and application or to conceptualize the subject on an interdisciplinary basis. Our intent in developing this book was to address this need. Further, we have tried to emphasize new approaches in threshold development that offer exciting prospects for expanding and improving the usefulness of thresholds.

The book is arranged in three parts: Theory, which includes considerations of EILs, economic thresholds, their components, and relationships of decision making to IPM (Chapters 1 through 6); Methods, which examines approaches to threshold development for different pest types (Chapters 7 through 9); and New Developments, which includes material on novel approaches for threshold development and visions of where future threshold development may lead (Chapters 10 through 15). In the Afterword, we offer a summary of these views and present our perspective on future directions in pest management decision making. The Selected Bibliography provides a summary of the literature on economic decision levels through 1993.

We have been ably assisted in this project through the enthusiasm and support of the contributing authors. Additionally, editorial reviews and assistance in manuscript preparation were provided by Stephen Spomer and Peggy Mutchie-Whisler, Department of Entomology, University of Nebraska–Lincoln.

Part 1: Theory

Larry P. Pedigo and Leon G. Higley

1
Introduction to Pest Management and Thresholds

Probably no concept has influenced the direction of pest technology in the past three decades more than Integrated Pest Management (IPM). The first published reference to "management" in the context of pests seems to have been made by Australian entomologists P. W. Geier and L. R. Clark in 1961 (Geier and Clark 1961). Even preceding the environmental epic, *Silent Spring* (Carson 1962), these authors argued for a change in the strategy of dealing with pests in the interest of sustainability and environmental quality. Their focus was on ecological and economic reality in pest-control programs, and they emphasized the way tactics should be used, rather than the tactics themselves. They called their concept *protective population management,* and the idea came to be known initially as *pest management.*

Through the following decades, pest management was adopted widely in entomology and subsequently in other pest disciplines. It eventually assumed the name of integrated pest management, which reflected its interdisciplinary nature and use of multiple tactics. Today, IPM is the predominant strategy at all levels of U.S. agriculture and has become a cornerstone of U.S. presidential policy in pest technology, which calls for 75% adoption by farmers by the year 2000 (Pedigo 1995).

As important as IPM has become, it owes much of its theoretical basis to *integrated control,* a pest technology initiated in the late 1940s and expanded in the early 1950s. Integrated control was conceptualized most thoroughly by Stern et al. (1959), in their work with the spotted alfalfa aphid, *Therioaphis maculata* (Buckton), on California alfalfa. Biological control was attempted against this exotic pest with an introduction of three wasp parasitoids brought from Europe. However, economic suppression was not achieved, and applications of the insecticide, dimethoate, were added to supplement mortality of the parasitoids in achieving a successful program. This combination of tactics—natural enemies and selective pesticide use—was a little-used pest strategy at the time. But with this and other successes, the idea of integrated control gained popularity and formed an indispensable foundation for the development of pest management. Therefore, it seems probable that IPM resulted from an evolutionary change through the years, proceeding from biological control to integrated control and on to pest management, rather than from a sudden revolution in thinking (Cate and Hinkle 1993).

In any case, a major premise in integrated control has been to conserve natural enemies and their effects when attempting to suppress pests. This has been a particularly daunting task because pesticides have been a key element in most pest programs, and they are also destructive of the natural-enemy complex. Consequently, researchers, by necessity, were (and continue to be) preoccupied with ways of limiting pesticide use as a means of conserving natural enemies. An important breakthrough in this regard was the idea of tolerating pests, albeit at levels that cause no significant harm. It was reasoned that most biological species are not pests and most pest species do not cause significant harm at all times and in all locations. Furthermore, when a pest species is a problem, not all pests need to be killed to achieve the desired level of crop protection. Consequently, assessing pest status was proposed as a means of judiciously using pesticides, which could result in natural-enemy conservation and improved net profits.

Thus were born the ideas of *economic damage, economic injury level,* and *economic threshold.* Stern et al. (1959) used these ideas as a means of assessing pest status, and the ideas became the underlayment of their integrated control theory. Together, these biologically based terms, couched in the context of economics, constitute the *economic-injury-level* (EIL) *concept*—the primary theme of this book.

Today, the use of *thresholds* is commonplace in pest technology and sets state-of-the-art programs apart from the identify-and-spray approaches of the past. However, thresholds and the EIL concept are not without critics. Many authorities have criticized the original EIL concept because it is too simple, overlooking the influence of other factors within and outside the crop/pest system (Pedigo et al. 1986). Some have emphasized that important features such as interseasonal dynamics, interactions with other pests and natural enemies, pesticide resistance, and environmental costs are not included in the typical economic-threshold decision. Although such criticisms may have merit, it is paradoxical that the very simplicity that is criticized may be the reason the threshold-decision rule has remained popular for more than 35 years.

To understand the importance of the EIL concept to IPM and set the stage for most of the topics in this book, it is important to understand the nature of IPM. Much has been written and said about IPM. It has been idealized by entomologists, environmental groups, the food industry, politicians, and others. One group or another has viewed it as a philosophy, a theoretical background, a set of rules, and a basis for judicious pesticide use. As an ideal, it is commonly suggested as a cure-all, prompting many pest technologies to be identified as IPM when clearly they are not, e.g., lawn-care concerns referring to scheduled pesticide treatments as IPM (Pedigo 1995).

Although there is much discussion of IPM, there is little agreement on an exact definition. It is beyond our purpose here to argue for any specific definition, but we can suggest unique goals that, as a group, set IPM apart from other pest technologies and help characterize it. The major goals include reducing pest status (not simply killing pests), accepting the presence of a tolerable pest density, conserving environmental quality, and improving user profits. Not all bona fide IPM programs achieve all of these goals, but most set out to do so. Procedures to achieve these goals vary: in general, strategies involve determining pest status (including economics), intervening in the production system when necessary, and using several selective and compatible tactics (e.g., conserving natural enemies, planting pest-resistant seed, and selective applications of a pesticide). Of course, it is clear that determining pest status must precede the other activities, indicating the importance of the EIL concept to IPM.

In addition to IPM being distinguishable from older pest technologies, it is, most importantly, superior to them. Before the development and acceptance of IPM, pest technology focused on *control,* rather than *management.* Control refers to having power over something; the true meaning of control is best exemplified in the use of conventional pesticides; i.e., chemical control. By contrast, management refers to a judicious use of means to accomplish a desired end. The main objective of pest control is to reduce pest impact—usually obtained by killing pests. Ultimately, killing may

Theory

become a major control objective in itself, rather than crop protection, and a compulsion for 100 percent mortality is prompted. With this thinking, the greater the mortality (i.e., "percent control") the better the chemical and the control program. Such a mind-set has resulted in pest overkill and is responsible for considerable quantities of unnecessary and undesirable pesticide residues in the environment (Pedigo and Higley 1992).

In contrast, IPM stresses reducing or *modifying* impact of pests and *reducing injury* to tolerable levels. These objectives do not necessarily depend on pest mortality; they do require an assessment of pest status, most often supplied by pest sampling and EIL's. Additionally, many control strategies are not founded on economic, ecological, or environmental considerations and are therefore a simpler approach to pest problems than is IPM. However, the control approach is effective only in instances when benefit-to-cost ratios are high, and then only in the short term. Indeed, control (particularly chemical control) has failed as a stand-alone strategy because of its limited long-term effectiveness and the continued obsolescence of the tactics used.

IPM offers sustainable solutions to pest problems, because both economics and ecology are considered as part of the solution. Additionally, IPM is a sustainable practice because it focuses on how tactics are used, rather than on the tactics themselves; whereas the control strategy focuses on tactics and the never ending search for new ones as the old ones become obsolete.

Also, when single tactics are emphasized, there has always existed the fetish of finding an all-encompassing solution to pest problems. This could be termed the *silver-bullet fetish,* which in pest technology emphasizes finding and using the most powerful tactics for killing pests. Geier (1966) called this approach "bulldozing nature." The silver-bullet approach certainly has fostered new discoveries, such as modern synthetic poisons for more effective pest destruction; however, because the focus is on killing pests rather than proper use, effectiveness of the pesticide eventually fails; i.e., the silver bullet becomes obsolete, which instigates a new search for more silver bullets. The silver-bullet fetish is exemplified in the discovery and use of DDT and other pesticides. For a time, these now-obsolete pesticides were strikingly effective in killing pests, allowing the temporary avoidance of crop loss. However, in the long term, primarily because of the way in which they were used, they became ineffective and/or caused environmental problems and were abandoned in favor of new silver bullets.

Although many pesticides have become obsolete with use, the silver-bullet fetish in pest technology is alive and well. New pesticides for use in chemical control programs continue to be discovered, and the fetish has been extended to encompass new tactics, or to emphasize old ones. The theme continues to be that a single tactic can solve all pest problems and that, if only the "correct" tactic can be found, pest and environmental problems will be solved permanently. Some of these efforts are couched within the context of IPM, but attempts to address pest problems with single tactics are as likely to fail today as they were in the past. Biological control, insect mating disruption with pheromones, and, most recently, insect resistance from transgenic plants, all have been touted as final solutions to pest problems. Most often these claims ignore biological realities and fail to deal with the dynamic nature of pest populations. In particular, they ignore the potential of pest populations to overcome strong selective pressures, allowing them to circumvent a single control tactic; in insects, for example, resistance has been documented in all categories of tactics, when a tactic is effective (Pedigo 1989). Even worse, emphasizing the search for silver bullets misplaces priorities and diverts resources away from multifaceted IPM programs, which can supply sustainable solutions to pest problems.

We do not argue that new management tactics are unnecessary in developing IPM programs; in particular, alternatives to pesticides are important in minimizing environmental impacts. However, we believe that proper use must be understood (i.e., in an ecological context) and that the new tactics should be *integrated* with other ecologically sound tactics to arrive at lasting solutions.

Furthermore, we argue that, important as new alternatives are, they are not the only means of protecting crops and reducing environmental impacts. Indeed the EIL, as a basic and unifying principle of IPM, provides the potential for improved profits and reduced environmental impacts (Pedigo and Higley 1992). By combining elements of pest and host biology with economics, the EIL concept has become the very fabric of advanced pest management and the basis for limiting pesticide use. As has been discussed, the primary purpose of EIL development was to apply pesticides in a rational and judicious manner, thus helping to alleviate ecological problems within agroecosystems and associated habitats. Considering this, further development and use of standard EILs should be a priority in any serious agenda for environmental conservation and sustainable agriculture.

To move forward, we need to recognize the unifying nature of the EIL for IPM and broaden its environmental scope. Indeed, if the major elements (variables) of the EIL are considered carefully, most of them can be manipulated in some way to reduce pest-management impacts on environmental quality (Higley and Pedigo 1993). Our ability to develop improved EILs will depend on our competence in quantifying and including environmental costs, developing tolerant plant varieties, and discovering environmentally responsible management tactics. Therefore, we believe that gaining knowledge about EILs and improving the EIL concept are essential if IPM is to progress toward durable and environmentally acceptable solutions to pest problems. This book is dedicated to furthering these goals.

References

Carson, R. 1962. Silent spring. Houghton and Mifflin, Boston MA.

Cate, J. R., and M. K. Hinkle. 1993. Integrated pest management: The path of a paradigm. Special Rept., National Audubon Soc. Washington DC.

Geier, P. W. 1966. Management of insect pests. Annu. Rev. Entomol. 11:471–490.

Geier, P. W., and L. R. Clark. 1961. An ecological approach to pest control, p. 10–18. *In* Proc. Tech. Meeting Intern. Union Conser. Nature and Nat. Resources, 8th, 1960, Warsaw, Poland.

Higley, L. G., and L. P. Pedigo. 1993. Economic injury level concepts and their use in sustaining environmental quality. Agric. Ecosystems Environ. 46:233–243.

Pedigo, L. P. 1989. Entomology and pest management. Macmillan, New York.

Pedigo, L. P. 1995. Closing the gap between IPM theory and practice. J. Agric. Entomol. 12:171–181.

Pedigo, L. P., and L. G. Higley. 1992. The economic injury level concept and environmental quality. Am. Entomologist 38:12–21.

Pedigo, L. P., S. H. Hutchins, and L. G. Higley. 1986. Economic injury levels in theory and practice. Annu. Rev. Entomol. 31:341–368.

Stern, V. M., R. F. Smith, R. van den Bosch, and K. S Hagen. 1959. The integrated control concept. Hilgardia 29:81–101.

Leon G. Higley and Larry P. Pedigo

2
The EIL Concept

Pest management rests on the premise that not all pests require management; some levels of pests are tolerable. The economic injury level (EIL) is a cornerstone for pest management because it defines how much pest injury (and therefore how many pests) can be tolerated. This definition incorporates biological criteria about a pest and its host, as well as economic criteria regarding host value and management costs. As a theoretical construct, the EIL is crucial to pest management, because our very definitions of pests and their importance (pest status [Stern et al. 1959]) depend upon having an objective means to weigh the importance of a pest species. On this theoretical level, whether our management options are therapeutic (curative) or preventive doesn't matter (Pedigo 1995), nor does the biological classification of the pest (insect, pathogen, or weed) matter; in all cases, the EIL defines the pest and its importance.

The EIL also has an important practical aspect, in that it forms the basis for determining economic thresholds (ETs), the operational criteria for determining whether or not management action is needed against a pest. The practical importance of EILs is tied to the nature of management options and to the nature of the pest. Specifically, EILs are most applicable in the use of therapeutic practices. For pests with few therapeutic management options, such as many plant pathogens, the usefulness of EILs is similarly limited. The EIL is of some value even for pests managed preventively, but with preventive management most emphasis is on predicting pest occurrence and not on assessing pest status. The EIL may be buried deep in the assumptions made in preventive management, but it typically has little direct application in management decisions for preventive action. For preventive management or therapeutic management under conditions of great uncertainty, alternatives to the EIL-ET (cost-benefit) model often are used, including payoff matrices, investment models, cost functions,

Theory

discrimination analysis, and various optimization procedures (see Norton and Mumford [1993] for a comprehensive discussion).

The practical focus of the EIL and ET on therapeutic practices has limited their usefulness for some types of pests. Additionally, since it is a practical index, it is important that we can quantify pests or pest effects easily to allow for accurate decisions. Because disease management primarily involves preventive tactics and because disease assessment is difficult, EILs and ETs do not offer the same advantages in disease management as they do for insect management (Chapter 8). Similarly, the development of thresholds for weeds was delayed until effective postemergence herbicides became available (Chapter 7). Thus, most development of EILs and thresholds has occurred for insects, because they present pest situations allowing therapeutic action and ready assessment of pest attack (Chapter 9, Selected Bibliography).

The use of EILs and ETs are sometimes criticized as being a pesticide management approach, rather than pest management. Such criticisms ignore the theoretical importance of EILs and the practicalities of management. Theoretically, EILs are essential for defining pests. Practically, EILs and ETs are limited because they are most applicable to therapeutic tactics; given that few therapeutic tactics are available other than pesticides, most uses of thresholds are associated with pesticides. The limitation is not with the EIL concept but with our current technologies for therapeutic management. That the use of thresholds is tied to available technology is well illustrated in weed science. The growing development of weed thresholds only began after cost-effective, postemergent herbicides (a therapeutic tactic) became available. Conceptually, there is no difficulty in using thresholds for innundative releases of natural enemies or similar approaches, but such uses have not been forthcoming because of underlying problems in the efficacy and economics of alternatives to pesticides.

Elsewhere (Pedigo et al. 1986), with S. H. Hutchins, we review the history of EILs and ETs, so this chapter will not repeat that discussion, other than to highlight a few important points. The concepts of the EIL, ET, economic damage, and pest status were advanced in a seminal article by Stern et al. in 1959. This paper by Stern et al. and the important contribution of Geier and Clark (1961)—work that has largely failed to receive the recognition it deserves—provide the theoretical foundation of pest management. Unfortunately, there were some deficiencies in the original definitions provided by Stern et al., particularly regarding economic damage, that led to subsequent confusion and delayed the development of EILs. We make this point not to diminish the contributions of Stern et al. but rather to explain why—following the original statement of the concept—it took 13 years for the first calculated EIL to appear (Stone and Pedigo 1972) and why there continues to be confusion regarding the distinction between EILs and ETs. In Pedigo et al. (1986), we railed against the misuse of terminology and the coining of new terms (which typically have little or nothing to distinguish them from existing words). Sadly, much of the literature on EILs and ETs is confused and offers little conceptual advance over the original contributions of Stern et al. Although we and our colleagues are guilty of occasionally offering new terms or definitions, hopefully we have done so in an effort to refine, rather than supplant, the work of Stern et al. We are convinced that an end to the confusion in the literature on EILs will come only through use of Stern et al. as the touchstone for broadening our understanding.

Despite confusion in the literature, the concept of the EIL itself is very straightforward. The EIL represents the point where costs equal benefits. Costs are the losses associated with pest action and the costs of managing pests; benefits are the losses prevented by management. Hidden in this simple statement is considerable biological complexity, and the application of the EIL to specific pest problems adds further complexity. Nevertheless, as a decision criterion, the EIL is perhaps the simplest statement possible for defining pest impact. Although many other decision-making approaches have been advanced and are used (Norton and Mumford 1993), none have the longevity or impact of the EIL. Certainly, the simplicity of the EIL is central to its prominence.

Defining the EIL

Although Stern et al. defined the EIL in terms of pest densities, it is actually a level of injury (because losses from pests are a function of their impact, or injury, rather than just of their numbers). Defining the EIL requires a consideration of economic and biological variables, and various definitions have been forwarded. Stone and Pedigo (1972) provided an early approach to this question. They initially defined economic damage as the economic value of losses necessary to equal economic costs of management. Expressed as a ratio of management costs ($/ha) to crop value ($/kg), they obtained an expression of yield loss (kg/ha) necessary to justify management, which they called the gain threshold. Stone and Pedigo used published data on yield losses from leaf loss, related these to leaf loss per insect, and finally identified the number of insects necessary to produce losses equivalent to the gain threshold. This number was the EIL.

Norton (1976) proposed a more general model (based on actual data for potato cyst eelworm on potato):

$$\theta = C/PDK \qquad [1]$$

where θ = level of pest attack (the EIL) in grams of nematode eggs per gram of soil, C = management cost per hectare of pesticide application, P = price of produce per ton, D = loss in yield (tons per hectare) associated with one nematode egg per gram of soil, and K = reduction in pest attack (expressed as a proportion).

Pedigo et al. (1986) drew from the Norton definition and proposed a more general model for the EIL:

$$EIL = C/VDIK \qquad [2]$$

where EIL = the EIL in injury equivalents per production unit (e.g., insects/ha), C = management costs per production unit ($/ha), V = market value per unit of production ($/kg), D = damage per unit injury (kg reduction/ha/injury), I = injury per pest equivalent (injury/insect), and K = proportional reduction in injury with management (under certain circumstances K may be interpreted as the proportional reduction in damage rather than injury). It may not be possible to distinguish between injury and damage for all pests, therefore D and I may be replaced by a single variable, D' (= yield loss per pest) in these instances

$$EIL = C/VD'K \qquad [3]$$

In this case, K represents the proportional reduction in damage (i.e., yield loss). Also, in many instances K is assumed to be near one and may be omitted from the equation. We argue elsewhere that greater attention to K offers important opportunities for improving the environmental responsiveness of the EIL (Chapter 14, Higley and Pedigo 1993).

This latter model has proven robust both in practical calculation of EILs and in the development of modifications to the original EIL concept for

The EIL Concept

incorporating environmental costs (Chapter 14, Higley and Wintersteen 1992) and aesthetic costs (Chapter 12, Raupp et al. 1988). Unfortunately, the misconception exists that the *C/VDIK* expression of the EIL implies a linear relationship between injury and damage, which it does not, and that there is no need to distinguish between injury and damage (Onstad 1987, Onstad 1988). Beyond these issues, questions regarding past injury and preventable injury also have been raised relative to the EIL (Onstad 1988). Pedigo et al. (1988) address these criticisms and justify the *C/VDIK* model, and we will consider these issues further in this chapter. As these comments indicate, we do not accept as valid the critique offered by Onstad of the *C/VDIK* model, but as we are partisans in this debate, the skeptical reader, to form an independent judgment, might best be served by reviewing Onstad's (1988) critique and our response (Pedigo et al. 1988).

The Onstad–Pedigo et al. debate not withstanding, the *C/VDIK* model is robust and is widely accepted as a general statement of the EIL (e.g., Metcalf and Luckman 1994). Nevertheless, it is possible to make a more general statement of the EIL model and to address directly distinguishing injury and damage and assessing past injury. We should emphasize, however, that although these alternative expressions may offer some slight improvement in distinguishing the exact nature of the EIL variables, more complicated expressions of the EIL probably are not warranted in most situations.

Injury and Damage

An important refinement of the basic equation is to recognize that yield loss is a function of some level of injury, therefore the variable D may be better represented as $D(I_t)$ where I_t is some level of total injury (notice that I_t is *not* injury per individual). With this modification it is possible to compare costs and benefits as

$$C = VD(I_t) \qquad [4]$$

and rearranging

$$D(I_t) = C/V \qquad [5]$$

then solving for I_t. If $D(I_t)$ is a linear equation, the solution is obtained by simple division:

Theory

$$D(I_t) = bI_t \qquad [6]$$

therefore,

$$bI_t = C/V \qquad [7]$$

and

$$I_t = C/Vb \qquad [8]$$

where b is the regression coefficient (i.e., slope, yield loss per unit injury). This regression coefficient is variously described by other authors. Often, it is not expressed as injury, but rather as yield loss per pest (where D is a function of pest number rather than injury).

When $D(I_t)$ is not linear the solution is more complex; for example, if $D(I_t)$ is quadratic,

$$D(I_t) = b_1 I_t + b_2 I_t^2 \qquad [9]$$

$$bI_t + b_2 I_t^2 = C/V \qquad [10]$$

from the quadratic formula (and accepting only biologically meaningful roots)

$$I_t = (-b_1 + (b_1^2 + 4b_2 (C/V))^{0.5}) / 2b_2 \qquad [11]$$

The functional nature of D often is neglected because if the function is linear, the yield loss per unit injury remains constant. Only if the functional relationship between yield loss and injury is curvilinear does yield loss per unit injury change. In fact, for calculating EILs, only that portion of the damage curve between the damage boundary (where measurable yield losses first occur) and the EIL may be of interest. Thus, for situations in which the entire damage curve is curvilinear, D may still be defined as a linear function if that portion from the damage boundary to EIL is adequately described as a line. Given the frequent experimental limitations in estimating the damage curve, a linear fit often is adequate for the available data.

By solving $D(I_t)$ we are able to calculate an EIL in terms of some total injury. From total injury we can then move to an EIL defined in terms of individuals or injury equivalents by dividing injury at the EIL by injury per individual (I_i) or by injury per injury equivalent (I_{IE}). More commonly, we

The EIL Concept

determine $D(I_t)$ (yield loss per total injury at the EIL) and multiply by injury per individual or injury equivalents to give

$$\text{EIL} = C/VD(I_t)I_{IE}K \qquad [12]$$

This expression is little different from Eq. [2], although it makes the functional nature of D more explicit.

The entire discussion regarding the functional relationship of yield loss *(D)* with respect to injury *(I_t)* could be repeated for the relationship of yield loss to individual pests by merely substituting the word *pest* wherever the word *injury* appears. Indeed, in some EIL calculations the concept of injury is never addressed; weeds, insects, or insect-population densities function as an index, or proxy, of injury. Why, then, if EILs can be calculated without explicitly considering injury, should we calculate an EIL in terms of injury?

At least three reasons exist for considering injury. The first is philosophical more than practical. By ignoring injury, we develop equations relating yield loss to pest numbers. These equations say nothing of how plants physiologically respond to stress from pests. Describing yield loss in terms of injury does not solve this problem, but it does provide a necessary basis for a quantitative examination of how injury reduces yield. Further, yield loss equations are greatly influenced by environmental conditions. Without a precise quantification of injury, it is impossible to develop an understanding of the physiological effects of injury or how environment alters those effects.

A second argument for describing an EIL in terms of injury is that the rate of injury per individual pest is not constant. For example, to relate an EIL defined in terms of injury to insect numbers, the level of injury at the EIL must be divided by injury per insect. Weed competitiveness varies based on age of the weed (Chapter 7). Immature insects show dramatic changes in injury rates as they develop (Chapter 3). Also, at high population densities, insects may interfere with each other, thereby reducing the injury produced by a single insect as compared with injury per insect at a lower population density. If the EIL falls within a high population density, then the injury per insect should be modified to reflect any reduced injury per insect. Similarly, populations of young immatures have different injury rates than populations of older immatures. Thus, an EIL calculated without considering injury may be based on overestimates of injury per insect and may produce too low an EIL.

The third argument for considering injury in EILs is to provide a mechanism for considering multiple species. Because different pest stages and species produce different levels of injury, differentiating injury from damage provides a mechanism for accommodating these differences in injury associated by stage or species. Recognizing injury and establishing injury equivalencies (relationships among pest stages and species based on injury) offer a powerful mechanism for developing more accurate ETs and multiple species thresholds (Chapter 3, Harcourt 1954, Shelton et al. 1982, Pedigo et al. 1986, Hutchins et al. 1988, Hutchins and Funderburk 1991). This approach is a necessity for weed EILs given that weeds invariably occur as a complex of weed species (Chapter 7).

This discussion on distinguishing pests in EIL calculations focuses on recognizing the distinction between pests and their action (injury). In most management, action is taken against pests so the EIL and ET must be related to pest densities, but this is not always the case. For example, with stand loss caused by pests, it is possible to determine a stand-loss EIL for making replanting decisions. Such an EIL only considers injury (stand loss) and represents an example of an EIL based on a nonpesticidal action (in this case replanting). Also, stand-loss EILs and ETs are equivalent (because injury is not increasing). However, stand-loss EILs are dependent on time; specifically, the longer past the optimum plant date, the greater the loss necessary to justify replanting (because replant yield potential decreases the farther it is from the optimum planting date). Thus, stand-loss EILs increase through time: more stand loss is necessary to justify replanting the farther the planting date is past the optimum.

Higley and Hunt (1994) present a brief discussion of stand-loss EILs. From their analysis, a general formula for a replant EIL after stand loss is given as

Replant EIL = % stand loss justifying replanting

$$= \frac{(1 - ((RY*V) - C)))* 100\%}{(EY * V)} \quad [13]$$

where EY = expected yield (bu/a), RY = replant yield (bu/a), V = market value ($/bu), and C = replanting costs ($/a). To illustrate use of such an EIL for a soybean field, two weeks after the optimum planting date, the appropriate data are EY = 40 bu/a, RY = 36 bu/a (90% of EY because past

The EIL Concept

optimum planting date), V = $5/bu, and C = $21.63/a ($6.93/a for fuel, repairs, and labor plus $14.43/a for seed costs from UN-L economic guidelines). Thus, the replant EIL is calculated as

$$\text{Replant EIL} = 1 - \frac{((36bu/a * \$5/bu) - \$21.63/bu)}{(40bu/a * \$5/bu)} * 100\% = 20.8\%$$

Therefore, approximately a 21% stand reduction is necessary to justify replanting.

Assessing Past Injury

The premise of the EIL (costs equal benefits) requires that management prevents injury (the benefit). Consequently, the EIL is valid only for preventable injury. Largely, this is a question of thresholds rather than EILs, in that thresholds are used to indicate when management is necessary. Thus, thresholds must focus on preventable injury. Indeed, it is for this reason that thresholds expressed in terms of injury are suspect. An EIL of 50% defoliation may represent where economic injury occurs, but a threshold of 50% defoliation (as is sometimes presented in extension recommendations) is rather meaningless, because the loss has already occurred and is not preventable. The one instance where levels of injury are legitimate in thresholds is when there is a period of tolerance (no yield loss associated with injury) up to a point where yield losses do begin to occur (Fig. 2.1).

A further issue arises based on how thresholds are used and how D, the yield-loss relationship, is defined. Sound pest management should involve assessment before significant injury has occurred. With immature insects and weeds, relatively little injury occurs in early stages, so there is a longer interval to allow decision making before serious injury accumulates. Most EILs and ETs have the unstated premise that no significant previous injury has occurred. If injury has occurred, a conventional EIL may or may not provide an accurate assessment. If the yield loss relationship (D or D') is linear and there is no significant tolerance to injury, then losses are constant and no adjustment is necessary for prior injury. In contrast, if the yield-loss relationship is curvilinear, then previous injury must be included in the calculation of D to provide an accurate assessment of losses. Figure 2.2 illustrates how past injury may be incorporated into assessments of injury at the EIL. The modification to the EIL model is (from Eq. [5])

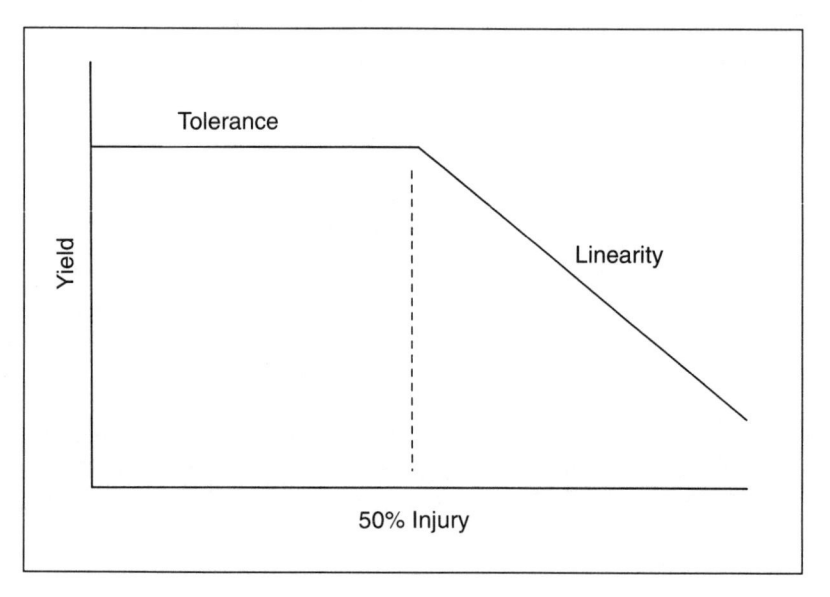

Figure 2.1. In this example, the inclusion of injury in the threshold is appropriate, because there is a period of tolerance, as shown on the damage curve.

$$D(I_t) = (C/V) + D(I_p) \quad [14]$$

where I_p = measured past injury. Substituting from Eq. [9],

$$D(I_t) = (C/V) + b_1 I_p + b_2 I_p^2 \quad [15]$$

This equation expresses the total yield losses from past injury and additional injury at the EIL. Consequently, I_t represents total injury, not injury to produce an EIL. Instead the injury at the EIL is

$$I_{EIL} = I_t - I_p \quad [16]$$

To further illustrate the complexity of reflecting past injury in the EIL, consider solving the EIL equation for I_{EIL} with a simple quadratic yield-loss relationship. From Eq. [11], [15], and [16], we see that

$$I_{EIL} = ((-b_1 + (b_1^2 + 4b_2 ((C/V) + b_1 I_p + b_2 I_p^2))^{0.5})/2b_2) - I_p \quad [17]$$

This is far from the simple expression of an EIL at the start of the chapter. Although it is possible to incorporate this level of complexity into

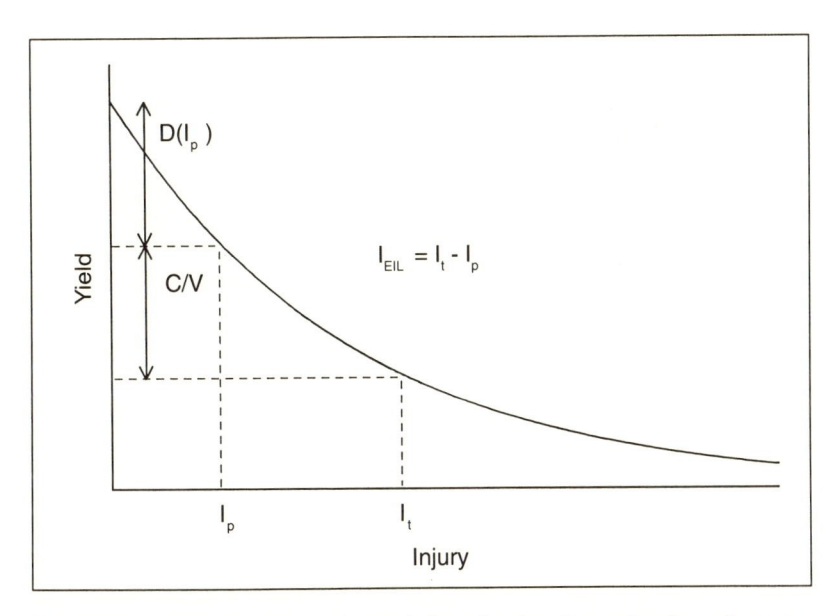

Figure 2.2. Incorporation of past injury in the determination of an EIL with a curvilinear damage function: D (I_p) = damage (= yield loss) arising from past injury, C/V = amount of damage (= yield loss) to justify management (this is the gain threshold), I_{EIL} = injury at the EIL, I_p = past injury, and I_t = total injury at an EIL after past injury (I_p) has occurred.

a conventional EIL, few circumstances seem likely to warrant such detail. As EILS and ETS are developed to consider stress interactions, it may be necessary to have some consideration of past injury for accurate yield-loss assessments. More generally, the premise of timely management seems a far better solution to the problem of making management decisions after significant injury has already occurred.

Expanding the EIL

As the preceding example illustrates, there are many possibilities for broadening the scope of the EIL. However, it is debatable whether or not benefits in such expansions offset the increased complexity or data requirements associated with their use. Among approaches that do seem to offer merit are considerations of perennial crops (Chapter 13), aesthetic pests (Chapter 12), and environmental costs (Chapter 14).

As a theoretical value for defining pest status, the EIL is fundamental and is unlikely to be supplanted. As a practical tool in the determination of thresholds, the EIL does have limitations, and alternative approaches to decision making may be appropriate. Nevertheless, conventional EILs and their associated ETs are the most common and useful decision tools for many management decisions. If alternatives to pesticides as therapeutic tactics can be developed, EILs will undoubtedly prove essential to their use. An expanded role for EILs in addressing key issues in pest management, particularly regarding environmental issues, has been proposed (Chapter 14, Pedigo and Higley 1992, Higley and Wintersteen 1992, Higley and Pedigo 1993). Seeking new applications or adaptations of EILs for resistance management or preventive management remain as unmet challenges.

References

Geier, P. W., and L. R. Clark. 1961. An ecological approach to pest control, p. 10–18. *In* Proc. Tech. Meeting Intern. Union Conser. Nature and Nat. Resources, 8th, 1960, Warsaw, Poland.

Harcourt, D. G. 1954. The biology and ecology of the diamondback moth, *Plutella maculipennis* (Curt.), in eastern Ontario. Ph.D. Diss., Cornell Univ., Ithaca NY.

Higley, L. G., and T. E. Hunt. 1994. Early-season soybean insects: Past problems and future risks, p. 91–99. *In* Proc. 1994 Integrated Crop Management Conference, Iowa State Univ., Ames IA.

Higley, L. G., and L. P. Pedigo. 1993. Economic injury level concepts and their use in sustaining environmental quality. Agric. Ecosystems Environ. 46:233–243.

Higley, L. G., and W. K. Wintersteen. 1992. A novel approach to environmental risk assessment of pesticides as a basis for incorporating environmental costs into economic injury levels. Am. Entomologist 38:34–39.

Hutchins, S. H., and J. E. Funderburk. 1991. Injury guilds: A practical approach for managing pest losses to soybean. Agri. Zool. Rev. 4:1–21.

Hutchins, S. H., L. G. Higley, and L. P. Pedigo. 1988. Injury equivalency as a basis for developing multiple-species economic injury levels. J. Econ. Entomol. 81:1–8.

Metcalf, R. L., and W. H. Luckmann (ed.). 1994. Introduction to insect pest management. John Wiley, New York.

Norton, G. A. 1976. Analysis of decision making in crop protection. Agro-Ecosystems 3:27–44.

Norton, G. A., and J. D. Mumford (ed.). 1993. Decision tools for pest management. CAB International, Oxford UK.

Onstad, D. W. 1987. Calculation of economic-injury levels and economic thresholds for pest management J. Econ. Entomol. 80:297–303.

Onstad, D. W. 1988. Letter to the editor. J. Econ. Entomol. 82:1–2.

Pedigo, L. P. 1995. Closing the gap between IPM theory and practice. J. Agric. Entomol. 12:171–181.

Pedigo, L. P., and L. G. Higley. 1992. The economic injury level concept and environmental quality: A new perspective. Am. Entomologist 38:12–21.

Pedigo, L. P., L. G. Higley, S. H. Hutchins, and K. R. Ostlie. 1988. Reply to Onstad letter to the editor. J. Econ. Entomol. 82:2–4.

Pedigo, L. P., S. H. Hutchins, and L. G. Higley. 1986. Economic injury levels in theory and practice. Annu. Rev. Entomol. 31:341–368.

Raupp, M. J., J. A. Davidson, C. S. Koehler, C. S. Sadof, and K. Reichelderfer. 1988. Decision-making considerations for aesthetic damage caused by pests. Bull. Entomol. Soc. Am. 34:27–32.

Shelton, A. M., J. T. Andaloro, and J. Barnard. 1982. Effects of cabbage looper, imported cabbageworm, and diamondback moth on fresh market and processing cabbage. J. Econ. Entomol. 75:742–745.

Stern, V. M., R. F. Smith, R. van den Bosch, and K. S. Hagen. 1959. The integrated control concept. Hilgardia 29:81–101.

Stone, J. D., and L. P. Pedigo. 1972. Development and economic-injury level of the green cloverworm on soybean in Iowa. J. Econ. Entomol. 65:197–201.

Leon G. Higley and Robert K. D. Peterson

3
The Biological Basis of the EIL

The EIL is an expression of both economic and biological parameters. Economic parameters include the market value of the crop and the cost of management used to prevent injury. The biological parameters include the nature of pest injury, the host response to pest injury, and the efficiency of the management tactic in reducing pest injury. Of these variables, the economic parameters and the efficiency of management tactics usually are derived most easily. In contrast, the biological variables of pest injury and host response may be very difficult to determine. However, it is these parameters that lie at the heart of the EIL's most fundamental purpose—to determine when pests can be tolerated.

Ironically, despite the crucial importance of pest/host relationships both for EILs and in defining pest status, surprisingly little attention has been given to these relationships as compared with other areas of pest management (especially tactics). The original development of calculated EILs was greatly delayed because of questions of pest/host relationships (specifically, determining methods to define economic damage), and this issue continues to be a major limitation in the development of EILs and ETs. Two important aspects of understanding pest/host relationships are important for pest management. First, basic understanding of how injury alters host physiology is needed. Second, clear mathematical expressions of injury and yield loss are necessary so these relationships can be incorporated into EILs. The question of pest-host relationships is equally important for animal and plant hosts; however, most of our comments for this chapter will be focused on pest-plant relationships. Chapter 11 addresses issues in pest-animal relationships, and Chapter 12 addresses the difficult question of

The Biological Basis of the EIL

pest-host relationships for situations where yield is defined in terms of aesthetics, rather than as a measurable plant product.

Characterizing pest-plant relationships is difficult for many reasons. One important limitation is that this area is not well researched. There is substantial work on what are called insect-plant relationships, but this research typically focuses on the impact of plants on insects, rather than the other way around (Welter 1993). Similarly, there is a large and growing literature on questions of pathogen-plant relationships, but much of this work focuses on the infection process, rather than on disease physiology.

What is needed is an understanding of how pests alter plant physiology and in the process reduce plant yield. Essentially, this is a question of defining biotic stress—plant stress produced by pests. The lack of research attention to biotic stress constrains our ability to develop EILs and represents an ongoing limitation to pest management. Although the literature on plant injury relationships is not extensive, for plant-insect relationships there are a number of key references, including Tammes (1961), Bardner and Fletcher (1974), Pedigo et al. (1986), Welter (1989), Higley et al (1993), Trumble et al. (1993), Peterson and Higley (1993), and Welter (1993). In ecology, understanding the role of herbivores (including insects) is an active area of inquiry and important reviews include papers by McNaughton (1983), Belsky (1986), Crawley (1989), and Whitham et al. (1991). Unfortunately, the ecological and agronomic literatures on herbivory have largely ignored each other; hence, there exists considerable need for synthesis. Chapter 7 provides a thorough review of literature on plant-weed relationships; Chapter 9 discusses experimental approaches to this issue for insects; and Chapter 13 explores interseasonal issues in characterizing plant responses to injury. Elsewhere, in joint efforts with others, we have tried to highlight the need for better understandings of biotic stress in general, and, in particular, of the impact of insects on plants (Pedigo et al. 1986, Higley et al. 1993, Peterson and Higley 1993).

Beyond questions of research emphasis, there are practical difficulties in understanding pest-plant relationships. Often the experimental techniques needed to explore these relationships require detailed knowledge both of the pest and of the plant, which can present challenging problems. Unless experimental procedures are employed with great care, it is easy to confound experiments and obtain equivocal results (see Chapter 7 and Chapter 9). Also, particularly in the entomological literature, there is a

Theory

tendency of relating pest levels only to yield—an endpoint—rather than trying to define the entirety of pests' effects upon the plant. This so-called black-box approach has not produced appreciable understanding of pest-plant relationships, and many have argued against such an approach (e.g., Bardner and Fletcher 1974, Fenemore 1982, Boote 1981, Pedigo et al. 1986, Higley et al. 1993).

Among the reasons it is difficult to establish simple relationships between pest numbers and yield is that these relationships are greatly influenced by environment (Welter 1993). Environmental conditions may alter pest numbers and may alter plant responses to pest attack. Plant pathogens are influenced by the environment to the extreme; indeed, without conducive environmental conditions, disease may not occur. Although environmental influences on insect numbers are not as profound as with pathogens, there are effects, nevertheless, particularly for species with short generation times, such as mites or aphids. Of the three pest types, weeds are least influenced by the environment, although weed numbers and competition do vary with abiotic conditions. Nor are numbers the only factor; the precise *nature* of plant response to pest injury also is conditioned by the environment. Again, this is perhaps more true for pathogens and insects than it is for weeds.

Environmental factors are so critical in the development of plant disease that the additional question, in using thresholds, of how disease broadly impacts plant physiology and reduces yield may be less emphasized. Indeed, one of the constraints in the use of thresholds for disease management involves precisely this issue. Because disease is highly dependent on environmental conditions, and because frequently we have few ways to impact disease after it is established in plants, thresholds for plant disease tend to focus on prevention, rather than cure. It follows that the biological parameters in determining these thresholds tend to address issues of epidemiology, more than plant stress. In the remainder of this chapter, therefore, we will focus most of our comments on insects and weeds, although many issues we discuss are also applicable to pathogens.

In describing the biological basis of the EIL, it is extremely important to distinguish between the direct effects of the pests and the plant response to those effects. This distinction acknowledges an underlying difference between the biology of the pest and the biology of the plant host. There are many reports in the literature that fail to draw this distinction and try to establish simple relationships between pest numbers and resulting yield loss, typically through regression procedures relating pest numbers directly to yield or yield loss. Usually such experiments are conducted over a limited range of pest numbers—typically at a range thought to be near the EIL. An alternative approach is to quantify more precisely the impact of pests and to consider effects of injury throughout plant growth, rather than only at harvest. Even in this more physiologically based approach, regression is still an important procedure for describing relationships mathematically. The distinction between physiology versus black-box approaches is that regressions in physiological analysis are used to identify pertinent physiological impacts of injury, which may provide more consistent explanations of yield loss. It has been argued that making distinctions between pest injury and plant responses is unnecessary (Onstad 1987); however, injury and plant response are distinctly different biological processes, and arbitrarily combining them obscures important phenomena and may impair the accuracy of the yield-loss component of the EIL. Chapter 2 provides additional discussion on this point; see also Onstad 1988, Pedigo et al. 1988, and Higley et al. 1993.

In distinguishing between pests and plant response to pest attack, it is important to use appropriate terminology. The distinction has long been made between injury, the action of a stressor on a plant, and damage, the plant's response to injury (Tammes 1961, Bardner and Fletcher 1974, Pedigo et al. 1986). In recognizing that plant stress occurs through changes in physiological processes, Higley et al. (1993) redefined these terms as follows:

injury—a stimulus producing an abnormal change in the physiological process;

damage—a measurable reduction in plant growth, development, or reproduction resulting from injury;

stress—a departure from optimal physiological conditions.

This terminology provides a common linkage for addressing all types of stress (not only biotic stress). The terms refer to a specific sequence of events: injury as a specific stimulus producing stress, stress as a deleterious alteration of physiological processes, and damage as a measure of how stress has impacted plant fitness (or, in agronomic words, has reduced yield).

Injury

In using economic injury levels, we express the individual impact of pests as injury in the I variable. Pedigo et al. (1986) and Higley et al. (1993) offer fundamental distinctions between pest types and the injury they produce. Differences in injury may be based on categories of physical impact, as well as on the duration and magnitude of injury. The issues of duration and magnitude of injury are particularly important for developing EILs, because they influence experimental approaches that quantify injury from individual pests. Two important types of injury are recognized: *acute injury*—injury occurring over a relatively short time in which each unit of injury is discrete and stress can occur from the effect of one or a few units of injury; and *chronic injury*—injury occurring over an extended time in which units of injury are indistinct and stress occurs only from the combined effect of many units of injury.

Many types of insects produce acute injury. Disease caused by plant pathogens is a classic example of chronic injury, where the impact of each pathogen occurs over a long time at a low level, ultimately to produce disease.

Other agents also produce chronic injury; for example, aphids, mites, and many other small sucking pests. Individually, these produce very low levels of injury; their ability to stress a plant is dependent both on an accumulation of many individual units of injury and the occurrence of injury over a period of time. Weeds are an example of agents that generally produce chronic injury, because their effects are expressed over a very long period of time. It must be recognized, however, that at specific points during crop/weed development, weeds also may actually produce injury over a short interval, more typical of acute than chronic injury—for example, by shading a crop at a critical period of plant development.

Other than for weeds, it is difficult to describe injury per pest (I) for agents producing chronic injury. With respect to insects, often the D and I variables are combined as a reflection of this experimental difficulty. The advantage of distinguishing between acute and chronic injuries is that it provides a mechanism that looks beyond biological pest classifications; instead, attention is focused on how injury impacts the plant, as a means to find commonalities. Another approach is to identify the precise physiological impact of injuries and to use this as a basis for combining injuries from different types of stressors (Hutchins et al. 1988).

Injury Guilds

Injury from pests typically induces one or more physiological responses from a host. Several pest species may induce similar physiological responses. Consequently, similar pests can be grouped into injury types. Before the development of injury types, insect pests were grouped into feeding guilds based on taxonomic relationships. This approach is not particularly useful, because pest species within a family can injure plants in distinct ways, producing different physiological responses. In the 1960s and 1970s, arthropods were placed in guilds based on the physical appearance of injury (Metcalf et al. 1962, Bardner and Fletcher 1974). Types of injury included leaf mining, leaf skeletonizing, stem boring, and fruit scarring.

In an important conceptual paper, Boote (1981) emphasized physiological responses of plants to different injury types. Based on physiological responses, he categorized arthropods as belonging to five injury types: stand reducers, leaf-mass consumers, assimilate sappers, turgor reducers, and fruit feeders. Pedigo et al. (1986) recognized the importance of Boote's categories and suggested another: plant architecture modification. Higley et al. (1993) modified the six injury types and incorporated several others to form categories of physiological impact. These categories include: population or stand reduction, leaf-mass reduction, leaf photosynthetic-rate reduction, leaf senescence alteration, light reduction, assimilate removal, water-balance disruption, seed or fruit destruction, architecture modification, and phenological disruption. Further discrimination of injury types is likely with a better understanding of physiological impact (Higley et al. 1993).

Similarities in physiological response have been identified for different pest species (Welter 1989, Welter 1991, Peterson et al. 1992, Higley et al. 1993, Peterson and Higley 1993, Welter 1993). This has been demonstrated most often with leaf-mass consumers (Hutchins et al. 1988, Hutchins and Funderburk 1991, Peterson et al. 1992, L. G. Higley, unpublished data).

If pests can be placed into injury guilds based on homogeneities of host response, then management procedures can be developed for a complex of pests. To date, work in this area has focused on the identification of injury types, the construction of injury guilds (different species producing a common injury) (Hutchins et al. 1988, Higley et al. 1993), and the development of multiple-species EILs (Hutchins et al. 1988, Hutchins and Funderburk 1991).

Injury guilds and injury equivalents (injury by one species expressed in terms of another species) have been developed for weeds (Wilkerson et al. 1991) and defoliating caterpillars (Hutchins et al. 1988) in soybean. Hutchins et al. (1988), Hutchins and Funderburk (1991), and Higley et al. (1993) provide requirements for establishing injury guilds based on homogeneity in physiological response to injury: pest species must (1) produce a similar type of injury, (2) produce injury within the same phenological time frame of the host, (3) produce injury of a similar intensity, and (4) affect the same plant part (Hutchins and Funderburk 1991).

Injury Rates

In addition to the physiological nature of injury, another important variable is the rate at which injury occurs. The question of injury rates actually has two components: the first is the rate of injury for individual pests; the second is the rate of injury for an entire pest population. Ostlie's (1984) work on insect defoliation of soybean demonstrated that plant response to injury is a function of the duration of the injury, not only the amount of injury. This finding has the important implication that studies attempting to relate injury imposed on a single day will not produce the same plant response as occurs when the same amount of injury is imposed over a longer period. On an individual level, the importance of describing injury rates relates to providing an appropriate measure of the I variable for EIL calculations.

For disease, this issue is relatively trivial, in that it is not possible to quantify the injury produced by individual pathogens. Similarly, identifying injury rates for other agents producing chronic injury, particularly small sucking insects, is a long-standing problem. As discussed in Chapter 7, establishing injury rates, called competitive indices, for weeds is challenging, because these rates differ among weed species and may be highly dependent upon the physical environment, including plant and weed population levels. Despite these difficulties, there has been considerable progress made on this question, and competitive indices have been determined for many weed species in many crops.

Establishing injury rates is relatively straightforward for insects producing acute injury. Commonly, insect injury to plants occurs from feeding, so injury is a function of consumption. Another type of injury sometimes produced by insects is through oviposition, either directly or as a function of larval feeding after oviposition. Although measuring acute injury by insects does not present the problems as measuring chronic injury, there are still complexities to be addressed. Insect consumption rates are a function of the insect developmental stage. Typically, consumption rates will show a logarithmic growth by instar. Early instars may consume only a few percent of total consumption, whereas later instars may consume between 85 and 95% of the total consumption. Because consumption rates differ by instars, it is not surprising that these rates would be temperature dependent (just as insect development is temperature dependent), but even within a stage, consumption rates may be temperature dependent (Hammond et al. 1979). In addition to a thermotemporal influence on insect consumption rates, specific differences in consumption rates of insects commonly occur. This is true even for insect species whose larvae are of similar size and have similar habitats (Boldt et al. 1975, Hammond et al. 1979, Hutchins et al. 1988).

Other factors—not often recognized, particularly in EIL calculations—may also influence consumption rate. For example, the action of natural enemies, such as parasitoids, can alter consumption rates (Armbrust et al. 1970, Duodu and Davis 1974). Also, at high densities of insects (where competition for food and interference between individuals become significant) consumption rates per insect may decrease (Hill et al. 1943, Jaques 1962, Bardner and Fletcher 1974, Pedigo et al. 1977, Pedigo et al. 1986). Typically, crowding effects occur only at densities well beyond the EIL, so changes in I based on pest density need not be included in EIL calculations.

Another issue in establishing rates of injury caused by insects depends not only on the insect itself but also on the host. Insects that defoliate newly expanding leaves have a final impact that includes both the direct tissue loss and a loss in expanding leaf area. Hunt et al. (1995) examined this question in detail for bean leaf beetle feeding on soybean seedlings. They discovered that injury on newly expanding leaves caused as much as an eightfold increase in final tissue-loss relative to tissue removed by the beetles. This is an important issue for injury to seedling plants—plants at a stage when most of the leaves are not fully expanded. On older plants, this question is of less importance; then, expanding leaves may represent only 5 or 10% of total leaf area. Hunt et al. present a scheme for reflecting the impact of leaf expansion on insect injury rates, and they determined appropriate EILs based upon these relationships.

Damage

Although the characterization of injury presents some difficulties, the greatest degree of complexity in the biological components of the EIL is associated with damage. This is not surprising, given the complicated relationships underlying plant stress and its expression through yield reductions. Explicit considerations of damage within the pest disciplines occur through the areas of disease physiology for pathogens (Chapter 8), weed ecology for weeds (Chapter 7), and plant-insect interactions for insects (Chapter 9). The concept of stress provides a unifying theme for considering the impact of various types of biotic agents, but relatively little integration across pest types has occurred (Higley et al. 1993). Efforts to develop comprehensive understandings of stress (Chapin 1991, Mooney et al. 1991) suggest that integration is possible, but so far these efforts have addressed only abiotic stress, largely ignoring biotic agents (Welter 1993).

Nevertheless, the theoretical relationship between injury and yield—the damage curve—does provide an important basis for characterizing the impact of different types of stressors. Moreover, a mathematical description for the damage curve (or parts of the curve) is essential in estimating the D variable for EIL calculations.

The Damage Curve

Although the theoretical and empirical basis for the damage curve was established by Tammes more than 30 years ago (Tammes 1961), terminology for specific portions of the curve was not developed until 1986. Pedigo et al. (1986) provided names for segments of the curve that represent unique types of response between injury and yield, or other parameters of interest (Fig. 3.1). Terms and definitions for the portions of the damage curve are

> *Tolerance:* No damage per unit injury; yield with injury = yield without injury;

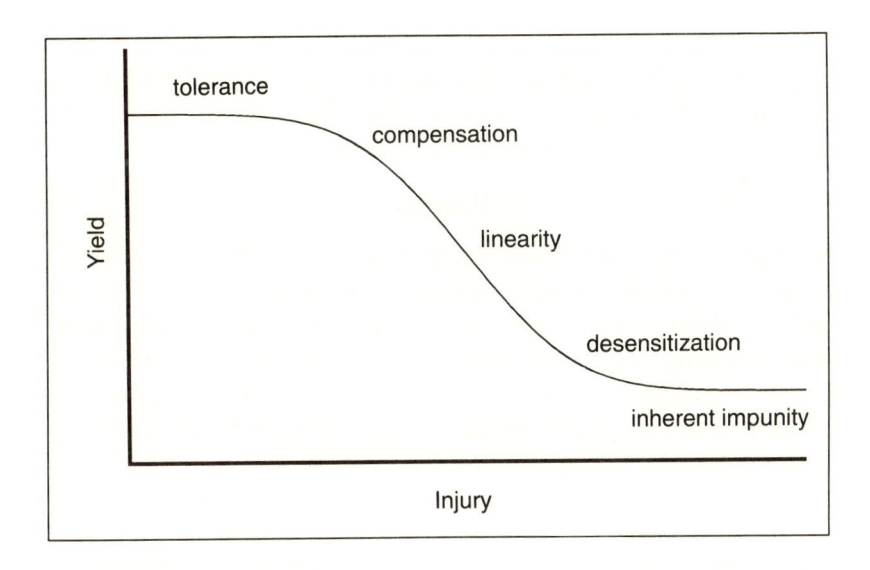

Figure 3.1. The damage curve—the relationship between yield and injury.

Overcompensation: Negative damage (yield increase) per unit injury; curvilinear relationship, positive slope;

Compensation: Increasing damage per unit injury; curvilinear relationship, negative slope;

Linearity: Maximum (constant) damage per unit injury; linear relationship, negative slope;

Desensitization: Decreasing damage per unit injury; curvilinear relationship, negative slope;

Inherent Impunity: No damage per unit injury; yield with injury < yield with no injury; constant slope.

Higley et al. (1993) note that "not all plants display the entire array of responses, but all potential responses are encompassed by the damage curve and its components." Portions of the curve also have been described as producing *tolerant, susceptive,* and *hypersusceptive* responses (Poston et al. 1983, and Chapter 9).

The damage curve is important from a practical standpoint because it is used to establish the yield-loss relationships necessary for calculating EILS

(Pedigo et al. 1986, Higley et al. 1993). The relationship between injury and yield has been characterized in hundreds of studies, although many studies that describe yield-loss functions do not include EIL calculations (Chapter 10).

Previous studies have provided examples of each portion of the damage curve, although it is unusual for all responses to be observed during one study. Nevertheless, in characterizing the relationship between rice yield and African white stem borer, *Maliarpha separatella,* Delucchi (1989) observed tolerance, compensation, linearity, desensitization, and inherent impunity. Shelton et al. (1990) observed all responses except overcompensation on cabbage yield after simulated lepidopteran defoliation. Hill et al. (1943) observed compensation, linearity, desensitization, and inherent impunity on wheat in response to Hessian fly injury.

Most of the studies that have characterized yield-loss functions have documented the linear (negative slope) relationship between yield and increasing injury. This is to be expected, because economic injury generally occurs in the linear portion of the damage curve and this region is of particular interest to researchers. Linear associations between injury and yield have been observed with insects on plants (e.g., Bechinski and Hescock 1990, Cuperus et al. 1982, Yencho et al. 1986, Michaud et al. 1989, and Peterson et al. 1993), insects on livestock (e.g., Campbell et al. 1987), weeds (e.g., Bauer et al. 1991, Weaver 1986, and Weaver 1991), and nematodes on plants (Noe 1993).

Many of the studies that have documented the linear portion of the damage curve also have observed one to three other portions. Tolerance and linearity have been observed with greenbug injury and wheat (Burton et al. 1985). Tolerance, compensation, and linearity have been observed with *Heliothis* injury and cotton (Hopkins et al. 1982), and with sugarbeet root maggot injury and sugarbeets (Bechinski et al. 1989). Desensitization and linearity have been observed with Colorado potato beetle injury and potato (Senanayake and Holliday 1990) and with grasshoppers on rangeland (Torrell et al. 1989). Desensitization, inherent impunity, and linearity have been observed with fall armyworm injury and sorghum (Martin et al. 1980), with nematode injury on cotton and soybean (Noe 1993), and with several broadleaf weed species and soybean (Weaver 1991).

There is also evidence of inappropriate curve-fitting in previous studies. Hopkins et al. (1982) fit a quadratic regression to cotton lint yield and percentage of terminal bud destruction. They observed an increase in yield at greater than 70% terminal bud destruction. Although their regression incorporates the increase in yield, Hopkins et al. (1982) state that the increase may be from a lack of data points. A more reasonable explanation is the recognition that a quadratic relationship makes little biological sense in describing injury/yield-loss responses. Jackai et al. (1989) indicate that the relationship between a coreid bug, *Clavigralla tomentosicollis,* population density, and cowpea yield "corresponded best with a third-degree polynomial fit." According to the polynomial relationship, at populations of 30 to 50 bugs per five plants, yield increased before decreasing again. As these and other studies illustrate, curve-fitting may indicate the best mathematical relationship between injury and yield, but this relationship may not reflect biological reality. The reciprocal problem, fitting lines through data that are actually curvilinear, also occurs in the yield-loss literature. In defining the linear portion of a damage curve, linear regressions are appropriate, but in many instances the actual relationship is obscured by focusing only on linear regressions. As a fundamental principle, researchers must be aware of the potential plant responses to injury (*sensu* Tammes 1961) when characterizing yield-loss functions, and these functions should make sense biologically.

Factors Influencing the Damage Curve

Many factors can alter the relationship between yield and injury, resulting in a new damage curve. It is important to recognize these differences, for many reasons. As an experimental problem, it is important to define injury/yield-loss relationships for unique circumstances. Similarly, it is important to recognize factors influencing these relationships in defining injury yields. It may be necessary to calculate different EILs to reflect different circumstances that alter injury yield relationships. An example of where this commonly occurs is in the calculation of different EILs for different plant phenological stages (which will have different injury-yield loss relationships). Pedigo et al. (1986) discussed parameters influencing the damage curve in some detail, so we will treat these issues only briefly.

Yield and injury have important implications in defining damage curves. For most crops, yield is defined as seed yield or fruit yield; for other crops, particularly forages, yield is defined as plant biomass. It is no surprise that injury/yield-loss relationships are substantially different depending upon whether yield is based on biomass or seed. Generally, seed and fruit yields are more sensitive to injury than is biomass.

A related issue arises with respect to injury type. In addition to classifying injury by physiological type or as chronic or acute, injury can be described based on its relationship to yield-producing structures. *Direct injury*—injury to yield-producing structures—is far less easily tolerated than is *indirect injury*—injury to non-yield-producing structures. In fact, we can anticipate a linear damage curve for most forms of direct injury, whereas a far more variable array of injury responses is typically produced by indirect injury.

Another factor influencing injury-yield loss relationships is intensity of injury. Injury intensity refers both to the quantity of injury and its duration. For a few types of injury, there is no clear relationship between quantity and ultimate yield loss; for this injury, what is important in producing loss is the occurrence of a single event of injury. We define this as *quantal injury*—injury in which the impact upon yield is independent of quantity of injury. (By definition, *quantal* designates an all-or-none response.) An example of quantal injury might be cosmetic injuries to fruits and vegetables; another is insect transmission of some plant pathogens. More generally, there is a strong relationship between the quantity of injury and the resulting yield loss; indeed, this is what is defined by the damage curve.

The issue of duration is less obvious. As we previously mentioned, it is established that plant responses to injury are not only a function of how much injury occurs but also of over what length of time the injury is imposed. Because most types of injury naturally occur over a given interval, it usually is not necessary explicitly to consider duration independent of injury quantity. However, when injury is simulated or imposed artificially, duration may be a confounding factor that must be addressed.

Various considerations regarding the host also impact the nature of the damage curve. Particularly important in this regard are plant phenology and plant anatomy. The nature of the damage curve is very dependent on when injury occurs relative to plant phenology. Early in their development, plants may be more susceptible to injury, but they have a longer time to compensate if they survive the injury. In contrast, during the late developmental stages, particularly reproductive stages, plants tend to be very sensitive to injury, and there may be less opportunity for compensation.

Phenology also is important as a response to injury itself. For many types of injury, a key physiological impact is an alteration of plant phenology, which has various implications.

Another important consideration is which plant parts are injured. Previously, we mentioned the importance of distinguishing between direct and indirect injury; now a further distinction is necessary: not all forms of indirect injury within a species may be the same and plant responses to injury frequently differ depending on where injury occurs. For gross differences in plant tissues, roots versus shoots, for instance, this distinction is obvious. But even within a given tissue type, there may be subtle differences in plant response. For example, defoliation impacting upper levels of a plant canopy often produce different responses than defoliation to lower portions of a plant canopy. Thus, in establishing injury guilds, it is important that species within the guild are all attacking comparable plant tissues.

One of the most important factors influencing damage is the impact of the biotic and abiotic environment. Plants stressed by biotic or abiotic agents may have different responses to subsequent stressors. The presence of an initial stress may alter the incidence of a subsequent stressor. The most common instance of such interaction is that of water stress. It is well recognized that the impact of injury from many agents is more severe for plants under water stress than those under nonstress conditions. Similarly, the impact of many stressors, particularly weeds, will be different if the nutritional status of the plant is impaired. Generally, separate EILs have not been calculated to reflect different environmental influences; however, there has been discussion of producing EILs for normal versus drought circumstances. At least one example of differing EILs for drought and nondrought conditions does exist (Hammond and Pedigo 1982, Ostlie and Pedigo 1985). A difficulty in better representing the impact of the environment on damage and in EILs relates to a relative lack of understanding of many of these relationships. Higley et al. (1993) argued for a more physiologically based approach to these questions, and it is hoped that as our fundamental understanding of stress and stress interactions improves we may be better able to incorporate the influence of the environment into EILs and ETs.

Future Issues

An understanding of the biological relationships between pests and their hosts is fundamental to pest management. Arriving at descriptions of these biological parameters is a long-standing challenge in development of EILs, and lack of understanding remains a serious impediment to expanding both the number of EILs and improving the usefulness of existing EILs. Given the subject's importance, it is sobering to recognize our relatively limited understanding of biotic stress. However, advances in instrumentation and new insights into plant physiology offer some promise for significant improvements in characterizing pest-plant relationships over the next decade. Ultimately, as we are able to build general models regarding the impact of biotic stressors on plants, it will become easier for us to develop EILs and thresholds for these stressors. Moreover, we may hope that better physiological insights into biotic stress may lead to new opportunities in pest management beyond EIL development.

Pest management is one aspect of an area we can call stress management, which mitigates the impact of biotic and abiotic stresses on plants. As our understandings improve, we may look for a more integrative understanding of stress and more integrative approaches to managing that stress.

References

Armbrust, E. J., S. J. Roberts, and C. E. White. 1970. Feeding behaviour of alfalfa weevil larvae parasitized by *Bathyplectes curculionis*. J. Econ. Entomol. 63: 1689–1690.

Bardner, R., and K. E. Fletcher. 1974. Insect infestations and their effects on the growth and yield of field crops: A review. Bull. Entomol. Res. 64:141–160.

Bauer, T. A., D. A. Mortensen, G. A. Wicks, T. A. Hayden, and A. R. Martin. 1991. Environmental variability associated with economic thresholds for soybeans. Weed Sci. 39:564–569.

Bechinski, E. J., and R. Hescock. 1990. Bioeconomics of the alfalfa snout beetle (Coleoptera: Curculionidae). J. Econ. Entomol. 83:1612–1620.

Bechinski, E. J., C. D. McNeal, and J. J. Gallan. 1989. Development of action thresholds for the sugarbeet root maggot (Diptera: Otitidae). J. Econ. Entomol. 82:608–615.

Belsky, A. J. 1986. Does herbivory benefit plants? A review of the evidence. Am. Nat. 127:870–892.

Boldt, P. E., K. D. Biever, and C. M. Ignoffo. 1975. Lepidopterous pests of soybeans: Consumption of soybean foliage and pods and development time. J. Econ. Entomol. 68:480–482.

Boote, K. J. 1981. Concepts for modeling crop response to pest damage. ASAE Pap. 81-4007. Amer. Soc. Agric. Eng., St. Joseph MI.

Burton, R. L., D. D. Simon, K. J. Starks, and R. D. Morrison. 1985. Seasonal damage by greenbugs (Homoptera: Aphididae) to a resistant and a susceptible variety of wheat. J. Econ. Entomol. 78:395–401.

Campbell, J. B., I. L. Berry, D. J. Boxler, R. L. Davis, D. C. Clanton, and G. H. Deutscher. 1987. Effects of stable flies (Diptera: Muscidae) on weight gain and feed efficiency of feedlot cattle. J. Econ. Entomol. 80:117–119.

Chapin, F. S. 1991. Integrated responses of plants to stress. Bioscience 41:29–36.

Crawley, M. J. 1989. Insect herbivores and plant population dynamics. Annu. Rev. Entomol. 34:531–564.

Cuperus, G. W., E. B. Radcliffe, D. K. Barnes, and G. C. Marten. 1982. Economic injury levels and economic thresholds for pea aphid, *Acyrthosiphon pisum* (Harris), on alfalfa. Crop Prot. 1:453–463.

Delucchi, V. 1989. Integrated pest management vs. systems management, p. 51–67. *In* J. S. Yaninek and H. R. Herren (ed.) Biological control: A sustainable solution to crop pest problems in Africa. International Institute of Tropical Agriculture. Ibadan, Nigeria.

Duodu, Y. A., and D. W. Davis. 1974. A comparison of growth, food consumption, and food utilization between unparasitized alfalfa weevil larvae and those parasitized by *Bathyplectes curculionis* (Thomson). Environ. Entomol. 3:705–710.

Fenemore, P. G. 1982. Plant pests and their control. Butterworths, Wellington, New Zealand.

Hammond, R. B., and L. P. Pedigo. 1982. Determination of yield-loss relationships for two soybean defoliators by using simulated insect-defoliation techniques. J. Econ. Entomol. 75:102–107.

Hammond, R. B., L. P. Pedigo, and F. L. Poston. 1979. Green cloverworm leaf consumption on greenhouse and field soybean leaves and development of a leaf consumption model. J. Econ. Entomol. 72:714–717.

Higley, L. G., J. A. Browde, and P. M. Higley. 1993. Moving towards new understandings of biotic stress and stress interactions, p. 749–754. *In* D. R. Buxton, R. Shibles, R. A. Forsberg, B. L. Blad, K. H. Asay, G. M. Paulson, and R. F. Wilson (ed.) International crop science I. Crop Science Soc. of America, Madison WI.

Hill, C. C., E. J. Udine, and J. S. Pinckney. 1943. A method of estimating reduction in yield of wheat caused by Hessian fly infestation. USDA, Circular No. 663.

Hopkins, A. R., R. F. Moore, and W. James. 1982. Economic injury level for

Heliothis spp. larvae on cotton plants in the four-true-leaf to pinhead-square stage. J. Econ. Entomol. 75:328–332.

Hunt, T. E., L. G. Higley, and J. F. Witkowski. 1995. Bean leaf beetle injury to seedling soybean: Consumption, effects on leaf expansion, and economic injury levels. Agron. J. 87:183–188.

Hutchins, S. H., and J. E. Funderburk. 1991. Injury guilds: A practical approach for managing pest losses to soybean. Agri. Zool. Rev. 4:1–21.

Hutchins, S. H., L. G. Higley, and L. P. Pedigo. 1988. Injury equivalency as a basis for developing multiple-species economic injury levels. J. Econ. Entomol. 81:1–8.

Jackai, L. E. N., P. K. Atropo, and J. A. Odebiyi. 1989. Use of the response of two growth stages of cowpea to different population densities of the coreid bug, *Clavigralla tomentosicollis* (Stål.) to determine action threshold levels. Crop Prot. 8:422–428.

Jaques, R. P. 1962. Stress and nuclear polyhedrosis in crowded populations of *Trichoplusia ni* (Hubner). J. Insect Pathol. 4:1–22.

Martin, P. B., B. R. Wiseman, and R. E. Lynch. 1980. Action thresholds for fall armyworm on grain sorghum and coastal bermudagrass. Florida Entomologist 63:375–405.

McNaughton, S. J. 1983. Physiological and ecological implications of herbivory. Encycl. Plant Physiol. New Ser. 12C:657–677.

Metcalf, C. L., W. P. Flint, and R. L. Metcalf. 1962. Destructive and useful insects: Their habits and control. 4th ed. McGraw-Hill, New York.

Michaud, O., R. K. Stewart, and G. Boivin. 1989. Economic injury levels and economic thresholds for the green apple bug, *Lygocoris communis* (Knight) (Hemiptera: Miridae), in Quebec apple orchards. Can. Entomol. 121:803–808.

Mooney, H. A., W. E. Winner, and E. J. Pell. 1991. Responses of plants to multiple stresses. Academic Press, San Diego CA.

Noe, J. P. 1993. Damage functions and population changes of *Hoplolaimus columbus* on cotton and soybean. J. Nematol. 25:440–445.

Onstad, D. W. 1987. Calculation of economic-injury levels and economic thresholds for pest management J. Econ. Entomol. 80:297–303.

Onstad, D. W. 1988. Letter to the editors. J. Econ. Entomol. 82:1–2.

Ostlie, K. R. 1984. Soybean transpiration, vegetative morphology, and yield components following actual and simulated insect defoliation. Ph.D. Diss., Iowa State University, Ames IA.

Ostlie, K. R., and L. P. Pedigo. 1985. Soybean response to simulated green cloverworm (Lepidoptera: Noctuidae) defoliation: Progress toward determining comprehensive economic injury levels. J. Econ. Entomol. 78:437–444.

Pedigo, L. P., R. B. Hammond, and F. L. Poston. 1977. Effects of green cloverworm larval intensity on consumption of soybean leaf tissue. J. Econ. Entomol. 70:159–162.

Pedigo, L. P., L. G. Higley, S. H. Hutchins, and K. R. Ostlie. 1988. Reply to Onstad letter to editor. J. Econ. Entomol. 82:2–4.

Pedigo, L. P., S. H. Hutchins, and L. G. Higley, 1986. Economic injury levels in theory and practice. Annu. Rev. Entomol. 31:341–368.

Peterson, R. K. D., S. D. Danielson, and L. G. Higley. 1992. Photosynthetic responses of alfalfa to actual and simulated alfalfa weevil (Coleoptera: Curculionidae) injury. Environ. Entomol. 21:501–507.

Peterson, R. K. D., S. D. Danielson, and L. G. Higley. 1993. Yield responses of alfalfa to simulated alfalfa weevil injury and development of economic injury levels. Agron. J. 85:595–601.

Peterson, R. K. D., and L. G. Higley. 1993. Arthropod injury and plant gas exchange: Current understandings and approaches for synthesis. Trends Agric. Sci. 1:93–100.

Poston, F. L., L. P. Pedigo, and S. M. Welsch. 1983. Economic injury levels: Reality and practicality. Bull. Entomol. Soc. Am. 29:49–53.

Senanayake, D. G., and N. J. Holliday. 1990. Economic injury levels for Colorado potato beetle (Coleoptera: Chrysomelidae) on 'Norland' potatoes in Manitoba. J. Econ. Entomol. 83:2058–2064.

Shelton, A. M., C. W. Hoy, and P. B. Baker. 1990. Response of cabbage head weight to simulated Lepidoptera defoliation. Entomol. exp. appl. 54:181–187.

Tammes, P. M. L. 1961. Studies of yield losses. II. Injury as a limiting factor of yield. Tijdschr. Plantenziekten. 67:257–263.

Torell, L. A., J. H. Davis, E. W. Huddleston, and D. C. Thompson. 1989. Economic injury levels for interseasonal control of rangeland insects. J. Econ. Entomol. 82:1289–1294.

Trumble, J. T., D. M. Kolodny-Hirsch, and I. P. Ting. 1993. Plant compensation for arthropod herbivory. Annu. Rev. Entomol. 38:93–119.

Weaver, S. E. 1986. Factors affecting threshold levels and seed production of jimsonweed (*Datura stramonium* L.) in soyabeans (*Glycine max* [L.] Merr.). Weed Res. 26:215–223.

Weaver, S. E. 1991. Size-dependent economic thresholds for three broadleaf weed species in soybeans. Weed Technol. 5:674–679.

Welter, S. C. 1989. Arthropod impact on plant gas exchange, p. 135–150. *In* E. A. Bernays (ed.) Insect-plant interactions. Vol. 1. CRC Press, Boca Raton FL.

Welter, S. C. 1991. Responses of tomato to simulated and real herbivory by tobacco hornworm (Lepidoptera: Sphingidae). Environ. Entomol. 20:1537–1541.

Welter, S. C. 1993. Responses of plants to insects: Eco-physiological insights, p. 773–778. *In* D. R. Buxton, R. Shibles, R. A. Forsberg, B. L. Blad, K. H. Asay, G. M. Paulson, and R. F. Wilson (ed.) International crop science I. Crop Science Soc. of America, Madison WI.

Whitham, T. G., J. Maschinski, K. C. Larson, and K. N. Paige. 1991. Plant responses to herbivory: The continuum from negative to positive and underlying physiological mechanisms, p. 227–256. *In* P. W. Price, T. M. Lewinsohn, G. Wilson Fernandes, and W. W. Bennson (ed.) Plant-animal interactions: Evolutionary ecology in tropical and temperate regions. Wiley, New York.

Wilkerson, G. G., S. A. Modena, and H. D. Coble. 1991. HERB : Decision model for postemergence weed control in soybean. Agron. J. 83:413–417.

Yencho, G. C., L. W. Getzin, and G. E. Long. 1986. Economic injury level, action threshold, and a yield-loss model for the pea aphid, *Acyrthosiphon pisum* (Homoptera: Aphididae), on green peas, *Pisum sativum*. J. Econ. Entomol. 79:1681–1687.

Larry P. Pedigo

4
General Models of Economic Thresholds

The EIL (discussed in Chapter 2) is the criterion for decision making in the economic injury level concept. This is because the EIL includes the basic damage potential of a pest population. Consequently, the EIL itself may serve to indicate the time to apply a management tactic. However, if this approach is followed, the risk of significant loss is present because losses accumulate as injury exceeds the damage boundary. Additionally, injury (pest population densities) may significantly surpass the EIL because of normal delays in the implementation of a tactic.

To minimize the risk of these losses, the EIL concept utilizes a safeguard instrument—the *economic threshold* (ET). As originally and conventionally described, the ET is the population density at which control measures should be initiated to prevent an increasing pest population from reaching the EIL. However, the ET actually represents the *time* for taking action against a pest; population density serves as a convenient index of that time.

The importance of the ET in the EIL concept cannot be overestimated, and some consider it an ideal operational decision rule (Mumford and Norton 1984). Certainly, the widespread use of ETs in IPM is testimony to its importance and usefulness. The essential value of the ET is as an operable decision criterion and a way of implementing the EIL. Consequently, management recommendations are expressed as ETs, and it is these that are actually used by the practitioner.

Practitioners use the ET most commonly where curative action is taken against a pest (Chapter 1); i.e., where a correction to the system is needed following the failure of preventive measures. Therefore, conducting a sampling program, estimating population densities, and comparing these estimates with the ET are the most common procedures followed in determining the destructive potential of a pest.

Theoretical Aspects of the ET

Although the economic threshold is key to implementing the EIL concept, probably no other aspect of Stern et al.'s paper (1959) has been more confused. Much of the confusion has been promoted by extension publications and other literature that incorrectly refers to the ET essentially as the break-even point; i.e., the EIL (see, for example, Headley 1972, Palti and Ausher 1986). In yet other interpretations (also at odds with Stern et al.'s usage), the ET is believed to be more indicative of a time to sample, with the EIL relegated to the role of time to implement a control tactic (Onstad 1987).

Other confusion over the ET involves the creation of new terminology such as "action threshold," "action level," "action threshold level," "control threshold," and more (Cancelado and Radcliffe 1979, Chant 1966, Sylven 1968). Unfortunately, many of these terms have been used to indicate entirely subjective decision criteria not based on the EIL. Although these newer terms may be more descriptive of the time to implement a management tactic, they have confused the intended meaning of the EIL concept. Consequently, advancements in ET development have been unnecessarily impeded: the new terms mean essentially the same thing as the ET. Future gains in decision making potentially are possible through standardization of the meaning and use of the term *economic threshold*. The abandonment of newer terminology and use of the ET term as originally defined by Stern et al. is indeed preferable. Subsequent clarifications of terms in the EIL concept can be made as required.

Relationships with the EIL and Damage Boundary

As originally envisioned by Stern et al. (1959), the ET is located at a level below the EIL, allowing action to be taken to prevent a growing population from exceeding the EIL and thereby preventing economic damage. This typical relationship is shown in Figure 4.1. As indicated in the figure,

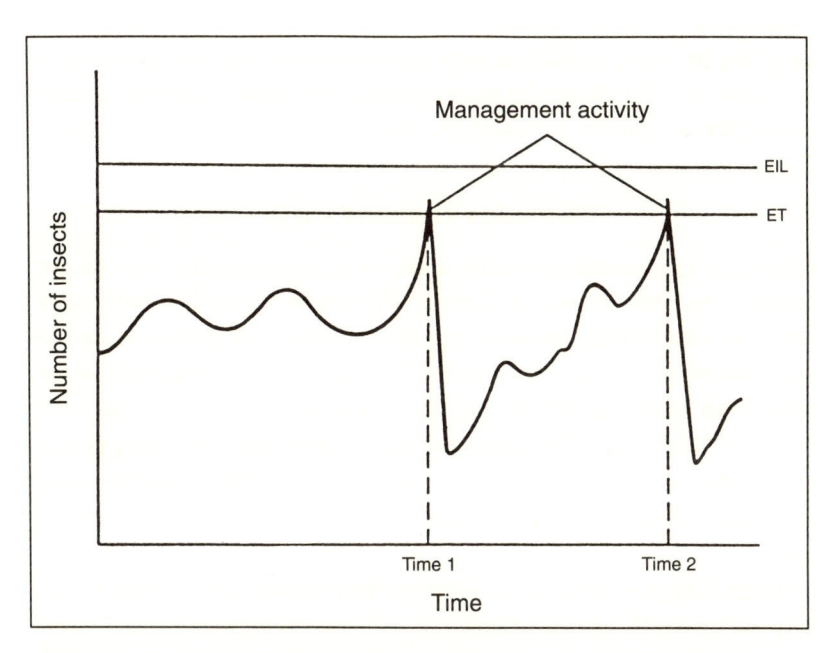

Figure 4.1. Diagram showing relationship of the ET to the EIL and time for management action. Reprinted with permission of Macmillan Publishing Company from *Entomology and Pest Management* by Larry P. Pedigo. Copyright © 1989 by Macmillan Publishing Company.

decisions are based on numbers of insects, and these numbers are an index of the time to apply a management tactic.

This conventional viewpoint of ET and EIL relationships assumes a growing population of pests, usually with continuous reproduction or recruitment and overlapping generations; e.g., many plant pathogens, spider mites, and aphids. Such relationships can be questioned when dealing with a pest for which decisions are made within a discrete generation (Chiang 1979, Andow and Kiritani 1983). For example, decision making may call for sampling leaf-feeding caterpillars at an early stage of development— that is, before they reach a stage where consumption rates are significant. Because substantial mortality may occur in the early stages, it may be prudent to set the ET above the EIL, thereby taking advantage of ecological factors that lower pest density. For example, the ET for *Spodoptera litura,* a caterpillar on soybean, is based on the number of first instars, but the EIL is

Theory

established for the number of sixth instars. For flowering-stage soybean, the ET is 215 to 1,698 larvae per plant when natural enemies are present, and the EIL for the same population is 3.2 (Andow and Kiritani 1983). Here, decisions made on action against higher numbers of small larvae take into account the potential of heavy mortality by natural enemies. Conversely, if the ET is expressed in units of pest injury, then the ET will always conform to the original relationship envisioned by Stern et al. (1959).

As discussed by Pedigo et al. (1986), the EIL also may be expressed as injury equivalents, and accompanying ETs also may be defined as injury equivalents. As discussed in Chapter 2, an injury equivalent is the amount of injury a single pest individual has the potential to cause throughout its complete life cycle; equivalency is the total accumulated injury equivalents for a population at a point in time.

An equivalency calculation of a pest population can be made by using data based on age (or life stage) of the pest and the relevant expectation of injury, then sampling the population to estimate density and age composition. Injury equivalency (IE) is given as:

$$IE = \sum_{i=1}^{n} e_i \cdot x_i \qquad [1]$$

where e = equivalency coefficient at stage i, x = number of damaging pests per sample in stage i, and n = total number of damaging stages for a pest. The injury equivalency at a point in time is a function of both the population density and the population's age structure.

An example of equivalency and its use has been given for the green cloverworm, *Plathypena scabra*, on soybean (Pedigo et al. 1986). Here, a table of equivalency coefficients was developed (Table 4.1) and referenced to determine sample equivalency. Using the table and a possible sample-unit of 12 second-stage, 6 fourth-stage, and 2 sixth-stage larvae, an injury equivalency of 2.86 would be calculated [IE = 12(0.0221) + 6(0.1000) + 2(1.000)]. By sampling over a period of time, injury equivalency for the green cloverworm population can be tracked and action taken when equivalency reaches the ET. Another example of injury equivalency for use with a multiple species feeding guild is given by Hutchins et al. (1988).

Although the approach of decision making through sample-unit equiva-

Table 4.1. Green cloverworm, *Plathypena scabra,* leaf consumption (cm²) and equivalence coefficients on soybean leaves for each larval stage (Pedigo et al. 1986).

Larval Stage[a]	Consumption[b] (cm²)	Fraction of total consumption per stage	Equivalence coefficients per stage
1	0.466	0.0086	0.0086
2	0.728	0.0135	0.0221
3	1.411	0.0262	0.0483
4	2.784	0.0517	0.1000
5	7.943	0.1474	0.2474
6	40.563	0.7526	1.0000
Total	53.895	1.0000	

[a] A seventh stage has been reported to occur 24.2% of the time. Therefore, the sixth stage includes any seventh-stage larvae that were found.
[b] Consumption of field-grown leaves.

lency has many advantages, determining pest age may be difficult for practitioners. In these instances, simple indices of age may be established (e.g., caterpillar length), or the injury itself (e.g., defoliation rates) may be estimated and used to track accumulation of injury caused by a pest population.

Factors Influencing the ET

As has been suggested, the ET is a complex value that depends on estimating and predicting several difficult parameters. The most significant of these include variable EILs, host phenologies, pest-population growth rates, and delays in management tactics.

Variable EILs. Because the ET is (or should be) based on the EIL, it varies as the EIL varies. Therefore, changes in EIL factors, such as management costs and commodity market values, cause proportional changes in the ET.

Theory

Inability to establish an objective EIL will also prevent the development of an objective ET.

Host Phenologies. The relationship of pest and host phenology is extremely important in causing changes in the ET. The EIL for a pest can change radically during the course of a growing season as crop plants change from vegetative to reproductive to senescent stages. These seasonal fluctuations in the EIL, as determined by host susceptibility to injury, cause direct and coinciding changes in the ET (Coggin and Dively 1980, Hammond and Pedigo 1976). It follows that pest colonization and developmental rate on a crop, relative to the crop's phenology, can substantially change pest status because of changing ETs. For instance, ETs for the alfalfa weevil, *Hypera postica,* can be three to four times lower for younger (= smaller) plants than ETs for older (= larger) plants of the first cutting (Wedberg et al. 1980). South of about 41° 30'N. latitude, *H. postica* frequently lays eggs in the fall, as well as the spring; consequently, the species is more likely to exceed the ET south of this latitude.

Pest Population Growth and Injury Rates. Pest population growth rates, accompanied by increased injury rates, greatly influence where the ET is set relative to the EIL. If population growth and injury rates are high, time between detection of a pest and occurrence of economic damage is correspondingly short. To prevent inadvertent losses from such pests, the ET is set relatively low. Conversely, if growth rates for a pest population and injury are low, ETs are set high, relative to the EIL, because proportionately more time is available to implement management tactics. Additionally, higher ETs can increase the chance for realization of pest mortality potentials from natural causes (Chiang 1979).

Delays in Management Tactics. Regardless of a practitioner's efficiency, there is always some delay between the time a pest population is assessed and the time when the full effect of tactic implementation is realized. The delay may be inconsequential (less than a day) in some instances, but substantial delays (up to several days) are more common. The length of delay also varies according to the tactic used.

Delays in implementing management tactics can be described by a series of separate delays, including decision delays, implementation delays, and suppression delays. To adequately establish ETs relative to EILs, some estimation must be made to determine an overall time-delay based on the total of the individual delays. As the delay period increases, the distance between the EIL and ET should be increased. This assures that accelerating injury from pests will not approach and exceed the EIL. Tactics with the shortest time delay, usually pesticides, are the least risky in this regard (Pedigo et al. 1986).

Risks and Trade-offs in Establishing the ET

Determining the ET represents an attempt to time the implementation of a management tactic adequately. In other words, because the ET takes into consideration proper timing, an implicit risk or uncertainty is involved in the assumption that pest-induced injury will reach or exceed the EIL (here, risk=known probability of occurrence; uncertainty=no known probability of occurrence). If that assumption is ill-founded, the cost of tactic implementation will not be completely offset and a net loss will be realized. Additionally, if the ET lies above the damage boundary, before the ET is reached a monetary loss that will not conform to the original break-even point (EIL) will accrue. In principle, this last problem can be avoided by basing the ET and EIL strictly on preventable injury and including an assessment of any past injury in calculations (a potentially difficult and expensive sampling problem). This level of refinement is not often used, and it seems to be impracticable in most situations; therefore, compromises are often made in estimating an ET.

The compromise falls somewhere between waiting too long to contain the pest population adequately, suffering economic loss, and taking action too soon, with the risk that uncontrolled pest populations will not reach the EIL. Because we can rarely be certain about pest-population growth rates, the ET is set to minimize such risks. Cautious producers might set the ET toward the lower end of the range to reduce the risk of catastrophic loss and thereby increase the risk of taking action when it is not needed. When pesticides are involved, environmentally conscious producers would tend to set the ET toward the upper end of the range and thereby reduce the risk of unnecessarily applying toxins.

These uncertainties highlight the predictive nature of the ET. Where predictions of pest impact are more uncertain, the use of the ET is correspondingly limited. Such a limitation in using ETs is strongest for pests requiring the use of preventive management; i.e., taking action before a pest occurs or without knowledge of current pest status. Many diseases and weeds, as well as some severe insect pests, are exclusively managed with preventive tactics, which mostly excludes the need for ETs.

Modes of ET Development

In establishing ETs, several approaches, representing different levels of sophistication, have been devised. The level of sophistication has been determined largely by existing data and the needs of particular management programs. Most of these approaches can be grouped into two broad classes, *subjective* and *objective*.

Subjective and Objective Determinations

Subjective determinations are the crudest approach to ET development. They are not based on a calculated EIL; rather, they are based on a practitioner's experience. These have been called nominal thresholds by Poston et al. (1983); they are not formulated from objective criteria. Nominal thresholds probably represent the majority of ETs found in extension publications and verbal recommendations. Although static, and possibly inaccurate, ETs based on nominal thresholds are still more progressive than using no ET at all, because they require pest-population assessment. Their use can result in reduced pesticide applications.

Objective determinations, on the other hand, are based on calculated EILs, and they change with changes in the primary variables of the EIL; viz., management costs, market values, injury per pest, and damage per unit of injury. With objective ETs, a current EIL is calculated, and estimates are made with regard to probability that the pest population will exceed the EIL. The final decision on action to be taken and timing is based on expected increases in injury and logistical delays, as well as activity rates of the tactics used. Considering the various types of objective ETs, at least three can by designated: fixed ETs, descriptive ETs, and dichotomous ETs (Pedigo et al. 1989).

The Fixed ET Category

The fixed ET, the most common type of objective ET, is set at a specified proportion or percentage of the EIL. For example, Hull et al. (1985) set the ET for the rosy apple aphid, *Dysaphis plantaginea*, at 50% of the EIL, with pest assessment set for the pink stage of tree phenology.

Use of the term *fixed* does not mean that these ETs are unchangeable; it signifies only that the percentage of the EIL is fixed. Fixed ETs change constantly with changes in the EIL.

The fixed ET ignores differences in population growth and injury rates. However, the percentages are usually set conservatively low to accommodate worst-case situations; i.e., highest injury rates. Therefore, a producer using fixed ETs is more likely to take unnecessary action than to fail to act when necessary.

Fixed ETs are crude, but they may be the best choice when pest-population dynamics are poorly understood. Certainly, they have found application across a wide variety of pests and crops, including pests on grapes, beans, soybean, sorghum, rice, and apples (Pedigo et al. 1986).

The Descriptive ET Category

Descriptive ETs are more sophisticated than fixed ETs. With descriptive ETs, a description of population growth is made. The need for action, as well as its timing, is based on expected future growth in injury rates. In developing these estimates, both stochastic and deterministic derivations have be used.

Stochastic Derivations. With stochastic derivations, action decisions and timing are derived from an understanding of previous pest-population growth rates and basing future rates on these. As an example, the green cloverworm can be sampled in soybean, beginning in late June, and an early growth curve can be established (Fig. 4.2). When larval numbers cause injury to reach the damage boundary, a statistical model based on sampling data can be applied to project future population growth. If these projections indicate that larval density will exceed the EIL during the susceptible plant-growth period, then action is taken; if not, incremental sampling usually would be continued to detect any unexpected population changes, until the crop is no longer susceptible.

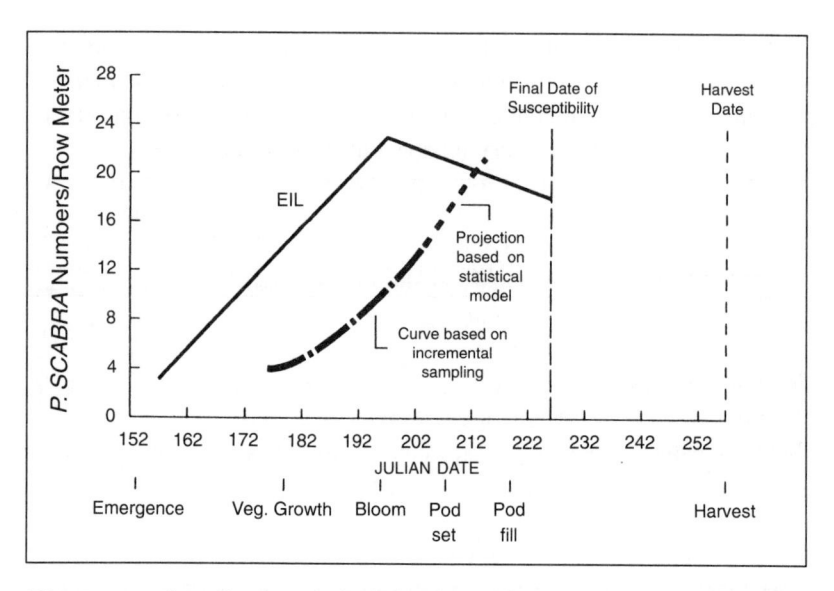

Figure 4.2. Growth of a green cloverworn, *Plathypena scabra,* population on soybean as indicated by incremental sampling, with projection of future population growth based on a statistical model.

Another example of stochastic determinations is found with spider mite, *Tetranychus* spp., management in cotton (Wilson 1985). Here, a computer model that considers the early rate of population growth and physiological time was developed. This model estimates damage trajectories and the date the ET will be reached. In this instance, the ET is the date on which 80% of the leaves of a cotton plant are infested.

The stochastic approach has the advantage of using current sampling data to keep track of the injuriousness of the pest population. Its greatest weakness is in making projections from earlier injury rates. If current and future rates do not show a strong relationship to past rates in a given season, there will be errors in decision making.

Deterministic Derivations. With deterministic derivations, estimates of future pest-population growth and injury rates are derived from age-specific parameters or processes. These growth estimates may be based on the probability of age-specific survival from life tables or on mechanistic models of a population process, such as predation or parasitization.

Table 4.2. Mortality-adjusted injury equivalence (IEQ), by generation, of green cloverworm instars during endemic (subeconomic) and outbreak (economic) population configurations. Calculations based on expectation of total injury (Ostlie and Pedigo 1987).

		Generation 2	
Larval Stage[a]	Generation 1[b]	Endemic	Outbreak
Small to medium	0.582	0.732	0.351
Medium to large	0.585	0.735	0.364
Large to pupa	0.822	0.981	0.712

[a] One (1.000) injury equivalent equals the total consumption of one green cloverworm larva living from egg hatch to pupation (53.9 cm^2).

[b] Stage-specific survivorship and, thus, injury equivalence does not differ significantly between population configuration during generation 1.

An example of the deterministic types can be seen in the work of Ostlie and Pedigo (1987) on the green cloverworm in soybean. In this work, stage-specific survival probabilities from life tables were used to compute mortality-adjusted injury equivalents (IEQs). An IEQ is the proportion of potential injury obtained by a damaging stage when premature mortality occurs. A different IEQ—which is specified for each larval stage—quantifies the amount of food consumed to date plus expected consumption based on the probability of survival. The calculated IEQs are applied to field samples. Here counts are made—by larval age (small, medium, and large larvae)—for a sampling unit and the number in each age class is multiplied by the appropriate IEQ for the stage (Table 4.2). By tallying the results, realized and future injuriousness of a pest population can be estimated and decisions are based on this estimated status. Furthermore, these look-ahead values can be used with conventional sequential sampling plans to determine whether or not to take action.

The Dichotomous ET Category

Dichotomous ETs can be developed by using a statistical procedure for classifying a pest population as economic or noneconomic from samples taken over time. The statistical procedure—termed *time-sequential sampling*—has been used to classify the adult population of the green cloverworm into outbreak or endemic classes (Pedigo and van Schaik 1984) and to determine ETs for the sorghum shoot fly, *Antherigona soccata,* on sorghum (Zongo et al. 1994). The procedure is based on the sequential probability ratio test of Wald (1947), as is conventional or spatial sequential sampling; however, in time-sequential sampling a time perspective, rather than a space perspective, is used to make decisions. In using time-sequential sampling with a damaging stage, in effect an objective ET is established.

To establish dichotomous ETs for the green cloverworm, larvae were sampled near Ames, Iowa, from 1977 through 1980. Population densities in 1977 and 1980 exceeded the calculated EIL (economic populations); densities in 1978 and 1979 did not exceed the EIL (subeconomic populations). The calculated EIL was based on a gain threshold (management cost/market value) of 28 kg/ha (0.5 bu/acre).

Subsequently, means, representing the economic populations and subeconomic populations, were calculated for each sample date in the season. These were plotted (Fig. 4.3) and the plots were used to select class limits, or critical densities, to describe a subeconomic population *(m_o)* and an economic population *(m_1)* during a seven-week period (Table 4.3). Subsequently, statistical dispersion models were fitted to the data, with the negative binomial model giving the best fit ($X^2 = 12.9$, $P > 0.05$, k = 1.164). By using the m_o and m_1 values, k from the negative binomial analysis, and alpha and beta error values of 0.1, a time-sequential sampling plan was calculated.

This sampling plan (Table 4.3) can be used by taking five 60-cm of row samples from a soybean field to calculate the mean, starting, in Iowa, the week of June 20. The mean count for a field is then multiplied by the weighting factor for the time period, and this value is compared with the upper and lower limits for making a decision. If a decision cannot be made on a date, the weighted value is accumulated over the next date. This accumulated value is compared on each date until a decision can be made or until seven sample periods have accumulated.

To test this dichotomous ET program, 12 computer simulations, 4 each of low-, medium-, and high-density populations, were run. Sample numbers, generated randomly from the simulations, were entered in the dichot-

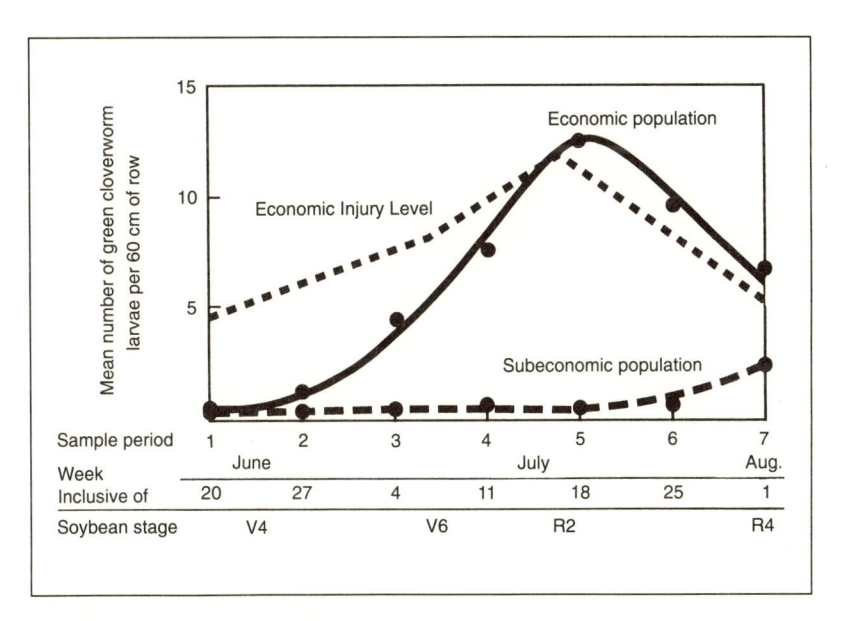

Figure 4.3. Economic populations (from 1977 and 1980) and subeconomic populations (from 1978 and 1979) of the green cloverworm, *Plathypena scabra,* in central Iowa. Data points represent mean number of larvae from six fields (three fields, each in two different years).

omous ET program for population classification. In all instances of high- and low-density populations, correct decisions were made, with average savings of ca. 25% over a fixed sample number. No decisions were made in two instances of moderate-density populations, and there were two incorrect decisions (treatment when unnecessary). In the instances of the incorrect decisions, the population peaked just under the EIL.

The dichotomous ET approach has several advantages. It is objective, easy to calculate, and simple to use. Its major disadvantage is that several years of population data at various densities are required to characterize pest population types. Moreover, some pests may not have distinctly different configurations from year to year; i.e., no distinct economic and subeconomic types. However, for many pests this ET approach deserves consideration.

Table 4.3. Time-sequential sampling form for green cloverworm larvae in Iowa soybean (Pedigo et al. 1989).

Grower		Beginning Date	
Field		Decision	
Scout			

Decision Criteria

a	b	c	d	e		
Date	Number Counted	Weighting Factor	Weighted Count	Lower Limit	Running Total of Weighted Count	Upper Limit
1	_____	× 1.7649	= _____	0	_____	2.82
2	_____	× 0.6197	= _____	0	_____	3.50
3	_____	× 0.3443	= _____	0	_____	4.38
4	_____	× 0.3371	= _____	1.35	_____	5.74
5	_____	× 0.3356	= _____	2.98	_____	7.37
6	_____	× 0.2994	= _____	4.34	_____	8.73
7	_____	× 0.1903	= _____	5.04	_____	9.43

DIRECTIONS: Sample field weekly beginning week of June 20. Take five 60-cm ground-cloth samples in an average-sized field and calculate a mean. Enter mean in column b and the date in column a. Multiply mean in b by number in c and record in d. Continue sampling once per week. Keep a running total of the numbers from d in column e. If the number in e exceeds upper limit, the population is economic, and insecticide treatment is recommended. If the number is below lower limit, it is subeconomic, and sampling activities can be discontinued. If no decision can be reached after seven sample periods, treat to reduce risk of loss.

Limitations of ETs and Future Outlook

A number of factors have limited both the development of ETs and the usefulness of existing thresholds. Many of these limitations are linked together. One major constraint is the absence of a thorough mathematical definition of the ET, although a working definition of the ET, as a level less than or equal to the EIL, often is used. Completely subjective ETs may not meet even the nonmathematical definition, and evaluating the validity of such thresholds can be as subjective as the thresholds themselves. Objective ETs do relate to an EIL, but even in this category, ETs may not be precisely defined. Among objective ETs, fixed ETs depend on a chosen level below the EIL; descriptive ETs depend on predictions of population trends, mortality, or injury rates; and dichotomous ETs depend on sampling errors and probability levels. None of the existing approaches for defining ETs is comparable to the precise, mathematical definition available for EILs.

A second constraint with economic thresholds comes from their relationship to EILs. Because any ET should be defined in reference to an EIL, those factors limiting EILs (see Chapters 2 and 5) also limit ETs.

Yet another constraint relates to evaluating injury by sampling pest populations. Some error is always associated with sampling, and although methods may improve, sampling error represents an absolute constraint in the use of ETs. The question of sampling error is particularly acute if ETs are used to optimize early management decisions.

Because the ET is predictive (i.e., it projects that pest numbers and resulting injury will reach the EIL), another constraint is created. An understanding of pest-population biology and prediction of population trends is one obvious requirement for accurate ETs, particularly for making early decisions. Similarly, the prediction of market values can be an important constraint for EILs, and subsequently for ETs. Weather is also important in determining population trends and many EIL variables. Because of the pervasive influence of weather factors on ETs and the inability to predict these factors in the long term, weather remains a major constraint on ETs.

For future development and improvement of ETs, these limitations must be pushed back. In three areas—implementation, theoretical development, and research—there is promise of improvement. In implementation, methods that allow for fluctuating variables, such as crop values, are desirable:

these approaches may include tabular ETs that provide an array of possibilities for different circumstances, interactive computer programs that produce calculations for specific conditions, ET templates for existing computer software (e.g., spreadsheets), simulation models that may improve predictions, and expert systems for situations that are more uncertain. In theoretical development, ETs will benefit particularly if more precise mathematical definitions can be formulated. Finally, in the research area, predictions of weather and pest-population dynamics will improve ETs.

Acknowledgments

Thanks are extended to P. Davis, Cornell University, L. Higley, University of Nebraska, and S. Hutchins, DowElanco, for their ideas and significant inputs into this chapter. Journal paper J-15307 of the Iowa Agriculture and Home Economics Experiment Station, Ames, Iowa; Project No. 2580 and 2903.

References

Andow, D. A., and K. Kiritani. 1983. The economic injury level and the control threshold. Jpn. Pestic. Inf. 43:3–9.

Cancelado, R. E., and E. B. Radcliffe. 1979. Action thresholds for potato leafhopper on potatoes in Minnesota. J. Econ. Entomol. 72:566–569.

Chant, D. A. 1966. Integrated control systems, p. 193–198. *In* Scientific aspects of pest control. Washington DC, Natl. Acad. Sci., Nat. Res. Counc. Pub. 1402.

Chiang, H. C. 1979. A general model of the economic threshold level of pest populations. FAO Plant Prot. Bull. 27:71–73.

Coggin, D. L., and G. P. Dively. 1980. Effects of depodding and defoliation on the yield and quality of lima beans. J. Econ. Entomol. 73:609–614.

Hammond, R. B., and L. P. Pedigo. 1976. Sequential sampling plans for the green cloverworm in Iowa soybeans. J. Econ. Entomol. 69:181–185.

Headley, J. C. 1972. Defining the economic threshold, p. 100–108. *In* Pest control strategies for the future. Natl. Acad. Sci., Washington DC.

Hull, L. A., E. H. Beers, and J. W. Grimm. 1985. Action thresholds for arthropod pests of apple, p. 274–294. *In* R. E. Frisbie and P. L. Adkisson (ed.) Integrated pest management on major agricultural systems. Texas Agric. Exp. Sta. MP-1616, College Station TX.

Hutchins, S. H., L. G. Higley, and L. P. Pedigo. 1988. Injury equivalency as a basis for developing multiple-species economic injury levels. J. Econ. Entomol. 81:1–8.

Mumford, J. D., and G. A. Norton. 1984. Economics of decision making in pest management. Annu. Rev. Entomol. 29:157–174.

Onstad, D. W. 1987. Calculation of economic-injury levels and economic thresholds for pest management. J. Econ. Entomol. 80:297–303.

Ostlie, K. R., and L. P. Pedigo. 1987. Incorporating pest survivorship into economic thresholds. Bull. Entomol. Soc. Am. 33:99–102.

Palti, J., and R. Ausher. 1986. Advisory work in crop pest and disease management. Springer-Verlag, New York.

Pedigo, L. P., L. G. Higley, and P. M. Davis. 1989. Concepts and advances in economic thresholds for soybean entomology, p. 1487–1493. *In* A. J. Pascale (ed.) Proc. World Soybean Res. Conf. IV, Vol. 3, Asociación Argentina de la Soja, Buenos Aires, Argentina.

Pedigo, L. P., S. H. Hutchins, and L. G. Higley. 1986. Economic injury levels in theory and practice. Annu. Rev. Entomol. 31:341–368.

Pedigo, L. P., and J. W. van Schaik. 1984. Time-sequential sampling: A new use of the sequential probability ratio test for pest management decisions. Bull. Entomol. Soc. Am. 30:32–36.

Poston, F. L., L. P. Pedigo, and S. M. Welch. 1983. Economic injury levels: Reality and practicality. Bull. Entomol. Soc. Am. 29:49–53.

Stern, V. M., R. F. Smith, R. van den Bosch, and K. S. Hagen. 1959. The integrated control concept. Hilgardia 29:81–101.

Sylven, E. 1968. Threshold values in the economics of insect pest control in agriculture. Statens Vaxtskyddsanst. Medd. 14:65–79.

Wald, A. 1947. Sequential analysis. John Wiley, New York.

Wedberg, J. L., W. G. Ruesink, E. J. Armbrust, D. P. Bartell, and K. F. Steffey. 1980. Alfalfa weevil pest management program. Ill. Coop. Ext. Serv. Circ. 1136.

Wilson, L. T. 1985. Developing economic thresholds in cotton, p. 308–344. *In* R. E. Frisbie and P. L. Adkisson (ed.) Integrated pest management in major agricultural systems. Texas Agric. Exp. Sta. MP -1616, College Station TX.

Zongo, J. O., C. Vincent, and R. K. Stewart. 1994. Time-sequential sampling of sorghum shoot fly, *Atherigona soccata* Rondani (Diptera:Muscidae), in Burkina Faso. Trop. Pest Manage. *in press*

Ronald B. Hammond

5
Limitations to EILs and Thresholds

Economic injury levels and economic thresholds are the cornerstones of any integrated pest management (IPM) program; however, limitations hinder their acceptance and use by IPM practitioners and end users (Poston et al. 1983, Pedigo et al. 1986, Wearing 1988). These limitations are associated with the establishment and implementation of EILs and ETs (limitations with ETs are often related to those constraints associated with IPM programs in general). A lack of multiple-stress EILs and ETs prevents full utilization and acceptance by growers; there also are constraints in settings other than the typical agricultural environment in which EILs initially gained acceptance. Examples of the latter are in dealing with medical and veterinary pests, where aesthetics are of primary importance, and in urban and forest settings. Also there is a need to address environmental limitations of EILs and ETs.

Various limitations associated with the development and use of EILs and ETs will be briefly described. No attempt will be made to outline fully ways of circumventing all the limitations; much of that is covered in many of the other chapters of this book. This chapter will serve only to open discussion.

EIL Establishment

The Basic Need: Improved Understanding

Understanding how pest injury alters yield, the D and I components of the EIL equation EIL = $C/VIDK$, is central to the EIL concept (Pedigo et al. 1986) (where EIL = economic injury level, C = cost of management tactic, V = market value per production unit, I = injury units per pest, D = damage per injury unit, and K = proportional reduction in pest attack). However, quantitative relationships with yield necessary for EILs have not yet been established for numerous pests and pest types (Pedigo et al. 1986). This aspect has been improving, not only for insects but also for weeds, diseases, and nematodes.

Currently, most of these relationships have been developed only for annual crops, and work has been minimal on perennial crops (Chapter 13 contrasts EILs in perennial systems compared with annual systems and associated problems). Moreover, the lack of a full understanding of how injury alters not only yield but also plant physiology limits the usefulness of EILs—a limitation further confounded by a lack of comprehension on how environmental factors and grower management practices affect these relationships.

This lack of a complete understanding is perhaps the primary limitation, and need, in the development of accurate EILs within all pest disciplines. Most fundamentally, we need to move toward a new and better understanding of how plants respond to pest injury, so that we may develop general models for all types of pest stress (see Higley et al. [1993] for a fuller discussion).

Experimentation

Related to the general lack of knowledge are problems of experimental design and data analysis necessary to examine relationships between pests and crops. Design and analyses should be based on the objectives of the experiment; i.e., the development of crop/yield-loss relationships. Cousens (1985, 1991) discusses various aspects of experimental design in weed competition experiments and points out the usefulness of certain designs (e.g., additive designs) in the face of published criticism.

Constraints associated with experimentation often relate to how the experiments are conducted. Most research efforts have been done with limited environmental variability and for short durations. The crop-production practices most common to the region are employed, often for ease of conducting the research. The crop/yield-loss relationships that result might not be applicable over the entire range of grower management practices. To illustrate, we can examine insect defoliation studies on soybean conducted in many Midwest states. These studies were often conducted for two years, with soybean rows planted in 76-cm widths, using conventional tillage practices and midspring planting dates. However, the use of narrow rows, some degree of conservation tillage, and various planting times are increasing in the Midwest, and defoliation/yield-loss relationships need to be examined under the new cultural conditions. For EIL development, whether the yield-loss data from one set of cropping practices should be applied to a quite different set often is not examined.

Another constraint is economic: substantial research costs and time investments are necessary (Pedigo et al. 1986). Studies are often labor intensive and lengthy. Data comparison between similar studies often is limited because different techniques have been employed by numerous researchers. Yield-loss relationships also can vary a lot under different environmental conditions (Pedigo et al. 1986). Two Iowa studies found that the pattern and amount of precipitation during the growing season significantly affected soybean response to a given defoliation level. A system of two EILs for defoliation based on precipitation levels was suggested (Hammond and Pedigo 1982, Ostlie and Pedigo 1985).

Numerous approaches are possible to circumvent these limitations. A unified approach to research should be taken. Common experimental designs appropriate to the main objectives (i.e., an understanding, crop by crop, of crop/yield-loss relationships and subsequent calculations of EILs and ETs) should be used, as should common experimental methodologies, data collection, and evaluation techniques. Regional research projects would be an appropriate place to develop common goals and procedures, allowing for replication under a greater range of environmental conditions so that the results would have more value in the making of comparisons.

Better understanding of how injury affects plant physiology would allow researchers to focus on general models of plant response to stresses (Higley et al. 1993). Such an effort will require a more detailed understanding of the pests and a thorough understanding of plant physiology. We must have a much more interdisciplinary approach than has been taken in the past—an approach with more precise experimentation and using the newer equipment now available (e.g., infrared gas analyzers for photosynthesis, leaf-area meters, leaf-canopy analyzers, personal computers, and better statistical packages).

Prediction of Management Costs and Market Value

Factors affecting the establishment of EILs of EILs that are difficult to predict include the cost of the management tactic *(C)* and the market value of the productive unit *(V)*. Although the cost of the management tactic will usually remain the same during a growing season, the value of the productive unit often varies, according to local, regional, national, and international conditions. A drought, for example, can cause local prices to rise—and thus cause EILs to decrease. The reverse might occur should a large foreign supply reduce local prices. Such factors add to the constraints associated with establishing and implementing EILs.

The Pest-Survival Element

The original EIL, initially developed for use against insects, assumed 100% mortality of the pest with management. Norton (1976) expanded the early EIL model to account for the *expected* level of control, including the percentage reduction in pest attack *(K)* value into the equation and providing a means to evaluate the impact of different control measures (e.g., pesticides); however, natural mortality agents also remove a proportion of the population, lessening the need for therapeutic action.

Ostlie and Pedigo (1987) proposed a method that accounts for mortality by converting insect counts into "insect injury equivalents," based on injury expectations for the injurious life stages. However, reliable estimates are critical in calculating natural mortality and survivorship. These are best obtained from partial life-table studies.

If it were to be incorporated into control-decision calculations, information about pest survivorship would conserve both management resources and natural enemies.

The ET and Its Calculation

There are limitations in calculating an ET from an established EIL, because mathematical definitions of the ET are largely unavailable (Pedigo et al. 1989). None of the existing approaches compare with the more precise definition of EILs. ET levels often are set subjectively, and at best are fixed, based only on experience, at a certain percentage of the EILs.

One objective method utilizes time-sequential sampling (see Chapter 4); however, substantial pest-population data are required with this technique.

EIL Implementation via the ET

Sampling

Many obstacles limit the implementation and acceptance of EILs and ETs, but many of the constraints are in fact more to be associated with limitations of IPM programs in general. Foremost is making the decision for the need to take action; i.e., the evaluation of the current pest status or injury level. Procedures for sampling insects and nematodes, determining defoliation levels, monitoring diseases, evaluating the weed situation, and other forms of injury evaluation are imperfect at best; they contain inherent errors (Pedigo et al. 1988, Pedigo and Buntin 1994). If errors are made in classifying a population density as economic (i.e., above the ET) when it is not economic, pesticide may be applied unnecessarily (β error). The reverse situation, classifying a population density as subeconomic (i.e., below the ET) when it is economic (α error), presents the opposite problem.

Accuracy of a sampling technique will often be poor when pests (e.g., nematodes) are not randomly distributed across a field (Carlson and Headley 1987). Sampling and quantifying the levels of many insects and pathogens is often too costly and impractical. Sampling methods and their efficiency often vary regionally, and with varying plant conditions; e.g., different plant types (bushy vs. thin), height, and row spacing. Other variables enter by way of different personnel; for example, whether insect counts of samples taken with a sweep-net in alfalfa were made in the field or in the laboratory has had an influence on decision making (Fleischer and Allen 1982). In another study, interobserver variation affected corn insect-population estimates and management decisions (Shufran and Raney 1989). In the latter study, the authors suggest that interobserver variation be kept in mind when using ETs.

Sampling procedures used should be as standard as possible and be those least affected by individual users. As we begin to understand better how injury alters the physiology of the plant, we can perhaps develop alternative methods for determining the need for therapeutic action. For example, current thought on insect defoliation is changing from "how much leaf area is removed" (i.e., percent defoliation), to "how much leaf area remains per unit area" (i.e., the leaf-area index). With the introduction of leaf-canopy analyzers that instantaneously measure leaf-area index (LAI) in the field, we might see ETs changing from a percent defoliation level to the LAI present at the time of sampling. When combined with measures of insect-population levels, we can perhaps make more accurate decisions.

The use of the LAI also has been applied to weeds. Yield-loss estimates based upon the LAI of weeds relative to that of the crop's LAI showed the least variability, compared with other measurements (Weaver 1991). Weaver noted that the usefulness of LAI measures may be limited by the ability to estimate LAI quickly and accurately, a problem that is addressed by leaf-canopy analyzers.

Bechinski (1994) reviews some of the practical impediments to insect sampling. Many of the issues Bechinski explores emerge as concerns for implementation of thresholds, as well as for sampling programs. In particular, the dichotomy between sophisticated sampling programs versus methods that are practicable for farmers was identified as an ongoing problem. A similar difficulty exists with more complex management thresholds.

Taking Action

After the decision to act is made, there is often a delay in the application of the therapeutic tactic. This delay can be caused by several factors: the time necessary to calibrate and otherwise prepare equipment and to obtain the pesticide; weather conditions; nonavailability of sufficient personnel and/or equipment. The latter is especially important during severe outbreaks. Numerous fields may require concurrent applications, sometimes requiring custom applicators.

Predictive Capabilities

Researchers have attempted to provide alternative ETs to growers because of variation in factors that alter EILs. ETs for various situations that growers encounter in their decision making have been put in tabular form, allowing the grower to choose ETs for specific management costs and market values of the productive unit. As we begin to understand better the effect of weather conditions and crop production practices (e.g., crop varietal choice, tillage methods, planting dates) on crop/yield-loss relationships, EILs for more of these situations will become available.

An important limitation is that EILs are most appropriately used when a single, curative action can be taken. Determination of economic injury (or having reached related population levels) is of little value if the only control options are those that must be used before infection and/or damage (e.g., the case with many plant pathogens) (Pedigo et al. 1986). The EIL most relates to preventive tactics only in assessing the success of these tactics, not in the need for their use.

Predictive models must increasingly incorporate current conditions on crop growth, future pest-population trends, weather forecasts, potential market-value changes, and so on. Researchers are now incorporating much of this information into computer programs that aid growers in making management decisions on all pest types.

Risk Aversion

Another impediment to the implementation of EILs and ETs in pest-management programs is associated with the amount of economic risk a grower is willing to assume. Conventional theory assumes that pest-control strategies are designed and implemented to optimize a system; the concept of EILs and ETs embodies this idea (Szmedra et al. 1990). However, as Szmedra et al. point out, incorporating risk into decision making by considering the variability of income and farmers' risk preference has lessened the emphasis on optimization. Some growers choose strategies that offer consistent income with little variability (risk-averse); others choose approaches that have potential for larger returns some years and small or less return other years (risk-taking). Szmedra et al. refer to a study by Day (1978) that suggests farmers are often averse to risk and generally search for stability in the system, rather than striving for optimal outcomes; this is seen by Day as "satisfying" or "economizing" behavior. This group would be less willing to following IPM strategies. Putting work by Akerlof and Yellen (1985) into the framework of ETs, Szmedra et al. also argue that the timing of applications differing from that suggested to be optimal (via ETs) *may not* result in different outcomes, compared with an IPM program that strictly follows ET guidelines. Using a model, they found that a regime of prophylactic calendar sprays was generally most risk-efficient economically compared with strategies that followed strict or modified ET guidelines that involved more economic risk. Their results suggest that threshold guidelines should be reevaluated and more finely tuned to assure that risk reduction takes place, thus increasing the likelihood of EIL and ET acceptance by growers. Acceptance of IPM methods will be by those farmers most willing to assume risk, whereas risk-averse farmers will only slowly modify a control strategy proven successful in the past.

Multiple Pests

When first developed, EILs were developed for a single species; however, growers face a multitude of pests. During a growing season, a grower will make control decisions concerning weeds, insects, pathogens, and nematodes. The status of developing EILs (and corresponding IPM programs) for different pest types varies; most past effort has dealt with insects (Thill et al. 1991). However, as the understanding of crop/yield-loss relationships has improved, researchers have begun to develop EILs and ETs for the other pests, in forms usable by growers and IPM practitioners.

Pests That Cause Similar Injury

The idea of developing EILs for multiple pests was considered by Pedigo et al. (1986), who suggested that if injuries from different pests produce the same host response these injuries could be regarded as the same, and EILs for multiple pests could be developed.

Hutchins et al. (1988) developed a technique that grouped insects into injury guilds based on the plant's physiological response to the injury. They used an insect equivalency system for estimating crop injury to refine the accuracy of EIL decision making. In their example, they grouped soybean defoliators into an injury guild and developed EILs based on insect equivalents.

A similar approach has been used for weeds, with the use of a competitive index: weeds are ranked in order of competitiveness (Wilkerson et al. 1991). Wilkerson et al. rated 76 weed species in terms of their competitiveness with soybean, and estimated potential yield loss based on the weed competitiveness rankings, number of weeds of each species present in the field, and expected weed-free yield. This information was used in a computer program, HERB, that helps evaluate potential crop damage from multispecies weed complexes in soybean and determine the appropriate course of action.

Pests That Cause Different Injuries

Developing a single EIL for different pest types, or for all insects or all pathogens on a plant, is probably unlikely, because different pest species impact crop physiology and growth differently. Currently, such differences cannot be reflected in a conventional EIL. However, if we present a multitude of ETs for numerous pest types to the grower, the likelihood of EIL acceptance will be diminished. Researchers have addressed this problem with the use of crop-simulation computer models and expert systems that incorporate rules for decision making for many pest types. These models also include input about other production variables that affect the grower's decision (e.g., crop varietal choice, market values, management costs, weather conditions).

Much modeling work has been done in cotton; numerous crop and pest models have been developed (El-Zik and Frisbie 1991). An early computer-based pest-management system, combining an understanding of the crop's growth and development with pest-management decisions, was SIRATAC (Ives et al. 1984). A later effort resulted in CALEX/Cotton, an integrated, expert system for cotton production and management that includes rule-based advisors for managing various insects, diseases, nematodes, and weeds, along with making crop-management decisions (Goodell et al. 1990). Suszkiw (1994) has described a combination of two systems—the well-known cotton simulation model and expert system for management decisions, GOSSYM/COMAX, and RBWHIMS (*r*ule-*b*ased [*W*] *H*olistic *I*nsect *M*anagement *S*ystem), an expert system for insect management on cotton.

Soybeans provide another example. AUSIMM (*A*uburn *U*niversity *S*oybean *I*ntegrated *M*anagement *M*odel) incorporates soybean cultivar selection (and their related maturity grouping, productivity, and susceptibility to nematodes and diseases), field history, plant growth data, pest population dynamics, pesticide efficacy, crop-loss functions, and weather data into a comprehensive computer package that is available to soybean growers (Backman et al. 1988). In the future, new modeling efforts may aid growers in integrating management decisions about crops and pests throughout the growing season.

Stress Interactions

The preceding section describes integrating EILs developed for single pest species, or pest species producing similar injury. Currently, there is a lack of EILs for situations where multiple pests interact or where environmental conditions vary. The approaches already mentioned assume an additive effect of the pests; however, pest damage might have an *interactive* impact on crop physiology and yield. Some studies have examined the interactive effects of two dissimilar pest types on EILs (e.g., Higgins et al. 1984), but more work is needed on all crops. There is the additional factor of possible interactive impact of varying environmental conditions (e.g., variations in rainfall). As discussed earlier, the interaction between plant stress and the environment is often strong and leads to quite different EILs (Ostlie and Pedigo 1985).

Overcoming Multiple Stress Limitations

A better understanding of how injury affects plant physiology will allow researchers to focus on general models of plant response to all biotic stresses (Higley et al. 1993). It may be that even dissimilar pest types similarly affect the crop physiology; such a situation could thus benefit from a pest-equivalency system similar to that suggested for insects (Hutchins and Funderburk 1991, Higley et al. 1993). More efforts are needed toward truly integrated studies in which the effects of two or more pest types on crop physiology and yield are determined. Such studies will be more complex, both in design and effort, than single-species work. Higgins (1985) presented various approaches and items to consider when studying interactive stress caused by insects and weeds—two different pest types. However, the same requirements necessary for single-species studies would enhance the effort on pest interactions (i.e., using common experimental designs and similar experimental methodologies, data collection, and evaluation techniques, and doing the work across multiple environments).

EIL Development in Settings Other Than Agriculture

Basic to the development of EILs is the evidence of quantitative relationships between pest damage and injury (Pedigo et al. 1986). There are settings (e.g., with vectors, medical pests, and veterinary pests) where these relationships are not as evident as they are with many agricultural crops. Assigning monetary value will always severely limit the application of EILs and ETs to medical pests because a "market value" cannot be assigned to human health and life. It is possible to place a price on animals; thus, EILs and ETs can play a role for veterinary pests. Chapter 11 addresses the use of ETs for veterinary pests and should be referred to for the status of related thresholds.

Aesthetic Considerations

In some situations, aesthetic considerations are of primary importance—often in urban areas, where, for example, pests must be managed on landscape plants (Raupp et al. 1992) and turfgrass (Potter and Braman 1991). The placing of monetary value on such products is usually very subjective, however, varying with the objectives of the pest manager and end user. Often, a difference exists between wholesaler and retailer. For example, retail nurserymen may have an extremely low EIL for pests that cause aesthetic damage on landscape plants since they sell directly to the homeowner; the wholesale nurseryman, on the other hand, might tolerate higher levels of pest injury before initiating control (Raupp et al. 1987).

The term "aesthetic injury level" (AIL), first used by Olkowski (1974), has been suggested for use in decision making in many of these situations. AILs would be considered equivalent to the EIL, except that aesthetic considerations come to be of primary importance in decision making. As efforts are directed toward this area, the need to increase the reliability of sampling and monitoring techniques, and then to reach a consensus on management-action thresholds for pests that cause aesthetic damage, is paramount (Raupp et al. 1992). Chapter 12 discusses the development and application of aesthetic thresholds in greater detail.

Forest Pests

Applying EIL and ET concepts to forest pests is problematic. When considering pests on short-term commodities such as Christmas trees, the situation becomes something of an aesthetic problem, similar to those discussed above.

In forest ecosystems, where pests are of great concern (witness the problem of gypsy moth), other factors enter the equation. The injury/crop-response relationship is difficult to establish because the growth of the tree spans many years. Other components necessary for the EIL calculations are difficult to establish because market values are hard to assess, both in the short and long term. Often the value is more aesthetic than monetary. Management costs often are large, especially when vast forested areas need spraying. Added to these problems are the environmental and social costs associated with the forest ecosystem. These constraints make it difficult to establish usable EILs and ETs for forest pests. The need exists, nonetheless. The use of EILs and ETs in interseasonal situations is discussed in Chapter 13.

Environmental Risk and EILs

A final limitation associated with the use of EILs and ETs involves the environmental costs associated with the use of pesticides. Of course, EILs were developed with the expressed intent of reducing pesticide use, and it was assumed that EILs would lead to benefits for the environment (less chemical residues in soil, less contamination of water supplies affecting wildlife and humans, and less adverse effects on operators, bystanders, and consumers [Thorton et al. 1990]). And using pesticides only when the ET is reached does help in improving environmental quality. Thus, the development of EILs and ETs has been seen as advantageous to the environment. As more research on EILs is conducted, greater availability and use of EILs and ETs will result.

However, more direct benefits for the environment might occur. EILs could be improved from an environmental standpoint by directly entering environmental costs into the EIL equation, through consideration of environmental risks associated with individual pesticides (Higley and Wintersteen 1992; Pedigo and Higley 1992). To consider environmental risks in

EILs, one must identify the risks, rank their relative importance, and assess a monetary estimate of the value of avoiding these risks. Higley and Wintersteen (1992), using contingent valuation techniques, determined these variables for numerous insecticides. They calculated environmental EILs for them by modifying the EIL equation: environmental EIL = *(PC + EC)/ VDIK,* where *PC* is the pesticide and application costs and *EC* is the environmental costs (the others variables remain the same). If a producer is unwilling to include the environmental costs directly into the decision, the grower could use the environmental-cost information to select the least environmentally damaging pesticide.

Hutchins and Gehring (1993) have questioned the general concept of environmental EILs. They argue that the use of contingent valuation incorporates incorrect health and environmental perceptions into a decision-making tool predicted by biological knowledge and economic circumstances. Hutchins and Gehring argue for complete disclosure and education on how to limit environmental risks with all tactics and believe that environmental EILs might reduce the objectivity in decision making, diminishing the implementation of IPM.

Kovach et al. (1992) conducted similar work on fruit and vegetables, developing environmental-impact quotients (EIQ) for many common pesticides. Their approach provides a method of incorporating environmental effects into the pesticide decision-making process, although they did not specifically develop environmental EILs. Chapter 14 addresses the concept of environmental EILs.

While the development of EILs has been one of the most important and useful concepts in pest management, limitations prevent their full implementation. Researchers need to refine existing EILs and ETs and develop new ones for a wider range of situations. Thus will researchers overcome resistance to EILs by growers and other end users.

References

Akerlof, G. A., and J. L. Yellen. 1985. Can small deviations from rationality make significant differences to economic equilibria? Am. Econ. Rev. 75:708–720.

Backman, P. A., T. P. Mack, R. Rodriquez-Kabana, and D. A. Herbert. 1988. A computerized integrated pest management model (AUSIMM) for soybeans,

p. 1494–1499. *In* A. J. Pascale, ed. Proc. World Soybean Res. Conf. IV, Vol 3. Asociación Argentina de la Soja, Buenos Aires, Argentina.

Bechinski, E. J. 1994. Designing and delivering in-the-field scouting programs, p. 683–706. *In* L. P. Pedigo and G. D. Buntin (ed.) Handbook of sampling methods for arthropods in agriculture. CRC Press, Boca Raton FL.

Carlson, G. A., and J. C. Headley. 1987. Economic aspects of integrated pest management threshold determination. Plant Dis. 71:459–462.

Cousens, R. 1985. A simple model relating yield loss to weed density. Ann. Appl. Biol. 107:239–252.

Cousens, R. 1991. Aspects of the design and interpretation of competition (interference) experiments. Weed Tech. 5:664–673.

Day, R. H. 1978. Adaptive economics and natural resource policy. Am. J. Agric. Econ. 60:276–283.

El-Zik, K. M., and R. E. Frisbie. 1991. Integrated crop management systems for pest control, p. 3–104. *In* D. Pimental, ed. CRC handbook of pest management in agriculture, Vol III, 2nd ed., Boca Raton FL.

Fleischer, S. J., and W. A. Allen. 1982. Field counting efficiency of sweep-net samples of adult potato leafhopper (Homoptera: Cicadellidae) in alfalfa. J. Econ. Entomol. 75:837–840.

Goodell, P. B., R. E. Plant, T. A. Kerby, J. F. Strand, L. T. Wilson, L. Zelinski, J. A. Young, A. Corbett, R. D. Horrocks, and R. N. Vargas. 1990. CALEX/Cotton: An integrated expert system for cotton production and management. Calif. Agric. 44:18–21.

Hammond, R. B., and L. P. Pedigo. 1982. Determination of yield-loss relationships for two soybean defoliators by using simulated insect-defoliation techniques. J. Econ. Entomol. 75:102–107.

Higgins, R. A. 1985. Approaches to studying interactive stresses caused by insects and weeds, p. 641–649. *In* R. Shibles, ed. Proc. World Soybean Res. Conf. III, Ames IA.

Higgins, R. A., L. P. Pedigo, and D. W. Staniforth. 1984. Effects of velvetleaf competition and defoliation simulating a green cloverworm (Lepidoptera: Noctuidae) outbreak in Iowa on indeterminate soybean yield, yield components, and economic decision levels. Environ. Entomol. 13:917–925.

Higley, L. G., J. A. Browde, and P. M. Higley. 1993. Moving towards new understandings of biotic stress and stress interactions, p. 749–754. *In* D. R. Buxton, R. Shibles, R. A. Forsberg, B. L. Blad, K. H. Asay, G. M. Paulson, and R. F. Wilson (ed.) International crop science I. Crop Science Soc. of America, Madison WI.

Higley, L. G., and W. K. Wintersteen. 1992. A novel approach to environmental risk assessment of pesticides as a basis for incorporating environmental costs into economic injury levels. Am. Entomol. 38:34–39.

Hutchins, S. H., and J. E. Funderburk. 1991. Injury guilds: A practical approach for managing losses to soybean. Agric. Zool. Rev. 4:1–21.

Hutchins, S. H., and P. J. Gehring. 1993. Perspective on the value, regulation, and objective utilization of pest control technology. Am. Entomologist 39:12–15.

Hutchins, S. H., L. G. Higley, and L. P. Pedigo. 1988. Injury equivalency as a basis for developing multiple-species economic injury levels. J. Econ. Entomol. 81:1–8.

Ives, P. M., L. T. Wilson, P. O. Cull, W. A. Palmer, C. Haywood, N. J. Thomson, A. B. Hearn, and A. G. L. Wilson. 1984. Field use of SIRATAC: An Australian computer-based pest management system for cotton. Prot. Ecol. 6:1–21.

Kovach, J., C. Petzoldt, J. Dengi, and J. Tette. 1992. A method to measure the environmental impact of pesticides. NY Food Life Sci. Bull. 139.

Norton, G. A. 1976. Analysis of decision making in crop protection. AgroEcosystems 3:27–44.

Olkowski, W. 1974. A model ecosystem management program. Proc. Tall Timbers Conf. Ecol. Anim. Control Habitat Manage. 5:103–117.

Ostlie, K. R., and L. P. Pedigo. 1985. Soybean response to simulated green cloverworm (Lepidoptera: Noctuidae) defoliation: Progress toward determining comprehensive economic injury levels. J. Econ. Entomol. 78:437–444.

Ostlie, K. R., and L. P. Pedigo. 1987. Incorporating pest survivorship into economic thresholds. Bull. Entomol. Soc. Am. 33:98–102.

Pedigo, L. P., and, G. D. Buntin (ed.). 1994. Handbook of sampling methods for arthropods in agriculture. CRC Press, Boca Raton FL.

Pedigo, L. P., L. G. Higley, and P. M. Davis. 1989. Concepts and advances in economic injury thresholds for soybean entomology, p. 1487–1493. In A. J. Pascale, ed. Proc. World Soybean Res. Conf. IV, Vol. 3. Associación Argentina de la Soja, Buenos Aires, Argentina.

Pedigo, L. P., and L. G. Higley. 1992. The economic injury level concept and environmental quality. Am. Entomologist 38:12–21.

Pedigo, L. P., S. H. Hutchins, and L. G. Higley. 1986. Economic injury levels in theory and practice. Annu. Rev. Entomol. 31:341–368.

Poston, F. L., L. P. Pedigo, and S. M. Welch. 1983. Economic injury levels: Reality and practicality. Bull. Entomol. Soc. Am. 29:49–53.

Potter, D. A., and S. K. Braman. 1991. Ecology and management of turfgrass insects. Annu. Rev. Entomol. 36:383–406.

Raupp, M. J., J. A. Davidson, C. S. Koehler, C. S. Sadof, and K. Reichelderfer. 1987. Decision-making considerations for aesthetic damage caused by pests. Bull. Entomol. Soc. Am. 33:27–32.

Raupp, M. J., C. S. Koehler, and J. A. Davidson. 1992. Advances in implementing integrated pest management for woody landscape plants. Annu. Rev. Entomol. 37:561–585.

Shufran, K. A., and H. G. Raney. 1989. Influence of inter-observer variation on insect scouting observations and management decisions. J. Econ. Entomol. 82: 180–185.

Suszkiw, J. 1994. Cotton advice you can bank on. Agric. Res. 42:18–19.

Szmerda, P. I., M. E. Wetzsten, and R. W. McClendon. 1990. Economic threshold under risk: A case study of soybean production. J. Econ. Entomol. 83:641–646.

Thill, D. C., J. M. Lish, R. H. Callihan, and E. J. Bechinski. 1991. Integrated weed management—a component of integrated pest management: A critical review. Weed. Tech. 5:648–656.

Thorton, P. K., R. H. Fawcett, J. B. Dent, and T. J. Perkins. 1990. Spatial weed distribution and economic thresholds for weed control. Crop Prot. 9:337–342.

Wearing, C. H. 1988. Evaluating the IPM implementation process. Annu. Rev. Entomol. 33:17–38.

Weaver, S. E. 1991. Size-dependent economic thresholds for three broadleaf weed species in soybeans. Weed Tech. 5:674–679.

Wilkerson, G. G., S. A. Modena, and H. D. Coble. 1991. HERB : Decision model for postemergence weed control in soybean. Agron. J. 83:413–417.

Michael D. Duffy

6
Alternative Approaches to Decision Making

This chapter takes a different approach than others in this book. In part, this reflects different outlooks associated with different disciplines, but the subject itself—alternative approaches in decision making—involves a wider set of considerations.

A discussion on alternative approaches to decision making could fill volumes. The focus here is restricted to pest-management decisions, broadened to include integrated farm-management decisions. It is important to note that the emphasis here will be on the variables considered in the decision making: specific methods and techniques will be presented, but mostly only for illustrative purposes.

Before examining alternatives, it is necessary to have a clear understanding of what we are discussing. Although some points may appear to be merely semantic, differences of opinion have great influence on accepting or rejecting an idea. A little history can also help avoid misunderstandings and allow an evaluation of alternatives.

Pests can be thought of as any living species that has a deleterious effect on humans and their activities. Pests can spread diseases, cause economic hardship, or create aesthetic damage. A pest also can be viewed as an organism out of natural balance or an organism that favors the environment created by humans.

"Out of natural balance" implies that the population levels that would exist in equilibrium or a steady state are disrupted. Populations can be greatly reduced or eliminated. They can expand when natural enemies are removed. Also, populations can greatly expand when they are introduced into a new environment.

Pests that develop out of preference for the environment created by humans have been around since the first domestication of crops and animals. A field that is constantly in row crops, planted at a certain density, and maturing at a given rate will favor some plants over others. The favored noncrop plants become pests. Similarly, constantly having the same species especially in enclosed areas can favor development of pests.

Pests, then, can be thought of as agents with a deleterious effect that must be controlled, or as aberrations brought on by some disruption of the natural ecosystem. This is not a trivial distinction. If one views pests only as an evil, management options are limited. On the other hand, from an environmental perspective, there are a large number of options to choose from when managing pests.

This distinction is related to approaches used to deal with pests. Specifically, management and control are not the same thing: control suggests a final solution to a problem, whereas management suggests alternatives that may be different, depending on the circumstances.

Management is a process and the most important key in the process is knowing the goals or objectives. Once the goals are known, the process flows more easily. Calkins and DiPietre (1983) discuss five functions of management: formulating goals, data compilation, planning, implementation, and evaluation. Thinking of management in this way can help guide the evaluation of alternative approaches. Before illustrating and discussing the alternatives, it will be helpful to examine our history and current situation.

IPM is a term coined some 30 years ago. Major emphasis was placed in the late 1960s and early 1970s on finding ways to reduce the potential damage from pesticide use and still remain profitable. IPM is a concept that favors using the best tool available for a particular problem. In some cases it was simply scheduling pesticide use; other applications favored a truly integrated or environmental approach.

Scouting is a key concept to IPM. Scouting information provides the basis for assessing population levels, development progression, and stage, and scouting lets us examine environmental conditions. One of the first people to hire scouts was Dwight Isley in 1925 in Arkansas.

Pest-management options were fairly limited until the last half of the 1800s, then for the next century or so the options were limited to the cultural techniques and insecticides. It was not until after World War II that chemical pesticides became widely available, and for more than just insects.

Laws affecting pest management reflect this development pattern. The first pesticide regulations were aimed at ensuring that the consumer got what was advertised. In the late 1940s, this emphasis shifted to product labeling and ensuring that the materials were safe to use.

Pesticide use was adopted so quickly that in just 40 or 50 years it went from almost nothing to being used on almost every acre of the major crops—corn, soybeans, cotton, and others. Pesticides have evolved and changed over time, and for some decision makers they today are the sole pest-management consideration.

The predominance of pesticide use led to secondary problems: pest resistance, for example, is a problem today. Concerns about pesticides go back almost to the start of the pesticide-use boom. In the mid-1950s, Congress passed the first food safety act that dealt with pesticide residues and presence in food. In the 1960s, Rachel Carson raised the issue of, and concern about, pesticides in the food chain. A current level of concern focuses on long-term implications of the presence of pesticides in rainwater and ground and surface water.

In the next section, we will focus on the factors that influence the pest-management decision; a subsequent section will look at the techniques used to evaluate the decisions today; and a final section discusses the alternatives that are available.

Factors that Influence the Pest-Management Decision

Obviously, a major factor in the pest-management decision is the pest population present. Not so obvious are a myriad of other factors that often are overlooked or not fully understood. Factors such as the life cycle of the pest, the stage of population development present at the time of decision making, the parasites or predators present, weather conditions, and so forth have a bearing on the decision.

As noted, scouting is the means to acquire this information. Good, systematically attained scouting data are the only way to incorporate the microenvironment fully into the decisions.

Another key in the decision making is the tools that are available. As noted earlier, the current situation involves too much reliance on the pesticide alternatives. An extreme version of this situation has pest-management decision makers focusing solely on which chemical to use, ignoring other options.

Pest-management options can be classified loosely into four categories: chemical, biological, cultural, and mechanical. There could be other categorizations and discussions about which technique fits into which category, but the point here is that there are options available, with different costs, efficacies, and results.

Chemical options are fairly straightforward. In some cases many alternatives are available; in other cases the options are limited. Regardless of the number of chemicals available, the choices should be evaluated on their efficacy, cost, toxicity, carryover, impact on nontarget species, and so forth. Too often, the decision is based only on the question: Will it work and how much money is exchanged?

Biological options can be defined very broadly. In general, naturally occurring or augmented parasites and predators, release of sterile males, sex pheromones, and other pest-specific techniques are the biologicals considered. Most biological controls are undertaken on a large scale for use against introduced pests.

Cultural techniques also can be defined broadly. These techniques include such things as cultivator selection, crop rotation, cleanliness, and other management practices. Cultural practices can be both individual-field or large-scale, depending on the practice and application.

Mechanical techniques involve some form of power. These sources have progressed over time from human to animal to fossil fuel. Today fossil fuels are the dominant power source, but all forms are still used.

One can quibble over classifications and what should or should not be included; likewise, there can be debates on the value of such classifications schemes. The point is that options are available—each with their own costs, benefits, and features. Too often today, sometimes even under the auspices of IPM, the only real consideration is at what point to initiate a pesticide treatment. An argument could be made that the concepts of the ET and the EIL are in themselves inherently pesticide-use biased. Have we, in fact, developed IPM or pesticide-use scheduling? This old argument is still relevant: IPM has to consider as many options as possible.

The Individual Farmer

One aspect that IPM strategies do not include, or include only in a limited manner, is the farmer—the decision maker—as an individual. In economics, the theory of the firm generally assumes profit maximization, or some variation. The theories on consumer behavior generally focus on maximizing individual utility or satisfaction. From a profit perspective, an individual could appear to have irrational behavior, whereas from a utility perspective, the behavior could be perfectly rational. There are many examples of this. Why would someone buy a car for $50,000 when all of the exact same features could be found elsewhere for less? Why do we buy different brands? We do so because, once basic requirements are met, other factors such as trust, perception of value, origin, and others come into play. If I only want to buy American, it really does not matter what else is available.

The analogy with the farmer as a firm decision maker and an individual is essential to IPM. A farm has to be profitable or have access to outside capital to stay in business. But, once a certain level of profit is reached, a variety of factors other than profits can enter the decision choice. A farmer may favor a particular pesticide, tractor, or whatever because of the dealer. Farmers as individuals can make decisions based on leisure-time impacts, environmental considerations, size of their farm, longevity of the farm, and so on.

Farmers as individuals also face resource constraints; each has a different bundle of resources available. The resources available can be thought of as the land, labor and capital availability, and managerial skills.

Resource quality and availability can influence the IPM decision. What may be appropriate for one farm may be wholly inappropriate for another. Similarly, as noted, one technique may be better suited to large-scale fields, another to a different field, depending on the resource-ownership patterns.

Profitability

Such recognition of other factors besides profitability is not to downplay the dominance of profits in the IPM decision making. Indeed, many studies by economists have recognized the farmer's dual role but then dismissed it by assuming that profit was the only argument in the utility function.

Alternative Approaches to Decision Making

Assuming profit maximization as the sole individual farm goal is one means to simplify the problem to allow research. But many times, profitability is so narrowly defined that it precludes any IPM decision making: the level of profits is just one factor; variability of the profits is also important in the decision. Several years of low or negative profits with one boom year is not the same as a consistently modest profit level. Moreover, the level of profits from this year's decision can impact next year's profit potential.

Other Dimensions

Economists have been working toward adding risk as an argument of the decision-making function in a variety of ways (Klemme 1985, Rook and Carlson 1985). There is also work showing how to incorporate the time dimension into the IPM decision process (Zacharias and Grube 1986). Such approaches are laudable and must be continued and expanded. One obvious weakness is data availability, especially interaction data.

Externalities

Before turning to currently used techniques and possible alternative approaches, a final factor must be considered. The choice of IPM practice will have nonmarket impacts in the environment, on food safety, on resource depletion, and in other ways. Nonmarket impacts are for items that do not bear a price, e.g., clean air, water quality, aesthetics, and so forth. These impacts can be either positive or negative.

The economists' term for nonmarket goods and services is "externality." Externalities have been a part of economic theory for at least 40 years, and yet many economists and most other disciplines continue to ignore them or treat them as a separate issue.

Consideration of externalities is extremely important in looking at IPM decision making. From a societal perspective, externalities must be considered for efficient allocation of resources. Profit from the macroperspective has to include all the benefits and costs from the IPM decision. Individual decision makers (the microperspective) can exclude the externalities if they choose. However, society can exert influence in the microdecision by altering the market price. Taxes, regulations, and other regulatory devices all influence the market price and the IPM decision.

Theory

Externalities also vary with the characteristics of individuals. The farmer who has a high level of concern about soil erosion, wildlife, condition of soil, and similar factors will arrive at a much different IPM decision than someone who looks only at the short term.

The opening sections of this chapter have tried to illustrate the depth and breadth to which true IPM decision making should go. Too often the approach is one-technique dimensional (pesticides) and looks only at a single-period monetary return to the decision. IPM decision making has to include the farmer's particularities, monetary returns, and externalities. The problem becomes one of what is or is not important or relevant to the decision and the perspective, micro or macro, from which the decision is viewed.

Current Techniques

To add realism to the modeling of the decision process, the tools and mathematical techniques used have been refined. Some of this work has already been mentioned. This section will briefly discuss some of the additional techniques.

Mumford and Norton (1984) described four alternative economic models. These were the economic-threshold, marginal-analysis, decision-theory, and behavioral-decision models.

The *economic-threshold* work is for the most part break-even analysis. Monetary benefits and costs are assigned to different practices at different population levels and times. Addition and subtraction will produce the break-even point. This is the most widely used and simplest procedure, but it requires carefully going into detail in estimating the benefits and costs.

Marginal analysis is simply looking for the optimal solution. What is optimal depends on the objective function. It could be profit maximization, cost minimization, or any number of related objectives. Marginal analysis is limited by a variety of factors: not all of the IPM functions are continuous, there is a variety of data needs, and many factors are assumed.

Mumford and Norton described the use of *decision-making theory* in great detail with numerous examples. This is a Baysian statistical approach, using expected values for various states of the world. These techniques allow inclusion of many factors that would otherwise be held constant.

The fourth model presented briefly by Mumford and Norton was called a *behavioral-decision* model. For reasons much the same as some of those suggested earlier in this chapter, they propose a two-step process involving a static decision model and a dynamic one, noting that farmers' control decisions "reflect their perceptions of the (insect) problem, not necessarily the actual situation" (p. 170). Mumford and Norton go on to say: "Whether the farmer thinks this is good, bad, or satisfactory will depend on his evaluation of this and other outcomes, in terms of his own personal objectives" (p. 170).

Complexities

As noted, many economists have been trying to expand and add more complexity to the model. Zacharias and Grube (1986) used a dynamic programming model to analyze the corn rootworm-soybean cyst nematode decision. The dynamic programming approach allowed not only a multiple period evaluation; it also allowed expansion of the factors considered—for example, rotation. As Zacharias and Grube noted: "Farmers generally do not make pest control decisions independently of other crop production decisions."

Recognition of the interdependence of the decisions was also central to the approach outlined by Farmer et al. (1986). They noted: "Pest management is usually implemented on a pest-by-pest basis, ignoring the probability that the methods used to control each pest can impact the means necessary to control the others." They took this fact and built a theoretical model of the decision-making process based on a series of pest, crop-growth, and other systems models related to potato production.

Babcock et al. (1992) took into account not only the quantity but also the quality dimension in their work on apple production. They used a damage-control approach that allowed "estimation of the distinct effects of pruning on potential." This allowed consideration of more than one control technique.

Among many other innovative approaches to adding realistic complexity to decision making, Klemme (1985) used a stochastic dominance to evaluate the expected value of alternative tillage strategies. Rook and Carlson (1985) examined pest-management decisions from a group-dynamic decision-making process. Such work on new and alternative evaluation techniques must continue. Diligence is necessary to ensure that the added sophistication of the modeling is matched by still more considerations. Some of the new and alternative approaches are the subject of the next section.

Alternative Approaches

The first thing necessary for true IPM decision making is an interdisciplinary approach. In the past, too often only lip service has been paid to this need.

Savory (1988) in his book on resource management used a color metaphor to describe interdisciplinary research. He suggested that each discipline could be seen as a color: say agronomy was blue, entomology red, economics green. The most common interdisciplinary work, he argued, was similar to a rainbow: each profession or discipline was clearly recognizable. Savory argued that instead of a rainbow, interdisciplinary work should be like studying grey; i.e., all of the colors mixed together. Each discipline contributes its share of knowledge to the study of the grey interdisciplinary problems.

Others have tried to capture their interdisciplinary notion in different ways. Dr. Lee Kolmer, former dean of the College of Agriculture at Iowa State University, once said: "Universities have departments and farmers have problems." To receive continued support, IPM decision making has to progress. IPM teams should be leading the way in finding more sophisticated, acceptable models.

The Systems Approach

The concept of IPM focuses on a systems approach. Evaluating the decision by a systems approach allows for more interaction and complexity. Farmer et al. (1986) describe a systems approach to potato IPM, basically looking at combining various process models into a single IPM model.

The systems approach has definite possibilities, but it will need refinement for use in IPM. A major problem with the systems approach is the lack of data. Systems work also does not have the same appeal to many researchers as component research. The primary reason for this is because a systems study will tell you if one system is better than another, but often can't tell you why.

There are other names for the systems approach. Regardless of the

Alternative Approaches to Decision Making

name, the idea is to capture as much of the interaction as possible. Within the systems approach, there is the pest system, the crop system, the human system, and so forth.

As noted above, the systems approach suffers from data limitations, it requires an alternative process, and it is site specific. In spite of its limitation, the systems approach is a more comprehensive means of looking at the IPM decision (Norton and Mumford 1993).

The Expert System

Another alternative approach to the decision making is the expert system (Mumford and Norton 1993). Examples of the use of such a system are not widespread.

An expert system could incorporate the abovementioned systems approach. To be truly an expert system, it would have to involve a multidisciplinary approach. The advantage to an expert system is that the user can quickly eliminate possibilities that are not feasible or acceptable to the farmer or decision maker.

Another approach to IPM, which actually combines the system and expert system approaches, is integrated farm management (IFM). Under this approach, pest-management decisions are a subset of decisions for the whole farm. IFM includes fertility, crop and animal selection, and other farm-management decisions. As noted, too often IPM was on a "pest-by-pest basis," ignoring pest-decision interdependency. IFM takes this argument a step further, saying that all the farm-management decisions are interrelated. The planting of a certain crop will influence the pest spectrum faced. Similarly, a different fertility program may influence pest populations.

The Need for More Data

The approaches discussed as alternatives to IPM decision making are very advanced. The major drawback to them at this time is data availability. Regardless of the approach there are several factors that have not always been considered in IPM. One of the more significant steps forward in IPM will be to continually bring these factors into consideration.

One of the major—often overlooked or assumed—factors is the decision maker. Mumford and Norton (1984) noted the importance of considering the decision maker. Earlier in this chapter, it was noted that the farmer–decision maker assumes many different roles: firm manager, consumer, individual, and environmentalist. Too often in IPM decision frameworks, only one objective, profit maximization, is considered. Profits are a driving force, but health concerns, environmental concerns, and other factors can influence the decision. This means the IPM strategy has to be consistent with goals and objectives of the farmer.

The resource endowment of the farmer is another important consideration. Some IPM decisions focus on trade-offs between capital and labor. For farmers with limited capital, the decision may look quite different than for someone with surplus capital.

Although most IPM decisions are individual, more attention must be paid to the societal and environmental impacts. Rightly or wrongly, some IPM techniques have been negatively stigmatized: food safety, water quality, resource depletion, and so forth are issues for everyone, and they must be considered in making an IPM decision.

Other factors, such as government programs, should be added to the IPM framework. In addition to expanding the variables included, it will be necessary in the future to broaden the techniques used. This will require new data.

IPM has evolved. At each step, new models and techniques are tried to bring the decision-making process closer to the realities. The evolution of IPM must continue, if it is to remain relevant.

There are several directions this evolution could take, and this chapter has identified some of the available options. IPM must be more than a decision framework of whether or not to use a pesticide. More nonchemical management alternatives must be included. There also has to be a more concerted effort to include a truly interdisciplinary approach. This does not mean individuals cannot do IPM research, but it means that at various stages in the research other disciplines should be consulted. This is especially true in formulating the research and interpreting the results.

There has been an increase in the sophistication of the models used. This work should be expanded. IPM is a systems problem and should be dealt with in the larger context of the whole farm. Finally, there has to be more recognition of the role of the individual decision maker. Attitudes, beliefs, resources, constraints, and other factors are relevant in the decision making.

But too much interaction will make estimation almost impossible. As the number of choice variables, interactions, and constraints increase, so too will our ability to capture the true essence of the decision. The problem becomes one of collecting enough data and knowing what is and is not relevant. We must broaden our horizons and abandon the status quo.

References

Babcock, B. A., E. Lichtenberg, and D. Zilberman. 1992. Impact of damage control and quality of output: Estimating pest control effectiveness. Am. J. Agric. Econ. 74:164–172.

Calkins, P. H., and D. D. DiPietre. 1983. Farm business management, Macmillan, New York.

Farmer, G. S., G. B. White, and D. A. Haith. 1986. Systems analysis for integrated management of four major potato pests. Cornell University, A. E. Res. 86–126.

Klemme, R. M. 1985. A stochastic dominance comparison of reduced tillage systems in corn and soybean production under risk. Am. J. Agric. Econ. 67:550–557.

Mumford, J. D., and G. A. Norton. 1984. Economics of decision making in pest management. Annu. Rev. Entomol. 29:157–174.

Mumford, J. D., and G. A. Norton. 1993. Expert systems, p. 167–179. *In* G. A. Norton, and J. D. Mumford (ed.) Decision tools for pest management. CAB International, Oxford UK.

Norton, G. A., and J. D. Mumford (ed.). 1993. Decision tools for pest management. CAB International, Oxford UK.

Rook, S. P., and G. A. Carlson. 1985. Participation in pest management groups. Am. J. Agric. Econ. 67:564–574.

Savory A. 1988. Holistic resource management. Island Press, Washington DC.

Zacharias, T. P., and A. H. Grube. 1986. Integrated pest management strategies for approximately optimal control of corn rootworm and soybean cyst nematode. Am. J. Agric. Econ. 68:704–715.

Part 2: Methods

David A. Mortensen and Harold D. Coble

7
Economic Thresholds for Weed Management

In recent years, U.S. grain producers have been confronted with drastic reductions in profit margins for their crops. The reasons for reduced profitability include the increased strength of the dollar in international markets, leading to reduced exports, increased competition from abroad, and the relatively higher costs associated with crop production in the United States. The decrease in absolute value of grain crops has forced growers to look carefully at ways in which their farm operations can increase production efficiency and net profitability. The realization that maximum yield may not be synonymous with maximum profit has fostered an attitude of openness toward innovative research concepts aimed at increasing the net profit of farm operations.

Weed scientists play an important role in determining ways that weed-management systems could be made more cost effective, thus increasing the net profitability of crop production. Such research includes reducing rates of preemergence and postemergence herbicides, crop rotations, and continued work on crop-management practices to reduce weed problems.

Recently, Cousens et al. (1986), Auld and Tisdell (1986), King et al. (1986), Coble (1985), Baldwin and Stantelman (1982), and McWhorter and Shaw (1982) emphasized the need to integrate the results of weed-interference studies with weed-control-decision models, as well as interactive crop-growth models. The primary goal of such studies is to foster the notion of managing populations by using weed-control decisions that are based on a knowledge of population density and the economics of control. These efforts are of practical significance when designed to aid the farmer in making sound weed-management decisions that increase net profitability and enhance or maintain environmental quality; they are also of academic interest, adding to our understanding of the basic biology of weed-crop interactions.

Review of Weed-Management Strategies

Three approaches may be taken for weed management in crops: eradication, prophylaxis, and remedial or containment practices.

Eradication is the elimination of all weeds of a particular species from a field. In general, such a management strategy is limited to extremely problematic weeds that are limited in distribution. Eradication programs are expensive, unprofitable in the short term, and usually successful only on small areas of land.

Prophylaxis is an insurance strategy that involves some application of a preemergence herbicide to the soil. Because the herbicide is applied before weeds germinate, the weed-infestation potential, both in amount and species, is difficult to assess. Such management approaches may result in excessive preemergence application, reducing environmental quality, and wasting chemicals and money; however, from the perspective of profit maximization, such a strategy is justified when preemergence herbicides are inexpensive, and if yield losses occur in most years (Matthews 1984).

Remedial or containment practices are used to keep a weed population at or below a specific level (Coble 1985, Cousens et al. 1985). Such a strategy involves the acceptance of some population of weeds. It results in weed control based on a known "threshold" weed population. Because decisions are based on a knowledge of the weed populations present, remedial weed-control practices have the potential of being the most cost effective and environmentally sound of the three management strategies just outlined.

In recent years, weed scientists and economists have begun to develop models for determining the cost-effectiveness of available weed-control practices for both prophylactic (King et al. 1986, Schweizer and Zimdahl 1984) and remedial control of weeds that escape preemergence control practices (Cousens et al. 1985, Wilkerson et al. 1987, Marra and Carlson 1983, Bloomberg et al. 1982, and Niemann 1986). Initial results of Schweizer and Zimdahl (1984) with prophylactic weed-management strategies were encouraging, indicating that control decisions based on some knowledge of the presence of weed seed in a field can help to reduce preemergence herbicide rates and frequency of application. While this research approach helped to identify cost-effective preemergence management approaches, practical application will be difficult, because herbicide rates are based on determination of the weed seedbank in the soil. Until fast and accurate methods of seed identification and counting are developed, this research approach will have limited practical application.

The development of effective postemergence herbicides with a high degree of selectivity allows the grower to make remedial treatments to fields that exceed economically damaging levels (Coble 1985). Postemergence or remedial weed-control models are based on the principle that a certain weed population can be tolerated in a field crop as long as the value of the predicted crop loss is at or less than the cost of control. Weed scientists refer to this principle as the economic threshold. The use of *economic threshold* in this way is not entirely consistent with the original use described by Stern et al. (1959). In their work, the break-even point (where costs of yield loss and control are equal) is termed the economic-injury level (EIL); the economic threshold is the pest-population level, based on the EIL, at which management action should be initiated. Because weed populations are essentially constant through a season, the EIL and economic threshold are equivalent. The economic threshold for a particular weed species is a reflection of its interference effects on yield and the cost and efficacy of control treatments.

Uniqueness of Weeds as Pests

Before going further with a discussion of economic threshold theory and application for weed management, it will be useful to outline inherent differences between weeds and other pests. These differences have influenced the development of pest-management theory in weed science as well as the implementation of threshold-based management on the farm. First, in contrast with insects and disease, weed-species diversity in a single field can be very high. It is not uncommon to have 15 to 40 species in a field.

This diversity varies with region. In the United States, diversity is greatest in the Southern states, where we find both New and Old World tropical species as well as temperate species. In Southern states, as many as 75 species may infest soybean, whereas only 20 to 25 species commonly occur in the North. In addition to higher species diversity, higher infestation levels, coupled with more-difficult-to-control weeds, result in higher percent yield reduction in Southern states (Bridges 1993). Because of the greater diversity and densities, the potential for multiple species interaction is also greater.

The probability that some weed species will be present in all fields is high. The ubiquitous nature of weeds is due to the high species diversity and the fact that weed seed and vegetative propagules can persist in soils for long periods. Seed-burial studies, leaving seeds undisturbed, have documented seed longevities of several years to several hundred years (Burnside et al. 1981). Even in studies with high seedbank and seedling mortality, seedbanks have persisted for the duration of the studies—from 6 to 10 years (Burnside et al. 1986).

Because of persistent propagule banks, the temporal and spatial variation in weed presence is smaller than that to be observed for insects. Prediction is facilitated if there is a stationary pest that, once emerged, is easily classified and counted.

Research is underway to characterize the stability of weed spatial patterns (Mortensen et al. 1995). Repeated assessments of a weed population in which the spatial location of the weed is recorded could lead to the development of a comprehensive understanding of infestation level and distribution. Such information could be linked to a number of management-decision aids.

Damage and Period Thresholds

The term *damage threshold* has been used to describe the weed population at which a crop response can first be measured (Coble et al. 1981), which is equivalent to the damage boundary as described by Pedigo et al. (1986). The damage threshold most often is used to identify the first point on the damage function curve that is different from zero, as determined by statistical procedures. The term has little practical meaning outside of academic circles.

Dawson (1986) described the *period threshold* as the time period early in the crop season before any crop losses occur from weed interference. It is important to gain understanding of this critical period, during which remedial control action may be taken to avoid crop loss. For most crops, this period varies from around 2 weeks to 8 weeks after crop emergence. It varies with crop, weed-species assemblage, and environment (Coble et al. 1981). The term period threshold also is used to define the time, early in the crop cycle, during which weed control must be maintained by prophylactic measures to avoid crop loss from weeds that may emerge later in the season (Martin and Field 1988, Oliver 1979). This period is also referred to as the *critical duration*.

Economic and Economic-Optimum Thresholds

Economic return and sustainability are significant concerns of producers. Because both biological and economic effects and costs are considered, an *economic threshold* (ET) offers a method by which profitable and sustainable weed-management decisions can be made. An ET for weeds may be defined as the weed population at which the cost of control is equal to the value of crop yield attributable to that control. Mathematically, the ET for an individual crop is defined as

$$t_E = (Ch + Ca) / (YPLH_E) \qquad [1]$$

where t_E is the ET weed population, Ch is the herbicide cost, Ca is the application cost, Y is the weed-free crop yield, P is the value of the crop per harvested unit, L is the proportional loss per unit weed density, and H_E is the proportional reduction in weed density resulting from the weed control (chemical, mechanical, or biological) treatment (Cousens et al. 1985).

Increases in herbicide or application cost will increase ET weed populations. Conversely, increases in crop yield, value, treatment efficacy, or crop loss per unit of weed density will lower economic thresholds. Three of the factors involved in ET calculations—herbicide cost, application cost, and crop value—can be estimated fairly accurately. Potential crop yield, proportional loss per unit of weed density, and treatment efficacy are more difficult to estimate because of variation in the effects of weather, weed-species composition, weed size, and cropping systems on these variables (Auld and Tisdell 1987, Bauer et al. 1991, Forcella 1990, Legere and Schreiber 1989, Mortensen and Coble 1989).

At present, ET models base in-season economic decisions in the current crop year and do not include costs associated with increases in the weed seedbank resulting from uncontrolled plants. Cousens et al. (1986) applied the term *economic optimum threshold* (EOT) to include the impact of seedbank dynamics on long-term profitability of weed-management decisions. Inclusion of the seedbank cost modifies Eq. [1] to include a proportional cost of the seedbank, H_{SB}:

$$t_E = (Cl + Ca) / (YPLH_E H_{SB}) \qquad [2]$$

The seedbank cost is high for weed species with long-lived seeds, large dispersal distances, and large per-plant seed production.

ETs in Perspective: Where They Fit In

The principle of estimating the pest population prior to some control action is a basic premise or foundation for any IPM program. Therefore, ET theory applies to any field production system in which something less than complete control is accepted as a management philosophy. Research in developing crop-loss functions, seedbank biology, spatial and numerical distribution, and the development of practical sampling methods is actively underway, and several ET computer software applications are in use (Mortensen and Coble 1991). However, it would be misleading to position ET management as a panacea: although it represents a sound management strategy, ET management should be seen as one component of an integrated system.

Postemergence-induced seedling mortality is but one mortality event working on one stage in the life cycle of weedy plants. With no other mortality or plant-fitness-reduction events working on the weed population, weed populations in most fields would exceed the ET, and computer software applications would be reduced to herbicide-recommendation programs. The cropping system must be integrated to enhance mortality of weed seeds during winter; to place weed seeds in sites unsafe for their establishment; and, by crop notation, to make temporal shifts in the occurrence and nature of crop, tillage, and herbicidal mortality.

In addition, management methods that effectively increase the EOT population in a cropping system are needed. Increasing the EOT could be accomplished by enhancing the weed-suppressive ability of the crop (competitive cultivars, planting geometries, allelopathic residue) or by decreasing the fitness of the weed population (sublethal rates of herbicides, mechanical disturbance, biological control agents).

Economic-threshold management is sensible for most weed species, though there will be instances where difficult-to-control weeds, recently established infestations, or herbicide-resistant weed populations will require other strategies.

Estimating Crop Losses

Weed Competition and Weed Interference

Some confusion exists concerning the use of the terms *competition* and *interference*. Competition is best defined as an interaction in which the presence of one plant reduces the availability of a growth-limiting resource to its neighbors (Harper 1983). Work in the area of allelopathy (Rice 1984, Muller 1969) indicates that plant growth can also be limited by the production of phytoinhibitory substances by neighboring plants. Harper (1983) uses the term interference to account for the competitive and allelopathic effects of one plant on another.

Weed scientists have studied the interference effects of many weeds in a variety of crops. Results of studies on weed-density and weed-duration have been reviewed by Zimdahl (1980) and Stoller et al. (1987). Weed-density studies are a form of additive design—a design in which a uniform crop population is grown in the presence of a range of weed densities. The advantage of this design is that it accurately simulates the field condition—a crop at uniform density infested with weeds. However, a problem inherent to the design is that total weed–plus–crop plant population density is not uniform across weed densities (Silvertown 1982). At some low densities of weeds, the weeds do not interfere with one another; at a higher weed density, they begin to interfere with one another. This departure from a linear relationship of weed density is not uniform; hence, it is not possible to distinguish between the effect of high densities of weeds and high plant-population density (i.e., weed plus crop). In order to differentiate between the two effects, the proportion of weeds must be varied in the crop while the total plant density is held constant. Although such research would aid in our understanding of many proposed mathematical relationships between weed density and crop yield (Cousens 1985), for practical reasons the weed scientist is concerned with the effects of low densities of weeds arranged in an additive design. It is at low densities of weeds that the decision about whether or not to invoke weed control must be made.

Dew (1972) was the first to introduce the concept of ranking the competitiveness of agronomic weeds. Although the term *competitive index* was used by Grime (1973) in undisturbed plant communities (to quantify the success of several plant species when grown in competition with one another), it was Dew (1972) who introduced the *index of competition*, a methodology that permits estimation of crop losses due to weeds from the results of weed-density studies. Wilson (1986) termed such competitive rankings *crop equivalents*; Aerts and de Visser (1985) used the term *standard weed units*. In a study of the effect of wild oat density on barley yield, Dew defined the index of competition as

$$b' = b/a \qquad [3]$$

where b is the regression coefficient of weed density on yield, a the weed-free yield, and b' the index of competition. In this work, Dew proposed the square-root model as the best-fit regression equation. Since then, many models have been proposed (Schweizer 1973, Marra and Carlson 1983, Wilson and Cussans 1983, Cousens 1985). Cousens (1985), reviewing the proposals, rejects the sigmoidal (square-root) model as the best-fit relationship between weed density and crop yield for the following reasons: the model approaches an infinite slope at low density of weeds and an infinite upper limit at high density of weeds. He goes on to state that the sigmoidal model makes adequate predictions of weed effects on yield at some poorly defined intermediate weed density only, and makes unreasonable predictions at low and high densities. Keisling et al. (1984) state that some of the confusion over the mathematical relationship between weed density and crop yield arises from the fact that most weed-density studies are conducted at weed populations that far exceed the ET.

Because of this, high densities of weed incorrectly support the sigmoidal model. Cousens (1985) argues that the best density series for estimating weed-density effects is a combination of geometric and arithmetic series, where weed density increases in a geometric progression at lower densities and in an arithmetic progression at higher densities. In an extensive review of weed-density effect studies, Cousens et al. (1984) conclude that the rectangular hyperbolic model is the best-fit model, citing the following reason: at low densities of weed the model suggests one order of competitiveness and at higher densities another. The hyperbolic model proposed by Cousens et al. (1984) requires that some effect of weed competition be measured even at low weed-population densities, thus it rejects the sigmoidal response proposed by Zimdahl (1980). However, most weed-interference experiments reviewed by Cousens (1985) were conducted with particularly troublesome weeds (i.e., weeds that are usually very competitive), and where the lowest density is constrained by rather small plot sizes (12–40 m^{-2}). Almost certainly, at very low population densities, weeds would have no measurable effect and may result in a yield enhancement.

Initially, weed-density studies were used to establish damage thresholds for individual weed species, the damage threshold being the minimum weed density at which a statistically significant yield reduction is detected (Wetala 1976). More recently, results from weed-density studies have been used in ET modeling. The methodology used was outlined by Buchanan (1977). Buchanan points out that while the data obtained with such studies is biologically meaningful, the research approach is limited by high labor costs, large land-area requirements, and large expenditures of research time. These constraints make determination of competitive indices from weed-density studies impracticable for, say, the broad spectrum of weeds infesting a soybean crop.

The Issue of Weed Size. An additional issue under study is determining the most appropriate independent variable for estimating crop-loss functions from a weed infestation. While much of the aforementioned crop-loss research was based on weed density, it is clear that variation in the relationship between weed density and crop loss is in part because of the variation in weed size. Ranges in weed size resulting from a continuum of weed-emergence times, growing conditions, growth characteristics, and fitness of individual biotypes can result in a range of interference effects. Density is a reliable predictor of yield for competitive grain row-crops (Bauer et al. 1991), particularly when weeds emerge with, or shortly after, the crop. However, other weed-infestation parameters appear to explain a greater amount of the variation in crop yield for less competitive crops or crops where the period of weed control is lengthy. Kropff and Lotz (1992), in studying weed effects in sugar beets, found that time of weed emergence and density were the best predictors of crop loss. Gerhards (1993), too, studying winter barley and winter wheat, found that time of weed emergence was an important factor in the resulting crop losses. In both of these studies, weeds emerging over a range of 0 to 5 months had a significant effect on crop yield and quality; therefore, time of weed emergence is agronomically meaningful.

A more direct measure of the influence of density and emergence time is provided by the relative leaf area–yield-loss model. Spitters and Aerts (1983) suggested that a close relationship exists between crop-yield loss and the leaf area of the weeds (relative to crop leaf area) determined shortly after crop emergence. Kropff and Spitters (1991) developed a simple model using relative leaf area for weeds in sugar beets and rice and found that this plant-canopy characteristic improved the ability to predict crop-yield reduction. This method, factoring relative leaf cover, warrants further study.

As an assessment method, relative leaf cover is difficult to measure early in the season, and particularly so at low levels of weed infestation. Harvey and Wagner (1994) and Martin et al. (1993) described a method of visual assessment of weed pressure. From an in-field implementation perspective, this approach is very promising. More research is needed to couple theoretical models with manageable assessment methods that can be adopted by practitioners in the field.

Stability of Crop Loss Functions

For reliable estimates of crop losses from a weed infestation to be made, the relationship between crop-yield loss and weed density should be relatively consistent, or the variation should be quantifiable.

In any given season, more than one factor may be responsible for reductions in crop yield. A need for research exploring numerous factors governing yield was highlighted in the 1986 report on U.S. soybean production and utilization research. The environmental factor most often limiting soybean yields during most seasons is inadequate soil water (Boyer 1982), and even modest soybean water deficits have been shown to restrict plant processes (photosynthesis, leaf expansion, and nitrogen fixation) crucial to high yields. A number of weedy species that are serious pests in soybean fields are relatively efficient in their use of soil water—a reason frequently given for the comparatively greater competitiveness of these species (Black et al. 1969). For example, the water requirement of common cocklebur (*Xanthium strumarium* L.) (415g water/g dry matter) is much lower than that of soybean (646g water/g dry matter), when the plants are grown in a common environment (Shantz and Piemeisel 1927). Water use by weeds, like crops, is a species-specific characteristic. The water requirements of nine species of weeds ranged from 330 to 1,900 grams of water per gram of plant tissue produced (Davis et al. 1965). Large differences in per-unit area transpiration rates have been documented for a number of weed species (Patterson and Flint 1983, Mortensen and Coble 1985). In studying seven weeds and soybean, Patterson and Flint found a large range in physiologic water-use efficiency from a low of 3.62 mg CO_2/g H_2O for soybean to a high of 14.60 mg CO_2/g H_2O for smooth pigweed (*Amaranthus hybridus* L.). The weed-crop growth and physiologic response to drought stress is not known under field conditions.

In a greenhouse study, Wiese and Vandiver (1970) grew corn, sorghum, and eight weed species together in soil boxes. The soils were variously maintained, in a wet, a moderately moist, and a dry condition. They found that species-relative competitiveness was dependent on soil-moisture content. Specifically, they found that common cocklebur competitiveness was reduced in the dry soil. In a field study designed to evaluate water extraction and water-use efficiency of common cocklebur and soybean plants grown under intraspecific and interspecific competition, common cocklebur was able to exploit a greater volume of soil for water than was soybean (Geddes et al. 1979). The amount of water extracted by the mixed stand of soybeans and common cocklebur was intermediate between the values for the two intraspecific treatments and reflected the intense competition for water between these two species. While the results of these two studies are somewhat contradictory, results of other field studies suggest that environment may have a strong influence on the weed interference effects exerted on a crop and thus the competitive index.

In studying threshold levels of jimsonweed (*Datura stramonium* L.) in soybean, Weaver (1986) found that percentage soybean-yield losses attributable to jimsonweed were greater in two years with above-average rainfall than in the one dry year of a three-year study. In a study involving ten site-by-year environments (Bauer et al. 1991), soybean-yield losses resulting from weeds were greatest in environments most favorable for soybean yield. In these studies, confidence intervals were constructed around the crop-loss regression function. The risk of over- or underpredicting the crop loss associated with a weed infestation could be determined using this approach. Hahn (1986) studied the effects of giant foxtail (*Setaria faberi*

Herrm.) density on soybean and found that percentage yield losses from giant foxtail were higher in the year with greater rainfall. A commonality among these studies is that when growing conditions are favorable for the crop, the weeds exert the largest impact on crop yield. It is possible that this enhanced weed effect is largely a result of the fact that the genetic potential for soybean yield is greatest under favorable soil-moisture conditions, and the only growth-limiting effect results from the presence of the weed. When soil moisture is limiting, the potential for soybean yield is suppressed, thus reducing the relative effects of the weed.

Although the results of these studies suggest that weeds have a greater effect in years when moisture is not limited, several studies have found exactly the opposite. Young et al. (1983, 1984) found that quackgrass (*Agropyron repens* (L.) Beauv.) had a reduced effect on soybean and corn yields when irrigated. In both studies, the crops were infested with high populations of quackgrass. Young et al. speculated that quackgrass intercepted infiltrating soil water near the soil surface. Marose (personal communication) found that when moisture stress developed during soybean pod fill, common cocklebur had a greater effect on soybean yield in drier years and sites. The major problem with these studies, in which the results of a wet year are compared with those of a dry year, is that it is difficult to single out water as the limiting factor. It is likely that environmental variables other than water changed as well. In addition, these studies generally have been conducted at unreasonably high densities of weeds—densities well above those tolerated by most growers.

Clearly, in order to predict crop losses associated with weed populations accurately, a significant data base on weed-interference effects must be established. The confidence-interval method outlined by Bauer et al. (1991) is a meaningful approach to describing the variation in crop-loss functions in ET models. Quantification of the extent of this variation can then be used in making management predictions over a range of risk-tolerance strategies. As ET models are developed and implemented, the importance of local data for parameter estimation cannot be overestimated.

Estimating a Mean in a Spatially-Aggregated Population

Economic thresholds are used as decision criteria for weed-control treatments. The crop-loss function defining the relationship between weed density and crop yield is established through studying the influence of weed density, biomass, or projected canopy area on a crop.

A major assumption used in weed-vs.-crop competition experiments is that weeds are homogeneously distributed. In the field, however, weeds occur in multispecies assemblages and are heterogeneously distributed in time and space (Thornton et al. 1990, Van Groendael 1988, Navas 1991, Mortensen et al. 1993a, Johnson et al. 1995a). Bioeconomic models relying on crop-loss functions must address this spatial heterogeneity. It has been shown that, where weeds are spatially aggregated, their impact on crop yield is less than a uniformly distributed population (Auld and Tisdell 1987, Brain and Cousens 1990, Thornton et al. 1990, Wyse et al. 1994). Therefore it is necessary to describe a weed distribution to estimate crop loss accurately. Brain and Cousens (1990) point out that theoretical aspects of threshold modeling have been discussed extensively and that "very little attention has been given to practical aspects such as counting weed populations." Accurate assessment of weed-population density is extremely important: it directly affects the decision to implement a weed-control strategy.

Time and labor constraints often are the limiting factors in weed-population sampling. Marshall (1988) points out that to estimate grass-population densities on a field scale, with a mean close to the threshold density and an error of less that 30%, at least 18 locations ha^{-1} were required. Alternative methods are currently under study. Sequential sampling techniques are reliable and rapid. Sequential sampling offers the possibility of increasing the reliability of the mean estimate while reducing the number of observations in individual fields (Fowler and Lynch 1987). Clearly, more research is needed in this important area.

It is conceivable that future weed sampling will be aided by geographic information systems (GIS), where spatial and temporal information on weed infestations can be managed. Using this approach, sampling histories would be saved and managed for individual fields. For example, data collection coupled with a specific location in the field (using global positioning or radio-frequency triangulation methods) would allow the assessor to collect pest information during planned scouting events, while cultivating, or combining. Numerous agriculturally oriented GIS mapping programs are under development or are available for this purpose.

Weed Seedbanks and Economic-Optimum Thresholds

At present, the economic threshold for a given weed population is based on decisions affecting the current crop. However, because of the tremendous capacity for production of weed seed (Aldrich 1984), a weed population at the ET one year may well affect whether or not the threshold is exceeded the following year (Cousens et al. 1985).

Cousens et al. (1986) constructed a model that simulated the management of wild oats (*Avena fatua* L.) in continuously cropped winter wheat. Weed seed production in winter wheat was studied and the results used to parameterize the model. In the simulation, when ET-based weed management was compared (single-year thresholds) with conventional weed management practices, ET management was more profitable. However, when seed production was taken into account, ET population densities for wild oats were reduced. Cousens et al. (1986) termed a threshold approach that considers weed seed production the *economic-optimum threshold* (EOT). Mortensen et al. (1993b) and Bauer and Mortensen (1992) found similar results for common sunflower (*Helianthus annuus* L.) and velvetleaf (*Abutilon theophrasti* Medik.) in continuously cropped soybean: the EOT was, respectively, 52% and 34% of the ET population. In both instances, seed production and seed longevity in the seedbank was high. Many weed species occurring in annual crops would have lower seed production and considerably shorter longevity in the soil.

In subsequent studies, Mortensen et al. (1993b) found that increasing rotational diversity from continuously cropped soybean to two and three crop-cycle rotations significantly reduced the seedbank cost. In a soybean–winter wheat–corn simulation, the EOT weed population approached the ET population, and likely resulted from the increased diversity of mortality events and temporal variation with respect to when the mortality event occurred. In another study, Lindquist et al. (1995) found seedbank costs reduced when *Verticillium* sp. infected and reduced the fitness of velvetleaf. Here, naturally occurring populations of *Verticillium* sp. decreased seedbank costs and increased the ET and EOT populations of the weed in corn and soybean by reducing velvetleaf fitness.

Although skeptics of the ET management approach argue that weed seed production at low population densities of weeds is unacceptable (Norris 1982), little data is available to support such claims. The skeptics propose the zero-threshold concept (Norris 1985) in which no weed seed production is acceptable. However, Norris has worked exclusively with high-cash-value vegetable crops in intensively managed systems, with noncompetitive crops. In a study evaluating the effect of a number of weed-management practices on decline of weed seed in corn, Burnside et al. (1986) found that even when intensive attempts were made to eliminate seed production over a period of five years (at unreasonably high levels of weed-control input) the weed seedbank in the soil remained very high.

The work of Cousens et al. (1986), Mortensen et al. (1993b), and Lindquist et al. (1995) is encouraging. In cropping systems where the frequency and variation in mortality events is increased and/or where weed fitness is reduced, the seedbank cost of the few species studied was largely eliminated. However, to date this subject has been studied through the use of theoretical simulation models; clearly, researching the long-term costs associated with inputs of weed seed will help identify the true costs associated with ET management. To do this, more research on the effects of low densities of weed on seed inputs and the fate of weed seed in the soil is needed.

Putting Economic Thresholds into Practice

Bioeconomic Models as Delivery Tools

ET-based models integrate an enormous amount of information. In addition to the crop-loss functions described above, information on costs of production and weed-control efficacy are integrated before management decisions are arrived at. In effect, these models offer a biological and economic analysis of the management options available to the farmer; the farmer then chooses how this information is to be used.

In recent years, several postemergence decision-making models based on economics, crop management, and weed biology have been proposed. Cousens et al. (1986) have developed a weed-control decision-making model for wild oat and black grass (*Alopecurus myosuroides* Huds.) control in winter wheat (*Triticum aestivum* L.); Coble (1987) and Wilkerson et al. (1987) have developed a model for broadleaf weed control in soybeans (*Glycine max* (L.) Merr.). In the Coble (1987) and Wilkerson et al. (1987) postemergence model, weeds are ranked in order of competitiveness. This relative competitive ranking has been termed the competitive index (Coble 1985). In the model, competitive-index and weed-population data are used to calculate the competitive load for a particular weed species. The total competitive load is then determined by summing across species. Coble (1985) analyzed the results of many weed-density studies, focusing on the effect of low densities of weed on soybean yield. The study found that the competitive-load units related to crop yield by a factor of 0.5%. A fundamental assumption made in the Wilkerson et al. (1987) model is that, at low densities of weed, the relationship between percentage crop reduction and weed density is linear (Cousens 1985).

In addition to the competitive effects of weeds, weed-management decisions are based on herbicide costs, application expenses, likely weed-free yield, and an estimate of the selling price of the crop to arrive at a "best" management recommendation.

Case Studies of ET Management in Practice

A number of ET-based models are currently publicly distributed (HERB and NEBHERB) and other, more comprehensive, decision aids are in advanced stages of field validation or under development.

HERB, a soybean ET software application, was evaluated in 76 separate trials in North Carolina in 1988 and 1989 (Coble and Mortensen 1992). These trials were conducted with a common format of three treatments: a weed-free check (cultivated and hand weeded), a weedy check (fields with near-threshold weed populations were selected), and a treatment recommended by the software application based on field scouting early in the crop season. Results from the validation trials are presented in Figure 7.1. HERB estimated the crop loss very accurately at low populations of weeds; the model appeared to overestimate crop loss at higher weed populations. Preliminary results with the validation of NEBHERB have found similar results, the model erring on the conservative side, at times recommending herbicide application when treatment was not justified, based on the observed crop loss in untreated plots. Presently, 270 copies of NEBHERB are in distribution to independent crop consultants, the principle user group.

Another threshold model developed for weed management in cereal grains has undergone extensive on-farm testing throughout Germany (Gerowitt and Heitefuss 1990). Performance of the model was assessed in 63

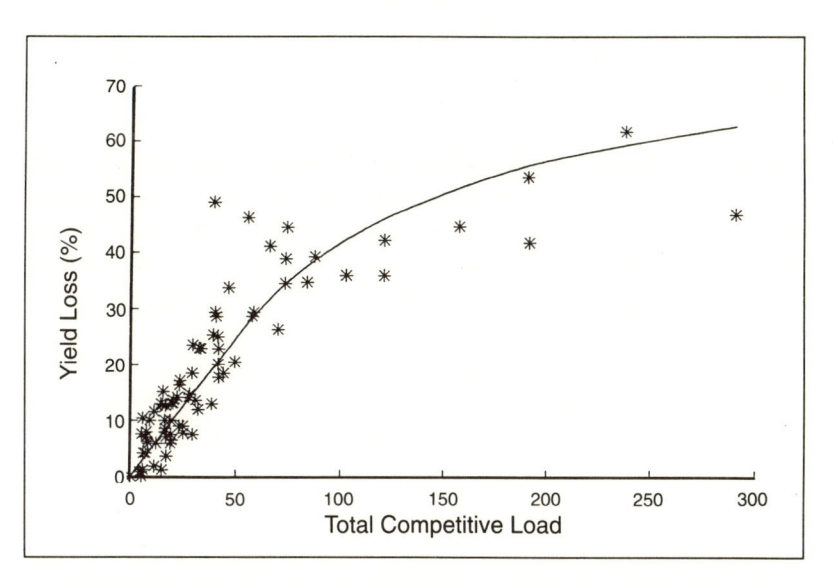

Figure 7.1. Soybean yield loss as a function of total weed-competitive load. Line is prediction for HERB and data points are yield losses calculated from field harvest data. Data are from North Carolina taken in 1988–1989.

validation trials in winter barley, 55 in winter wheat, and 30 in winter rye—148 in all. The utility of the model was assessed by comparing the yield and input costs for plots in which a model recommendation was implemented with test plots in which weeds were left uncontrolled or were maintained weed free. Model predictions were divided into four categories: no control, resulting in negative net return; no control, resulting in positive net return; control with negative net return; and control with positive net return. The following fixed threshold values were used in the validation experiments of 20 to 30 plants m^{-2} for grass weeds, 0.1 to 0.5 *Galium aparaine* plants m^{-2}, 2 *Fallopia convolvulus* plants m^{-2}, and 40 to 50 plants m^{-2} for all other broadleaf plants. It was determined that weed-management decisions based on a fixed threshold were more profitable than prophylactic herbicide application. In addition to the use of the fixed thresholds, an ET model was developed and used. The ET model further improved the net return for the fields surveyed. Consistently, the model and the fixed thresholds erred on the conservative side (i.e., recommending spraying when it was not cost effective). In 8% of the cases, a herbicide treatment was not recommended when it would have been profitable; in 33% of cases, herbicide application was recommended when it was not needed.

The overly conservative nature of these models was probably the result of a combination of factors. First, weed distributions occur as aggregates across fields, with the most distinct aggregation occurring at low infestation of weeds (Johnson et al. 1995b). Marshall (1988) demonstrated that when weeds are nonuniformly distributed, insufficiently sampled fields tend to result in overestimation of infestation. Second, weed fitness is often reduced by preemergent weed control. The influence of sublethal effects of preemergence herbicides (used prior to the postemergent decision) on the crop-loss function is poorly understood. The limited data available suggest that weed-competitive effects can be reduced from 10% to 30% by commonly used preemergence herbicides (Weaver 1991, Schmenk and Kells 1993). Third, to account for the uncertainty associated with weed seedbank costs, crop-loss functions in these models have been parameterized to err on the side of overcontrol.

Much research has centered on the development of bioeconomic models that go beyond postemergent weed-control decisions to comprehensive pre- and postemergent decisions (Swinton and King 1994, Lybecker et al. 1991, Forcella et al. 1993). The Lybecker et al. (1991) bioeconomic model was developed for weed-management decisions in continuous corn production. Through the model, seed germination is simulated and crop-yield loss projected for the resulting weed population. Use of this model has resulted in both a 20% reduction in pesticide use and a shift from pre- to postemergent weed control. It is important to point out that population assessment continues to be an obstacle to practical adoption of this approach. Variation in distribution of weed seed within and between subpopulations coupled with variation in seedling recruitment must be addressed before practical application of such models is realized.

Constraints on Implementing ET Management

To implement these integrated weed-management programs, it is important to recognize who is involved in the decision-making processes: who decides on treatment strategy and approach? Owner-operator farmers have the greatest flexibility in weed-management practices. The risk-management philosophy within this group will significantly influence the control strategies adopted. Risk aversion within this group is a more compelling basis for pest-management decisions than is profit maximization (Reichelderfer 1980). Risk-averse producers—whether they are averse by financial necessity or by choice—will be less likely to adopt ET decision-making criteria.

Land ownership also affects openness toward ET management. A landlord is less likely to adopt ET approaches to weed management but will rather prefer a risk-averse prophylactic approach. Producers perceived as "poor" managers (weedy fields being one criterion) will find it more difficult to lease cropped land. Whether land is privately owned or leased, the crop consultant/ land manager will significantly affect practices adopted by producers. In the Midwestern United States, approximately 30% of grain-crop acres are managed by crop consultants or land managers. The philosophy and logistics access of the consultant/land manager will affect the rate of adoption of new models for integrated weed management; for example, if a consultant can visit a field only once a month, it is unlikely that an ET approach will be feasible.

Social attitudes also influence adoption of threshold-based management. Cousens (1987) stresses that more research in the area of grower perceptions of weed problems is needed. At present, 80% of the soybean growers in North Carolina apply postemergence herbicides to their crops (personal communication, A. C. York). These control decisions are most often based on subjective visual thresholds that are usually well below the ET. Such excessive weed-control measures result from the grower's concerns about risk aversion and the cosmetic appearance of a field, as well as a lack of knowledge of quantitative weed effects on the crop. Agrichemical salespeople also make many recommendations. Adoption of progressive integrated models by this group could result in changes in pest-management practices on vast numbers of cropland acres.

As weed scientists conceptualize and develop weed-management programs, it is important to consider the targeted user. If not, we stand to miss a large percentage of those involved in weed-management decision-making. The advantages of ET management must be made clear. Enhanced information about availability, profitability, and more judicious use of herbicides must be communicated to the user.

Threshold management of weeds has been adopted by a limited number of row-crop producers in the United States. Although the benefits of reduced-spray frequency (reducing cost and herbicide input in the environment) are attractive to some producers using postemergence herbicides—and the expanding body of science on assessment may soon extend to preemergence decisions as well—there are still barriers to its adoption. There are concerns about seeds produced by uncontrolled weeds, the risk of failing to control weeds, spatially variable distributions, cosmetic effects, and collecting the information needed to make decisions. Research addressing these concerns must continue. It is clear that the practice of all-or-nothing weed control must be reconsidered; rather, a weed-management system in which weed mortality arises from a diversity of sources and that reduces weed fitness will be a big step forward.

References

Aerts, H. F. M., and C. L. M. de Visser. 1985. A management information system for weed control in winter wheat, p. 679–686. *In* Proc. 1985 Br. Crop Prot. Conf.–Weeds.

Aldrich, R. J. 1984. Weed-crop ecology. Breton Publishers MA.

Auld, B. A., and C. A. Tisdell. 1986. Economic threshold/critical density models in weed control, p. 261–268. *In* Proc. Eur. Weed Res. Soc. Symp. 1986, Economic Weed Control.

Auld, B. A., and C. A. Tisdell. 1987. Economic thresholds and response to uncertainty in weed control. Agric. Syst. 25:219–227.

Baldwin, F. L., and P. W. Stantelman. 1982. Weed science in integrated pest management. Bioscience 30:675–678.

Bauer, T. A., and D. A. Mortensen. 1992. A comparison of economic and economic optimum thresholds for two annual weeds in soybeans. Weed Tech. 6:228–235.

Bauer, T. A., D. A. Mortensen, G. A. Wicks, T. A. Hayden, and A. R. Martin. 1991. Environmental variability associated with economic thresholds for soybeans. Weed Sci. 39:564–569.

Black, C. C., T. M. Chen, and R. H. Brown. 1969. Biochemical basis for plant competition. Weed Sci. 17:338–344.

Bloomberg, J. R., B. L. Kirkpatrick, and L. M. Wax. 1982. Competition of common cocklebur *(Xanthium pensylvanicum)* with soybean *(Glycine max)*. Weed Sci. 30:507–513.

Boyer, J. S. 1982. Plant productivity and environment. Science 218:433–448.

Brain, P., and R. Cousens. 1990. The effect of weed distributions on prediction of yield loss. J. App. Ecol. 27:735–742.

Bridges, D. C. (ed.). 1993. Crop losses due to weeds in the United States. Weed Sci. Soc. Am., Champaign IL.

Buchanan, G. A. 1977. Weed biology and competition, p. 25–41. *In* B. Truelove (ed.) Research methods in science, 2nd edition. So. Weed Sci. Soc.

Burnside, O. C., C. R. Fenster, L. L. Evetts, and R. F. Mumm. 1981. Germination of exhumed weed seed in Nebraska. Weed Sci. 29:577–586.

Burnside, O. C., R. G. Wilson, G. A. Wicks, F. W. Roeth, and R. S. Moomaw. 1986. Weed seed decline and buildup in soils under various corn management systems across Nebraska. Agron. J. 78:451–453.

Coble, H. D. 1985. The development and implementation of economic thresholds for soybeans, p. 295–307. *In* R. E. Frisbie and P. L. Adkisson (ed.) Integrated pest management on major agricultural systems. Texas A&M University, College Station TX.

Coble, H. D. 1987. Using economic thresholds for weeds in soybeans. Abstr. Weed Sci. Soc. Am. 27:94.

Coble, H. D., and D. A. Mortensen. 1992. The threshold concept and its application to weed sciences. Weed Tech. 6:191–195.

Coble, H. D., F. M. Williams, and R. L. Ritter. 1981. Common ragweed *(Ambrosia artemisiifolia)* interference in soybeans *(Glycine max)*. Weed Sci. 29:339–342.

Cousens, R. 1985. A simple model relating yield loss to weed density. Ann. Appl. Biol. 107:239–252.

Cousens, R. 1987. Theory and reality of weed control thresholds. Conf. Australian Weed Sci. Soc., Feb. 1987. Plant Prot. Q. 2:13–20.

Cousens, R., C. J. Doyle, B. J. Wilson, and G. W. Cussans. 1986. Modelling the economics of controlling *Avena fatua* in winter wheat. Pestic. Sci. 17:1–12.

Cousens, R., N. C. B. Peters, and C. J. Marshall. 1984. Models of yield loss–weed density relationships, p. 367–374. *In* Proc. 7th International Colloquium on Weed Biology, Ecology and Systematics, Columa-EWRS, Paris.

Cousens, R., B. J. Wilson, and G. W. Cussans. 1985. To spray or not to spray: The theory behind the practice, p. 671–678. *In* Proc. 1985 Br. Crop Protection Conf.–Weeds.

Davis, R. G., A. F. Wiese, and J. L. Pafford. 1965. Root moisture extraction profiles of various weeds. Weeds 13:98–100.

Dawson, J. H. 1986. The concept of period thresholds, p. 327–331. *In* Proc. Eur. Weed Res. Soc. Symp. 1986, Economic Weed Control.

Dew, D. A. 1972. An index of competition for estimating crop loss due to weeds. Can. J. Plant Sci. 52:921–927.

Forcella, F. 1990. Breeding soybeans tolerant to weed competition. Abstr. Weed Sci. Soc. Am. 30:51.

Forcella, F., D. D. Buhler, S. M. Swinton, R. P. King, J. L. Gunsolus, and B. D.

Maxwell. 1993. Field evaluation of a bioeconomic weed management model for the Corn Belt, USA, p. 755–760. *In* Quantitative approaches in weed and herbicide research and their practical application, 8th Eur. Weed Res. Symp., Braunschweig, Germany.

Fowler, G. W., and A. M. Lynch. 1987. Sampling plans in insect pest management based on Wald's sequential probability ratio test. Environ. Entomol. 16:345–354.

Geddes, R. D., H. D. Scott, and L. R. Oliver. 1979. Growth and water use by common cocklebur *(Xanthium pensylvanicum)* and soybeans *(Glycine max)* under field conditions. Weed Sci. 27:206–212.

Gerhards, R. 1993. Alternative verfahren der unkrautkontrolle in wintergetreide mit hilfe digitaler bildverarbeitung. Ph.D. Diss., Rheinischen Freidrich-Wilhelms-Universitat, Bonn, Germany.

Gerowitt, B., and R. Heitefuss. 1990. Weed economic thresholds in cereals in the Federal Republic of Germany. J. Crop Prot. 9:323–331.

Grime, J. P. 1973. Competitive exclusion in herbaceous vegetation. Nature 242: 344–347.

Hahn, K. 1986. Effect of time of DPX-Y6202 application and giant foxtail *(Setaria faberi)* density on soybean *(Glycine max)* yield. M.S. Thesis, University of Illinois, Urbana IL.

Harper, J. L. 1983. Population biology of plants. Academic Press, New York.

Harvey, R. G., and C. R. Wagner. 1994. Using estimates of weed pressure to establish crop yield loss equations. Weed Tech. 8:114–118.

Johnson, G. A., D. A. Mortensen, and A. R. Martin. 1995a. A simulation of herbicide use based on weed spatial distribution. Weed Res. 35:197–205.

Johnson, G. A., D. A. Mortensen, L. J. Young, and A. R. Martin. 1995b. The stability of weed seedling population models and parameters in eastern Nebraska corn *(Zea mays)* and soybean *(Glycine max)* fields. Weed Sci. 43:604–611.

Keisling, T. C., L. R. Oliver, R. H. Crowley, and F. L. Baldwin. 1984. Potential use of response surface analyses for weed management in soybeans *(Glycine max)*. Weed Sci. 32:552–557.

King, R. P., D. W. Lybecker, E. E. Schweizer, and R. L. Zimdahl. 1986. Bioeconomic modeling to simulate weed control strategies for continuous corn *(Zea mays)*. Weed Sci. 34:972–979.

Kropff, M. J., and L. A. P. Lotz. 1992. Systems approaches to quantify crop-weed interactions and their application in weed management. Agric. Syst. 40:265–282.

Kropff, M. J., and C. J. T. Spitters. 1991. A simple model of crop loss by weed competition from early observations on relative leaf area of the weeds. Weed Res. 31:97–105.

Legere, A., and M. M. Schreiber. 1989. Competition and canopy architecture as affected by soybean *(Glycine max)* row width and density of redroot pigweed *(Amaranthus retroflexus)*. Weed Sci. 37:84–92.

Lindquist, J. L., B. D. Maxwell, D. D. Buhler, and J. L. Gunsolus. 1995. Modeling the population dynamics and economics of velvetleaf *(Abutilon theophrasti)* control in a corn *(Zea mays)*-soybean *(Glycine max)* rotation. Weed Sci. 43:269–275.

Lybecker, D. W., E. E. Schweizer, and R. P. King. 1991. Weed management decisions in corn based on bioeconomic modelling. Weed Sci. 39:124–129.

Marose, B. H. 1987. Personal communication. Integrated Pest Management Extension Specialist, Dept. of Crop Science, University of Maryland.

Marra, M. C., and G. A. Carlson. 1983. An economic threshold model for weeds in soybeans *(Glycine max)*. Weed Sci. 31:604–609.

Marshall, E. J. P. 1988. Field-scale estimates of grass populations in arable land. Weed. Res. 28:191–198.

Martin, A. R., D. A. Mortensen, and G. E. Meyer. 1993. Visual and photographic assessment of herbicide efficacy trials for use in bioeconomic modeling. Abst. Weed Sci. Soc. Am. 33:53.

Martin, M. P. L. D., and R. J. Field. 1988. Influence of time of emergence of wild oat on competition with wheat. Weed Res. 28:111–116.

Matthews, G. A. 1984. Pest management. Longman, London.

McWhorter, C. G., and W. C. Shaw. 1982. Research needs for integrated weed management systems. Weed Sci. 30 (Suppl. 1):40–45.

Mortensen, D. A., and H. D. Coble. 1985. Water use and leaf area development of weeds in soybeans. Proc. South. Weed Sci. Soc. 38:400.

Mortensen, D. A., and H. D. Coble. 1989. The influence of soil water content on common cocklebur *(Xanthium strumarium)* interference in soybeans *(Glycine max)*. Weed Sci. 37:76–83.

Mortensen, D. A., and H. D. Coble. 1991. Two approaches to weed control decision-aid software. Weed Tech. 5:445–452.

Mortensen, D. A., G. A. Johnson, D. Y. Wyse, and A. R. Martin. 1995. Managing spatially variable weed populations, p. 397–415. *In* P. C. Robert and R. H. Rust (ed.) Site specific crop management. Agron. Soc. Am. Press, Madison WI.

Mortensen, D. A., G. A. Johnson, and L. J. Young. 1993a. Weed distribution in agricultural fields. *In* P. C. Robert, R. H. Rust, and W. E. Larson (ed.) Soil specific crop management. Agron. Soc. Am. Press, Madison WI.

Mortensen, D. A., A. R. Martin, T. E. Harvill, and T. A. Bauer. 1993b. The influence of rotational diversity on economic optimum thresholds in soybean, p. 815–823. *In* Proc. Eur. Weed Res. Soc. Symp. Braunschweig.

Muller, C. H. 1969. Allelopathy as a factor in ecological processes. Vegetation, Haag 18:348–357.

Navas, M. L. 1991. Using plant population biology in weed research: A strategy to improve weed management. Weed Res. 31:171–179.

Niemann, P. 1986. Mehrjahrige anwendung des schadensschellen—prinzeps bei der unkrautbekampfung auf einem landwirtschaftlichen betrieb, p. 385–392. *In* Proc. Eur. Weed Res. Soc. Symp. 1986, Economic Weed Control.

Norris, R. F. 1982. Interactions between weeds and other pests in the agroecosystem, p. 343–406. *In* J. L. Hatfield and I. J. Thomason (ed.) Biometeorology in integrated pest management. Academic Press, New York.

Norris, R. F. 1985. Weed population dynamics and the concept of zero thresholds. Abst. Weed Sci. Soc. Amer. 25:58.

Oliver, L. R. 1979. Influence of soybean *(Glycine max)* planting date on velvetleaf *(Abutilon theophrasti)* competition. Weed Sci. 27:183–188.

Patterson, D. T., and E. P. Flint. 1983. Comparative water relations, photosynthesis, and growth of soybean *(Glycine max)* and seven associated weeds. Weed Sci. 31:318–323.

Pedigo, L. P., S. H. Hutchins, and L. G. Higley. 1986. Economic injury levels in theory and practice. Annu. Rev. Entomol. 31:341–368.

Rice, E. L. 1984. Allelopathy, 2nd edition. Academic Press, New York.

Reichelderfer, K. H. 1980. Economics of integrated pest management: Discussion. Am. J. Agric. Econ. 62:1012–1013.

Schmenk, R. E., and J. J. Kells. 1993. Velvetleaf *(Abutilon theophrasti* (Medik.)) competitiveness in corn as influenced by preemergence herbicides. Abstr. Weed Sci. Soc. Amer. 34:41.

Schweizer, E. E. 1973. Predicting sugarbeet root losses based on kochia densities. Weed Sci. 19:125–128.

Schweizer, R. E., and R. L. Zimdahl. 1984. Weed seed decline in irrigated soil after six years of continuous corn *(Zea mays)* and herbicides. Weed Sci. 32:76–83.

Shantz, H. L., and L. N. Piemeisel. 1927. The water requirements of plants at Akron, Colorado. J. Agric. Res. 34:1093–1189.

Silvertown, J. W. 1982. Introduction to plant population ecology. Longman, London.

Spitters, C. J. T., and R. Aerts. 1983. Simulation of competition for light and water in crop-weed associations. Aspects Appl. Biol. 4:467–484.

Stern, V. M., R. F. Smith, R. van den Bosch, and K. S. Hagen. 1959. The integrated control concept. Hilgardia 29:81–101.

Stoller, E. W., S. K. Harrison, L. M. Wax, E. E. Regnier, and E. D. Nafziger. 1987. Weed interference in soybeans *(Glycine max)*. Rev. Weed Sci. 3:155–182.

Swinton, S. M., and R. P. King. 1994. A bioeconomic model for weed management in corn and soybean. Agric. Syst. 44:313–335.

Thornton, P. K., R. H. Fawcett, J. B. Dent, and T. J. Perkins. 1990. Spatial weed distribution and economic thresholds for weed control. Crop Prot. 9:337–342.

Van Groendael, J. M. 1988. Patchy distribution of weeds and some implication for modeling population dynamics: A short literature review. Weed Res. 28:437–441.

Weaver, S. E. 1986. Factors affecting threshold levels and seed production of jimsonweed (*Datura stramonium* L.) in soybeans (*Glycine max* [L.] Merr.). Weed Res. 26:215–223.

Weaver, S. E. 1991. Size dependent economic thresholds for three broadleaf weed species in soybeans. Weed Tech. 5:674–679.

Wetala, M. P. E. 1976. The relationship between weed and soybean yields, p. 156–168. *In* Proc. 6th E. Afric. Weed Sci. Conf.

Wiese, A. F., and C. W. Vandiver. 1970. Soil moisture effects on competitive ability of weeds. Weed Sci. 18:518–519.

Wilkerson, G. G., H. D. Coble, and S. A. Modena. 1987. A postemergence herbicide decision model for soybeans. Abstr. Weed Sci. Soc. Am. 27:95.

Wilson, B. J. 1986. Yield responses of winter cereals to the control of broad-leaved weeds, p. 75–82. *In* Proc. Eur. Weed Res. Soc. Symp. 1986, Economic Weed Control.

Wilson, B. J., and G. W. Cussans. 1983. The effect of weeds on yield and quality of winter cereals in the UK, p. 121. *In* 10th Int. Cong. Plant Prot.

Wyse, D. Y., D. A. Mortensen, and T. E. Harvill. 1994. The influence of spatially variable weed populations on weed management decisions. Abstr. N. Cent. Weed Sci. Soc. 49.

Young, F. L., D. L. Wyse, and R. J. Jones. 1983. Effect of irrigation on quackgrass *(Agropyron repens)* interference in soybeans *(Glycine max)*. Weed Sci. 31:720–727.

Young, F. L., D. L. Wyse, and R. J. Jones. 1984. Quackgrass *(Agropyron repens)* interference on corn *(Zea mays)*. Weed Sci. 32:226–234.

Zimdahl, R. L. 1980. Weed-crop competition: A review. Int. Plant Prot. Center, Oregon State University, Corvallis.

Paul A. Backman and James C. Jacobi

8
Thresholds for Plant-Disease Management

The area of plant diseases is quite different from that of insects in the way that economic thresholds are determined. In the case of plant disease, typically the pest population cannot easily be quantified, because the infective propagules are microscopic and they occur in large numbers, of which only a fraction are ultimately infective. Because of this, predictive (rather than therapeutic) systems are often employed which rely on knowledge of severity in the previous crop or estimates of the conduciveness of the environment for the establishment of disease.

As mentioned in Chapter 2, two factors—the need for preventive management practices and difficulties in assessing pest impact—limit and alter the use of thresholds for plant-disease management. Even in the case of plant-parasitic nematodes, which have received considerable research attention regarding population assessment techniques and yield-loss relationships, information that would help in the practical use of ETs is limited (Barker and Noe 1987). Nevertheless, developing ETs is an active research direction in nematology; Barker and Noe (1987) review work in this area. For other types of plant pathogens, difficulties in assessment and management options continue to constrain the usefulness of thresholds. The making of decisions on where thresholds can be useful is very much a function of the disease-management strategy and the specifics of a given pathogen-host system.

Disease-management strategies consist of a reduction in the amount of primary inoculum or a reduction in the rate by which the inoculum can increase itself over time—or a combination of these approaches. Typically, soil-borne diseases are monocyclic: disease severity at the end of the season is directly related to the amount of inoculum present at the beginning of the season. Conversely, foliage diseases are generally polycyclic (i.e., they have more than one cycle of inoculum production during a season), and strategies relating to the suppression of secondary inoculum often must be implemented.

In this review, we have chosen to provide examples of host-pathogen systems in which thresholds have been developed and discuss how they work and the problems associated with them.

Management of Initial Inoculum or Initial Disease

Soil-borne diseases particularly, and a few foliage diseases, reach end-of-season disease severities in direct proportion to the amount of primary inoculum present in the field at the beginning of the cropping season. Numerous strategies have evolved for management of these populations of primary inoculum, with the primary focus being to reduce the numbers of infective propagules present at the field site before the season begins. If this is successful, midseason strategies are often unnecessary.

Tactics used for reduction of primary inoculum include

vertical resistance, which renders large portions of the inoculum noninfective on the cultivar selected;

rotation or fallowing, which depletes the populations of infective propagules during nonhost periods;

deep plowing, which buries the inoculum, rendering it more susceptible to microbial or chemical breakdown, and often separates the inoculum from its primary infection site;

clean seed or transplants, which limits primary inoculum;

flooding, which depletes oxygen, thereby killing inoculum;

chemical pesticides that reduce the infective propagules by direct toxicity.

Before any of the preceding tactics are used, it must be determined if there is a need for management. For the stem and root diseases caused by *Sclerotium rolfsii,* several strategies for doing this have evolved. The simplest of these is to determine numbers of infected plants in the previous cropping season. Rodríguez-Kábana et al. (1975) counted numbers of southern stem rot disease loci in peanut fields at several locations randomly distributed around the field. If average numbers exceeded a threshold, then chemical control strategies would be economical for the next crop. Similarly, and also in peanut, Bailey and Maytac (1989) developed a decision model for use of the fumigant metam sodium for control of Cylindrocladium black rot. This model uses disease severity in the previous crop of peanut, regardless of how long it preceded the present crop (because microsclerotia of *Cylindrocladium crotalariae* are very long lived). The other input into the model is the field yield in that crop to predict the dollar value of growing resistant-fumigated, resistant-nonfumigated, or susceptible peanut cultivars.

Direct measurements of infective propagules in the soil also can be useful for many plant pathogens. Leach and Davey (1938) developed a wet-sieving technique for estimating total sclerotia (live plus dead) of *S. rolfsii* in soil intended for planting of sugar beet, so that severely infested fields could be rotated and moderate fields could be managed with modified cultural practices. This forecasting method, coupled with the control practices, greatly reduced losses from *S. rolfsii* on sugar beets grown in California's Central Valley. However, the methods were labor intensive; they also did not indicate what percentage of the sclerotia were viable. A later paper by Backman et al. (1981) presented a method for monitoring viable sclerotia in sugar beets with fewer labor requirements: here, samples of soil removed from farm fields or from soil in the bottom of trucks (tare soil) delivering sugar beets for processing were used for analysis. These samples were air dried, screened to remove large soil clumps and stones, spread in a thin layer, wetted with 0.1% methanol solution, incubated at 30°C, and after 48 hours the number of white colonies of *S. rolfsii* were counted. Viable sclerotia populations from each of the two methods of obtaining soil were regressed against disease severity with high correlations. With this information, the processing company receiving the sugar beet could advise farmers on when to rotate to nonhost crops and which field could be planted to sugar beets with least damage from the disease. The methanol stimulation method was not sensitive enough to pick up the low populations of *S. rolfsii* sclerotia required to cause economic loss in peanuts (P. A. Backman, unpublished).

DeVay et al. (1974) used semiselective agar media for estimation of *Verticillium dahliae* microsclerotia in soil samples taken from cotton fields. In this technique, soil samples are finely ground, then weighed aliquots are distributed in small "micropiles" (400 small piles created with an Anderson sampler) of soil on the surface of an agar selective medium contained in a petri dish. The number of piles that produce growth of *V. dahliae* is directly related to the potential of the soil to produce disease.

The previous methods provide accurate procedures for predicting the potential of a field to produce disease, but, frequently, predictions based solely on counts of propagule numbers are inaccurate because of overriding environmental effects. Recently, a method for predicting severity of Sclerotinia blight of peanut caused by *Sclerotinia minor* was developed that used measurements of the conduciveness of the subcanopy environment to promote disease (Bailey et al. 1993). The environmental thresholds developed minimized application of fungicides for disease control during environments that were not supportive of disease development.

Virus-disease severity also can be forecast based on presence of primary inoculum. Hull (1958) and Watson et al. (1975) developed systems for predicting severity of beet yellows virus and beet mild yellows virus that relied on three factors: the number of aphid vectors present in the fall; the numbers of winter frost days; and the mean April temperature. Insecticides could then be applied when thresholds were reached or when secondary influxes of aphids occurred. Similarly, Stewart's wilt of corn caused by *Erwinia stewartii* could be predicted by summing the mean temperature for the months of December, January, and February (Stevens 1934). If the means totaled more than 100°F, then severe disease could develop, but if the sum was less than 90°F, disease would be unlikely to develop.

Management of Secondary Inoculum

Disease thresholds or forecasts are used to predict disease outbreaks and resulting yield losses of a wide variety of pathogens with repeating or secondary cycles of infection and disease development (Madden and Ellis 1988). The polycyclic nature of these diseases commonly requires that thresholds be used repeatedly during a growing season. A variety of methods have been used to develop economic thresholds designed to initiate management practices. They can be grouped into systems that have thresholds based on measuring disease severity, pathogen population, environmental conditions that favor pathogen increase, or a combination of these.

Thresholds Based on Direct Assessment of Pathogen Populations

In the literature, there are several examples of foliar diseases in which thresholds are based on measurement of pathogen populations (Berger 1969, Burleigh et al. 1969, Roelfs et al. 1970). Most examples measure the aerial densities of fungal spores. Spore collection may involve methods ranging from adhesive-coated glass slides to expensive mechanical spore-sampling devices. There is potential for two problems in the use of spore densities as ETs. First, the pathogen requires a favorable environment, which may or may not be present. Second, in some instances it may be difficult to interpret spore-trapping data where spores of similar fungi are present (Franc et al. 1988).

Systems that combine spore-trap information with environmental parameters have been used to overcome the potential for some of these problems. Berger (1969) developed a system to schedule fungicide sprays on celery to control *Cercospora apii* based on a combination of spore count and weather conditions favorable for disease development. The system was designed to identify periods favorable for pathogen development, allowing growers to modify their spraying schedule as needed. By monitoring the aerial spore load and weather, a grower could either reduce or increase the frequency of fungicide sprays: during periods of high spore loads and favorable environment, the grower may have to spray as many as seven times a week to maintain adequate disease control. Shaw and Royle (1987) developed a two-stage system to time fungicide applications for control of Septoria leaf blotch on wheat caused by *Septoria tritici*. In the first stage, spore numbers from traps are assessed. If the count is greater than the threshold amount, then the second stage, a reading on upward rain-splash, is taken. Upward rain-splash (or the "splashiness" of the rain) is measured with a device that detects the upward movement of an aqueous dye. When rainsplash has been recorded a fungicide application is needed. The system assumes that adequate leaf wetness for infection by *S. tritici* will be provided by rain that causes rain-splash.

Bacterial plant pathogen populations have traditionally been difficult and time consuming to monitor. Laborious methods have been used. Lindermann et al. (1984) developed a model to predict the incidence of bacterial brown spot of snap bean caused by *Pseudomonas syringae* pv. *syringae*. In each of two years, disease symptoms did not appear until epiphytic populations of *P. syringae* pv. *syringae* were > 10^4 colony forming units on individual leaves. Although the model was very effective in predicting disease incidence one week after full flower, practical application of the model is limited by the time-consuming dilution plating methods required to estimate bacterial populations. Recently, a rapid and reliable method that uses a tube nucleation test was developed to determine, indirectly, population sizes of the ice-nucleation-active bacteria (Hirano et al. 1987). They showed that the proportion of leaves frozen at temperatures between −2.0 to 2.5°C was highly correlated with the incidence of bacterial brown spot three and eight days later.

Thresholds Based on Indirect Assessment of Pathogen Populations

In many instances, a direct assessment of the pathogen population is not necessary, or not possible because of a lack of spore sampling or unavailability of equipment. Thresholds based on indirect assessment methods are particularly suited to quantifying disease problems in developing countries.

Rice blast caused by *Pyricularia oryzae* is a serious production constraint in many rice-growing areas of the world. Surin et al. (1992) reported the use of trap plants to monitor both the pathogen population and weather conditions in order to time fungicide applications more accurately. On several occasions during the growing season, boxes of rice trap plants, seven-day-old rice seedlings of the highly susceptible cultivar RD23, were placed in rice fields. These seedling boxes were left in growers' fields for a two- to three-day exposure period and then returned to the greenhouse. After five days, disease severity and spore production of *P. oryzae* were determined. High levels of disease in the trap crop indicated that systemic fungicide sprays were needed to prevent rapid increase in rice-blast severity. Spray advisories based on the trap-crop information allowed growers to reduce losses to the disease.

In Germany, a similar system (PHYTOPROG) is used to provide potato late blight warnings (Zadoks 1984). The system uses highly susceptible pilot plots of potatoes ("late blight gardens") to provide disease warnings.

Under this system, as long as the pilot plots remain free of disease, no fungicide treatments are recommended.

Thresholds Based on Meteorological Variables

Disease thresholds employing meteorological variables are used as a device for timing management decisions. Weather-based systems—in which decisions are based on how favorable the environment is for disease development—have been widely used to make fungicide applications more timely and to reduce unnecessary applications. Their acceptance has been best for relatively high-value crops—for example, apple, potato, and peanut (Backman et al. 1991, Cu and Phipps 1993, Jensen and Boyle 1966, Jones et al. 1984, Krause et al. 1975, Phipps and Powell 1984). These systems can be empirical; i.e., based on observation and analysis of current and/or historical weather data (Madden and Ellis 1988). The disease forecast may be based on weather information from either a region or a single field. As size and climatic variability of a region increase, the accuracy of the forecasts may decline, but overall, both regional and local forecasts have provided good information. Forecasts based on regional meteorological information have been used successfully for pathosystems of stripe rust in winter wheat (Coakley et al. 1982) and early leaf spot in peanuts (Jensen and Boyle 1966, Parvin et al. 1974, Phipps and Powell 1984).

With most polycyclic foliar diseases, a successful disease forecasting system must time not only one but several fungicide applications during the season. One of the most successful disease forecasting systems developed to date is BLITECAST (Krause et al. 1975, MacKensie 1981). Fungicide spray recommendations are made based on the accumulation of total weekly severity values and rainfall. Potato growers using the system record their own weather information (rainfall, temperature, relative humidity) and then call a BLITECAST computer to receive a recommendation. Several alternate delivery systems have been developed, including hand-held programmable calculators and on-site, microcomputer-based weather stations that collect and interpret data and provide forecasts to growers (MacKensie 1981).

Early leaf spot *(Cercospora arachidicola)* and late leaf spot *(Cercosporidium personatum)* are widespread and potentially damaging diseases wherever peanuts are grown (Porter et al. 1984). Peanut growers rely heavily on fungicides applied on a calendar (fixed) schedule for control of these diseases (Smith and Littrell 1980). However, the use of fixed schedules results in unnecessary fungicide applications during periods unfavorable for pathogen development. Jensen and Boyle (1966) developed a technique to schedule fungicide applications based on whether weather conditions are favorable for pathogen development. Another system, called the AU-PNUTS Advisory, uses a combination of recorded rainfall and National Weather Service five-day precipitation probabilities to schedule fungicide applications (Backman et al. 1991, Davis et al. 1993). The use of precipitation forecasts decreases the likelihood that a grower will be kept from making a fungicide application because of inability to get spray equipment into a wet or rainy field. Accurate predictions are particularly important when using protectant fungicides.

Backman et al. (1984) developed a meteorological timing system to control midseason foliage and stem diseases (frogeye leaf spot, brown spot, and anthracnose) of soybean. It is based on a required yield expectation and meteorological records of the number of days with measurable rain, extended periods of fog, or heavy dew. When bloom begins, if such events are recorded on two or more days within a four-day period and two more are expected within the next five days, an application of a systemic fungicide (benomyl) is recommended. Once an application is made, records need not be kept for the next eight days. After this period, a second application may be made if such events occur again on three days during a four-day period and are predicted for two of the next five days. In the same manner, a third application may also be made. This technique resulted in an average of 1.3 applications per crop-season, with significant economic improvements compared with the calendar-application method previously employed. Similar approaches, using point systems to evaluate the conduciveness of the environment with regard to disease development, also have been developed for soybeans (Jacobsen et al. 1987, TeKrony et al. 1985).

Thresholds Based on Disease Severity or Incidence

Thresholds based solely on the amount of disease at a specific time are infrequently used. Zadoks (1984) pointed out that, in many instances, it is too late to intervene when disease symptoms become visible. This is particularly true with explosive diseases such as late blight of potato, caused by *Phytothphora infestans*. In some other instances, long latent periods may not allow a wait-and-see approach. Disease forecasting systems using level of disease as the only input are also unpopular because of the need for frequent crop monitoring of disease levels.

Shoemaker and Lorbeer (1977) developed an ET for onions, timing the initial fungicide application when Botrytis leaf blight lesions averaged one per 10 leaves. This threshold was used to initiate a weekly fungicide spray program, resulting in higher yields of onions than spray programs begun before or after the threshold was reached.

Kucharek and Sanden (1976) developed a dynamic threshold based on disease severity and the age of peanut plants. Using this system, 50 peanut leaves are sampled from the midcanopy in each peanut field, and the total number of early and late leaf spot lesions is counted. The number of leaf spots in the sample and the age of the peanut crop are then used to determine the effectiveness of the grower's leaf spot control program. Growers are provided with a graph of acceptable disease severity by plant age and can adjust their spray schedules if measures deviate from this value.

Management of Multiple Pests

The ET concept is more complicated when multiple pests or pest complexes are taken into consideration. This is the real-life situation faced by most growers. Crop-monitoring systems are one approach to this problem, but a major limitation of such systems is that they are relatively time-consuming and expensive and may require immediate pesticide application; moreover, individual field counts most likely offer a more accurate assessment of pest numbers. Nevertheless, a number of systems for multiple pests have been developed.

The system called EPIPRE (Epidemiology, Prevention, and Prediction) is designed to monitor and control six fungal diseases and all aphid pests of wheat in the Netherlands (Zadoks 1984, 1990). The system has also been adopted in other European countries. After initial instruction on pest monitoring, the grower provides basic pest and crop information to a central computer by mail. The grower then receives a recommendation, with a financial justification in terms of yield based on whether the recommendation was to treat or not. The success of this system depends on rapid communication between the computer center and the grower; the turnover time is about three to four days. The growers who use the system gain a better understanding of potential pests and can reduce pesticide inputs (Zadoks 1984). Currently, the system does not include direct inputs of weather information (Zadoks 1990).

Another system, WDCA (Wheat Disease Control Advisory), is a computerized decision-support system or expert system for managing Septoria leaf blotch, leaf rust, and yellow rust in Israel (Shtienberg et al. 1990). The system was designed to give growers a method of determining when use of a foliar fungicide will be profitable. It takes into account economic, agronomic, and disease factors, as well as recorded and forecasted weather.

Another multiple–pest management system is AUSIMM (Auburn University Soybean Integrated Management Model). AUSIMM, a computerized IPM model in the southeastern United States (Backman et al. 1989), predicts the profitability of soybean pest-management practices. The model begins with a file that ranks ca. 120 cultivars according to maturity group, relative productivity, and susceptibility to nematodes and diseases. Potential yields of optimal cultivars are calculated, using location, soil type, planting date, rotation, and pest histories of the field. In the initial procedures, yields are adjusted for nematode species and populations. As the season progresses, submodels are used for management of plant disease and insect pests. These models integrate soybean plant-growth data, population dynamics, pesticide-efficacy data, weather data, effects of weather on pest populations and pesticide performance, and crop-loss functions. The effectiveness of any practice is compared with its cost. The submodels are linked by a core program that processes potential yield and crop value throughout the season, accumulating changes in yield potential, crop value, and management costs. AUSIMM improved grower net profits and decreased risks over local practices in 70% of 36 on-farm tests. Profit increases were primarily from reduced pesticide applications and the use of resistant cultivars. Because the model required a computer, acceptance by farmers in the region has been slow (Mack 1992).

Future Directions

The U.S. Department of Agriculture (USDA) has set a target for U.S. agriculture to have 75% of all acreage under IPM management by the year

2000. The USDA's intent is to reduce greatly the country's reliance on pesticides for producing crops and to develop an ecologically based management system for pests. There is also an effort by the U.S. Congress to foster development of alternative pest-control strategies, particularly biological control, by removing hindrances to its development and by passing legislation that will accelerate implementation.

Driving this rapid conversion of U.S. pest-control practices is an effort to reduce quantities of pesticides released into the environment, and particularly residues on foods. Pesticides targeted for likely earliest removal will be pesticides classified in the B_2 group, because of their oncogenicity rating. This list of potential carcinogens is dominated by fungicides. For the control of plant diseases, these policies mean that practices must not only be incorporated into a comprehensive IPM system, they will also require further modification because of losses of chemical strategies previously available.

In order to achieve these biointensive IPM systems for controlling plant pests, scientists (particularly plant pathologists) will be required to expand greatly the information base available for decision making. Key components will be the development of benefits-based ETs for plant diseases and nematodes—ETs that must be at once effective and simple enough for the end user to understand. These requirements may greatly expand the importance of the crop advisor, who will likely interface between complex decision-making aids and uncomprehending growers. As new management models are created, every effort should be made to develop systems that require simple inputs. We do not have to look far back in the literature to find numerous examples of effective management models that lie unused because growers are unwilling to invest the time or money to fulfill the training requirements and/or the hardware acquisition. For the next decade or more, scientists can expect to see not only increased funding for component research but also funding for research directed to the implementation of IPM that will link industry, extension, university researcher, and grower.

References

Backman, P. A., M. A. Crawford, and J. M. Hammond. 1984. Comparison of meteorological and standardized timings of fungicide applications for soybean disease control. Plant Dis. 68:44–46.

Backman, P. A., J. C. Jacobi, and D. P. Davis. 1991. Peanut leafspot control system cuts fungicide use, maintains disease control. Highlights of Agric. Res. Ala. Agric. Exp. Stn. 38:3.

Backman, P. A., T. P. Mack, R. Rodríguez-Kábana, and D. A. Herbert. 1989. A computerized integrated pest management model (AUSSIMM) for soybeans grown in the southeastern United States, p. 1494–1499. *In* A. J. Pascale (ed.) Proc. World Soybean Res. Conf. 5–9 March, 1989, Buenos Aires, Argentina.

Backman, P. A., R. Rodríguez-Kabána, M. C. Caulin, E. Baltramini, and N. Ziliani. 1981. Using the soil-tray technique to predict the incidence of Sclerotium rot in sugar beets. Plant Dis. 65:419–431.

Bailey, J. E., and C. A. Maytac. 1989. A decision model for use of fumigation and resistance to control Cylindrocladium black rot of peanuts. Plant Dis. 73:323–326.

Bailey, J. E., P. M. Phipps, T. A. Lee, and J. Damicone. 1993. Utilization of environmental thresholds to minimize fungicide applications for control of Sclerotinia blight of peanut. Proc. Am. Peanut Res. Educ. Soc. 24:46.

Barker, K. R., and J. P. Noe. 1987. Establishing and using threshold population levels, p. 75–81. *In* J. A. Veech and D. W. Dickson (ed.). Vistas on nematology: A commemoration of the twenty-fifth anniversary of the Society of Nematologists. Soc. of Nematologists, Hyattsville MD.

Berger, R. D. 1969. A celery early blight spray program based on disease forecasting. Proc. Fla. State Hort. Soc. 82:107–111.

Burleigh, J. R., R. W. Romig, and A. P. Roelfs. 1969. Characterization of wheat rust epidemics by numbers of uredia and numbers of urediospores. Phytopathol. 59:1224–1237.

Coakley, S. M., W. S. Boyd, and R. F. Line. 1982. Statistical models for predicting stripe rust on winter wheat in the Pacific Northwest. Phytopathol. 72:1539–1542.

Cu, R. M., and P. M. Phipps. 1993. Development of a pathogen growth response model for the Virginia peanut leafspot advisory program. Phytopathol. 83:195–201.

Davis, D. P., J. C. Jacobi, and P. A. Backman. 1993. Twenty-four-hour rainfall, a simple environmental variable for predicting peanut leaf spot epidemics. Plant Dis. 77:722–725.

DeVay, J. E., L. L. Forrester, R. H. Garber, and E. J. Butterfield. 1974. Characteristics and concentration of propagules of *Verticillium dahliae* in air-dried field soils in relation to the prevalence of Verticillium wilt in cotton. Phytopathol. 64:22–29.

Franc, G. D., M. D. Harrison, and L. K. Lahman. 1988. A simple degree-day model for initiating chemical control of potato early blight in Colorado. Plant Dis. 72:851–854.

Hirano, S. S., D. I. Rouse, and C. D. Upper. 1987. Bacterial ice nucleation as a predictor of bacterial brown spot disease on snap beans. Phytopathol. 77:1078–1084.

Hull, R. 1958. The spray warning schedule for control of sugar beet yellows in England. Summary of results from 1959–1966. Plant Pathol. 17:1–10.

Jacobsen, B. J., M. C. Shurtleff, H. W. Kirby, and T. A. Melton. 1987. Condensed plant disease management guide for field crops. Univ. Ill. Coop. Ext. Serv. Circ. 1231.

Jensen, R. E., and L. W. Boyle. 1966. A technique for accurately forecasting leafspot on peanut. Plant Dis. Rep. 50:810–814.

Jones, A. L., P. D. Fisher, R. C. Seem, J. C. Kroon, and P. J. Van de Motter. 1984. Development and commercialization of an in-field microcomputer delivery system for weather driven predictive models. Plant Dis. 68:458–463.

Krause, R. A., L. B. Massie, and R. A. Hyre. 1975. BLITECAST, a computerized forecast of potato late blight. Plant Dis. Rep. 59:95–98.

Kucharek, T., and G. Sanden. 1976. Peanut leafspot assessment and control. Extension Plant Pathology Report No. 19. University of Florida, Gainesville.

Leach, L. D., and A. E. Davey. 1938. Determining the Sclerotial population of *Sclerotium rolfsii* by soil analysis and predicting losses of sugar beets on the basis of these analyses. J. Agric. Res. 56:619–631.

Linderman, J., D. C. Arny, and C. D. Upper. 1984. Use of an apparent infection threshold population of *Pseudomonas syringae* to predict incidence and severity of brown spot of bean. Phytopathol. 74:1334–1339.

MacKensie, D. R. 1981. Scheduling fungicide applications for potato late blight with BLITECAST. Plant Dis. 65:394–399.

Mack, T. P. 1992. Implementing innovative insect management systems in soybean in the southeastern U.S., p. 36–45. *In* L. G. Copping, M. B. Green, and R. T. Rees (ed.) Pest management of soybean. Elsevier Applied Science, New York.

Madden, L. V., and M. A. Ellis. 1988. How to develop plant disease forecasters, p. 191–209. *In* J. Kranz and J. Rotem (ed.) Experimental techniques in plant disease epidemiology. Springer-Verlag. New York.

Parvin, D. W. Jr., D. H. Smith, and F. L. Crosby. 1974. Development and evaluation of a computerized forecasting method for Cercospora leafspot of peanuts. Phytopathol. 64:385–388.

Phipps, P. M., and N. L. Powell. 1984. Evaluation of criteria for utilization of peanut leafspot advisories in Virginia. Phytopathol. 74:1189–1193.

Porter, D. M., D. H. Smith, and R. Rodríguez-Kábana (ed.). 1984. Compendium of peanut diseases. APS Press, St. Paul MN.

Rodríguez-Kábana, R., P. A. Backman, and J. C. Williams. 1975. Determination of yield losses to *Sclerotium rolfsii* in peanut fields. Plant Dis. Rep. 59:855–858.

Roelfs, A. P., J. B. Rowell, and R. W. Romig. 1970. A comparison of rod and slide samplers used in cereal rust epidemiology. Phytopathol. 58:1150–1154.

Shaw, M. W., and D. J. Royle. 1987. Saving septoria sprays: The use of disease forecasts. 1986 Br. Crop Prot. Conf.–Pests and Diseases 8c-33:1193–1200.

Shoemaker, P. B., and J. W. Lorbeer. 1977. Timing initial fungicide application to control *Botrytis* leaf blight epidemics on onions. Phytopathol. 67:409–414.

Shtienberg, D., A. Dinoor, and A. Marani. 1990. Wheat Disease Control Advisory, a decision support system for management of foliar disease of wheat in Israel. Can. J. Plant Path. 12:195–203.

Smith, D. H., and R. H. Littrell. 1980. Management of peanut foliar diseases with fungicides. Plant Dis. 64:356–361.

Stevens, N. E. 1934. Stewart's wilt in relation to winter temperatures. Plant Dis. Rep. 25:152–157.

Surin, A., W. Rodjanahusdin, P. Arunyanart, and S. Disthaporn. 1992. A trap crop method to predict the occurrence of rice blast. Int. Rice Res. News. 17:4.

TeKrony, D. M., R. E. Stuckey, D. B. Egli, and L. Tomes. 1985. Effectiveness of a point system for scheduling foliar fungicides in soybean seed fields. Plant Dis. 69:962–965.

Watson, M. A., G. D. Heathcote, F. B. Lauckner, and P. A. Souray. 1975. The use of weather data and counts of aphids in the field to predict the incidence of yellowing viruses of sugar beet crops in England in relation to the use of insecticides. Ann. Appl. Biol. 40:38–59.

Zadoks, J. C. 1984. A quarter century of disease warning, 1958–1983. Plant Dis. 68:352–355.

Zadoks, J. C. 1990. Management of wheat disease in northwest Europe. Can. J. Plant Pathol. 12:117–122.

G. David Buntin

9
Economic Thresholds for Insect Management

The use of economic thresholds as a basis for decision making is a fundamental component of integrated pest management (IPM). Stern et al. (1959) proposed the concepts of an economic injury level (EIL) and economic threshold (ET) as a rational comparison of the economic costs and benefits of pesticide use. EIL is defined as the lowest population density (number) that will cause economic damage, where economic damage is the amount of damage that equals the cost of control (Stern et al. 1959, Pedigo et al. 1986). Implicit in the EIL concept is that not all damage is economically significant and that in many instances a certain level of insect injury may be tolerated. Stern et al. (1959) proposed the ET as the operational decision rule—the population density at which control measures should be initiated to prevent an increasing population from reaching the EIL. The ET is referred to as the action threshold by plant pathologists (e.g., Zadoks 1985) and some entomologists (e.g., Yencho et al. 1986).

In the discussion of EILs, it is useful to maintain a distinction between injury and damage. Injury can be defined as the effect of insect activities on host physiology, damage as the measurable loss of host utility, which is usually measured by reductions in commodity yield or quality (Bardner and Fletcher 1974, Pedigo et al. 1986). Consequently, not all injury causes damage, and a damage threshold, or boundary, defines the level of injury where damage occurs (Tammes 1961, Pedigo et al. 1986, and Chapter 3). Injury, per se, often is difficult to measure in the field; hence, typically, insect numbers are used as an index of injury.

This chapter will discuss how the components of the EIL are derived, with particular emphasis on procedures for relating insect numbers to crop damage. Calculation of the ET from the EIL will also be discussed.

Procedures for Assessing EIL Components

Determination of Insect-Number/Damage Relationships

Accurate quantification of the relationship between pest number and crop damage (loss of commodity utility) is crucial to the development of EILs and has been a major preoccupation of agricultural entomologists.

Experimental approaches to the assessment of crop response to insect attack can be grouped into four categories: observation of natural populations, establishment of artificial populations, modification of natural populations (usually with insecticides), and simulation of damage (Poston et al. 1983). The experimental approach to a specific insect-crop system often is dictated by the characteristics of the crop, nature and timing of injury, and feasibility of caging, collecting, or rearing a particular insect.

Observation of Natural Populations. Relating natural populations to commodity yield probably is the simplest, but least analytical, approach to describing damage-loss relationships. The approach typically involves taking insect counts with a given sampling technique at a certain stage of crop development and statistically relating counts with plant yield or productivity from a series of crop units such as plants, plots, or fields. Dutcher and All (1979) successfully used multiple-regression procedures to relate girdling and number of feeding sites of the grape root borer, *Vitacea polistiformis* (Harris), to berry yield of individual grapevines. By sampling 50 commercial fields of snap bean, Weinzierl et al. (1987) demonstrated a linear relationship between the number of western spotted cucumber beetles, *Diabrotica undecimpunctata undecimpunctata* Mannerheim, at two weeks before harvest and the percentage of beetle-scarred pods at harvest. In this study, pod quality rather than yield was the measure of plant utility.

Advantages to the use of natural populations are (1) simplicity; (2) lack of disturbance of "natural" pest phenology and damage; and (3) comprehensiveness, because the sampling of a large number of fields, representing a variety of growing conditions, may provide more robust estimates of the relationship between pest number and crop loss than will estimates derived from a few intense plot studies. Problems with the natural-population approach are: (1) natural infestations may be substantially above or below the range of densities needed for EIL calculations; (2) a large enough number of fields must be sampled to provide a range of potential population densities above and below the EIL and reasonable precision in statistical estimates—and this may require several seasons; (3) variations in agronomic and environmental conditions may mask damage-loss relationships; and (4)—probably the most important problem—this approach usually provides little information about insect biology and the underlying mechanism of insect injury and crop response to injury.

Establishment of Artificial Populations. The creation of artificial populations is a useful approach for occasional pests with which occurrence of infestation is not reliable and when a precise range of population densities is desired at a particular location. The procedure involves collecting or rearing the insect and artificially infesting plots with a known number of individuals. Depending on the mobility and activity of natural enemies of the damaging stage in a specific crop, pests may be restrained, through the use of cages or some other barrier, or left unrestrained. Generally, winged adults must be confined with cages (e.g., Mailloux and Bostanian 1988, Michels and Burkhardt 1981, Hall and Teetes 1982, Hutchins and Pedigo 1989, Ba-Angood and Stewart 1980). Wingless insects such as aphids or lepidopteran larvae usually do not require confinement (e.g., Bode and Calvin 1990, Breen and Teetes 1990) or can be confined with some sort of open ground barrier (e.g., Showers et al. 1983, Buntin and Pedigo 1985, Bechinski and Hescock 1990, Davis and Pedigo 1990). However, if natural enemy populations are present, nonmobile insects may need caging to reduce pest mortality that would occur in unrestrained settings. The mobility of a pest may not be the same in all crops, thus requiring confinement in one crop but not in another. For example, once introduced, second-stage larvae of the fall armyworm, *Spodoptera frugiperda* (J. E. Smith), usually remain in the whorl of infested corn plants and do not require confinement (Morrill and Greene 1974). However, in bermuda grass, fall armyworms are active on the soil surface. With this crop, Jamjanya and Quisenberry (1988) successfully used metal barriers to confine larvae in small plots at desired population densities to study their effect on forage-yield loss.

Difficulties can arise from creating artificial infestations. Genetic shifts may cause laboratory-reared populations to behave differently than feral populations, and artificial infestations may not mimic natural population phenology and dynamics. Furthermore, mortality from biotic and abiotic factors can be large; thus, infestations must be monitored carefully to document changes in pest number, without excessive disturbance of small plots. To avoid potentially high mortality rates of early instars, pests as immatures, such as lepidopteran larvae, often are infested as mid to late instars (Morrill and Greene 1974).

Unrestrained infestations are most desirable, because this avoids the added complications of disturbance caused by barriers and cages (i.e., cage effects). Establishing barriers or cages can disrupt and damage plants growing in a canopy inside and outside the plot area and cause soil compaction or affect soil properties if cages or barriers are buried in the soil. An impermeable barrier may affect water runoff and movement in the soil, causing an enclosed plot to fill with water after a heavy rain. Cages often alter the microclimate under the cage, reducing wind and photosynthetically active radiation (light) levels, increasing humidity, affecting (usually increasing) ambient and soil temperatures, and possibly reducing penetration of rain. Screen-mesh size should be as large as possible to minimize microclimatic changes while still confining the pest. Cages typically are used for a short time (1 to 2 weeks), and it usually is assumed that any cage effect is transient and not substantial over an entire season. Nevertheless, caged and uncaged controls should always be included as treatments in cage studies. Furthermore, data from caged controls rather than uncaged controls should be used in generating yield-loss equations.

Many of the problems of artificial infestations (e.g., phenology disruption and/or cage effects) can be minimized by infesting plots with eggs or manipulating adult populations to achieve a gradient of egg deposition in a series of plots. The yield response of corn to injury from larval corn rootworm, *Diabrotica* spp., has been studied by infesting corn plants with different densities of laboratory-reared eggs (Sutter and Branson 1980, Spike and Tollefson 1989). Parman and Wilson (1982) studied alfalfa crop response to feeding injury by *Philaenus spumarius* (L.) nymphs in the spring growth cycle by caging various densities of ovipositing adults in the fall. This created a gradient of egg deposition that produced a range of larval infestations the following spring. In these examples, artificial infestations were created with minimal environmental disruption from confinement and without affecting plant growth and insect biology and phenology.

Modification of Natural Infestations. A third approach is to modify natural infestations. This approach is most suitable for severe or perennial pests whose infestations almost always cause economic damage (Poston et al. 1983). Natural pest populations usually are modified by use of insecticides. This can either reduce pest numbers or enhance them by eliminating natural enemies. Populations also may be reduced by use of natural enemies (e.g., Hartstack et al. 1978); or they may be enhanced by attractant baits (Vea and Eckenrode 1976) or trap crops and genotypes of the same crop (Cancelado and Radcliffe 1979). Populations also can be modified by using a series of crop genotypes with varying degrees of pest resistance (Buntin and Raymer 1989).

The main criticism of such modification of natural infestations with insecticides is that the insecticide may alter plant physiology (Jones et al. 1986). Numerous pesticides affect (usually suppress) plant physiological processes such as leaf photosynthetic rates and stomatal conductance and adversely affect plant growth and yield (see reviews by Ferree 1979, Jones et al. 1986). Adverse effects of pesticides on plant health often occur with high dosages or after frequent use, such as weekly or biweekly applications, to eliminate a target pest (Toscano et al. 1982, LaPre et al. 1982). Although the effects are usually transient, even a single application within recommended dosages can reduce plant photosynthetic rates for several weeks (Jones et al. 1983, Trumble et al. 1988). Before a pesticide is used in studies to modify pest populations, potential effects on plant physiology should be evaluated.

Several approaches to such uses of insecticides have been employed. Serial dilutions from a standard rate of an insecticide can create a gradient of pest densities (Wilson et al. 1969, Hintz et al. 1976). Wilson et al. (1969) examined the yield response of oats to the cereal leaf beetle, *Oulema melanopus* (L.), using serial dilutions of malathion to create a gradient of larval densities (Table 9.1). Populations also can be modified using different insecticides of various efficacy to reduce pest numbers directly or induce population increases by disrupting natural enemies. Wilson et al. (1991) modified spider mite populations in cotton by using various rates of dicofol or methyl parathion to suppress mite numbers and used permethrin to eliminate predators, thereby enhancing mite numbers. This created a gradient of spider mite densities, making it possible to study cotton yield response to spider mite injury. A third approach is to compare pest numbers in untreated plots with the difference in yield between treated and untreated plots for a series of many paired plots (e.g., Walker et al. 1992).

Table 9.1. Yield response of oats to infestation of cereal leaf beetle, *Oulema melanopus* (L.), larvae created by serial dilution of malathion rates in two years (data from Wilson et al. 1969).

Malathion Rate (kg/ha)	1966 Larvae/stem	1966 Yield (kg/ha)	1968 Larvae/stem [a]	1968 Yield (kg/ha)
0	3.47	1,814	4.64	1,337
0.14	2.29	2,068	3.84	1,552
0.28	1.36	2,129	1.36	1,928
0.56	0.21	2,115	1.36	1,803
1.12	0.19	2,125	1.74	1,914

[a] Larval numbers adjusted for leaf-area damage [see Wilson et al. 1969].

Another commonly used approach is to select a series of target ETs in which the pest is controlled with an insecticide application whenever populations reach the target density (e.g., Archer and Bynum 1992). Crop production and marginal $ returns of each target threshold are compared for the best return. One advantage of this approach is that it mimics how control measures would be used in a commercial situation. A big disadvantage is that threshold selection is empirical; consequently, the optimal ET may not be selected. In other words, although one threshold may be the best alternative for the thresholds examined, it may not be the true ET. Evaluation of target thresholds often is used for severe pests where ETs are low, as in many vegetable crops, or in situations where normally multiple applications are applied during the season (Kirby and Slosser 1984, Durant 1991, Yencho et al. 1986). For example, Stewart and Sears (1988) compared a standard biweekly spray schedule with spraying at thresholds of 0.25, 0.5, or 1.0 CLEs (cabbage looper equivalents—one equivalent equals the feeding caused by one cabbage looper) per plant for control of lepidopterous caterpillars on cauliflower. They found that the 0.25 CLE threshold reduced the number of sprays by about 50% and increased net revenue without reducing yield, compared with the biweekly spray schedule (Table 9.2).

Table 9.2. Application number, yield, control costs, and revenue of managing lepidopterous larvae in cauliflower with spray schedules using different target economic thresholds for treatment (data from Stewart and Sears 1988).

Target threshold for treatment (CLE/plant/week)	Applications per season	Crop yield (kg/ha)	Pest control costs ($/ha)	Net Revenue ($/ha)
Biweekly	6	16,938	247.08	4,156.80
0.25	3	16,938	198.54	4,205.34
0.50	2	15,894	157.36	3,975.08
1.00	2	11,700	157.36	2,884.64
Untreated	0	594	0.00	154.44

Simulation of Insect Injury. Probably the most controversial loss-assessment method is damage simulation using surrogate injury techniques (Poston et al. 1983). In this approach, laboratory consumption data are used to relate injury to pest number, and insect injury is related to crop loss by manually removing or injuring plant tissue to mimic actual tissue injury caused by the pest. Damage simulation allows more flexibility in investigating damage-loss relationships than other procedures because the amount of injury can be precisely controlled, and injury per pest and damage per unit of injury (I and D of the EIL equation, respectively) can be separated experimentally. Damage simulation most often is used to study crop response to injury by mandibulate insects that chew on plant tissue, such as defoliators, seedling cutters (Showers et al. 1979), and flower or pod feeders (Todd and Morgan 1972, Brook et al. 1992), or by insects that damage apical terminals (Hopkins et al. 1982, Funderburk and Pedigo 1983). However, Rogers (1976) simulated flower-bud abortion in guar caused by larvae of the cecidomyiid *Contarinia texana* (Felt), by excising buds. Simulation is rarely used to mimic injury by insects with piercing/sucking mouthparts.

Damage simulation has been used most extensively to study crop response to insect defoliation. Daily or incremental leaf-mass consumption (usually expressed as leaf area consumed) per insect is measured in the laboratory. Crop response to defoliation is measured by removing a series of specific amounts or percentages of leaf mass at a growth stage when defoliation is likely to occur. Techniques include picking entire leaves or leaflets (e.g., Peterson et al. 1993), cutting a portion of a leaf with scissors (Stewart et al. 1990), or punching holes in the leaf, using a cork borer or paper punch (Shelton et al. 1990). For example, Hammond and Pedigo (1982) estimated EILs for the green cloverworm, *Plathypena scabra* (F.), in soybean by removing leaf tissue with a cork borer to achieve various percentage levels of defoliation. Foliage was removed according to a consumption model developed from laboratory measurements of daily development and consumption rates, in which a hypothetical cohort of larvae consumed a total of 54.3 cm^2 of leaf tissue per larva over about a 12-day period (Hammond et al. 1979).

Most criticisms of damage simulation regard questions concerning the fidelity of surrogate injury to simulate actual injury. Insect actions or saliva secreted while chewing potentially may induce plant chemical defenses or change in plant physiology processes that are not reproduced in the event of mechanical injury; however, Smith (1989) listed a number of examples where both insect and mechanical injury induce plant chemical defenses. Welter (1991) compared leaf-mass removal with scissors to defoliation by *Manduca sexta* (L.) in tomato and found no significant changes in photosynthesis per unit leaf area of remaining leaf tissue of damaged leaves and of undamaged leaves on the same plant. Likewise, Peterson et al. (1992) compared picking leaflets with defoliation by the alfalfa weevil, *Hypera postica* (Gyllenhal), in alfalfa and found that photosynthetic and transpiration rates of remaining leaf tissue were similar. In another study, net photosynthesis of remaining leaf tissue was not substantially different between soybean leaves with punched holes and leaves defoliated by *P. scabra* (Poston et al. 1976).

Ostlie and Pedigo (1984) compared water loss in soybean leaves after defoliation by *P. scabra* and cabbage looper, *Trichoplusia ni* (Hübner) with leaf picking and hole punching. They found that water loss increased in proportion to the amount of cut-leaf edge per leaf area removed, punched holes having more cut edge than defoliation caused by either insect, and picked leaves having much less. However, simulated and natural defoliation produced differences in whole-plant transpiration only during the first 16 hours after defoliation, with total water loss for 48 hours after defoliation not being significantly different between methods, indicating that both simulation methods adequately mimicked actual defoliation. The aforementioned studies support the contention that surrogate injury can simulate defoliation because the principal effect of defoliation is loss of quantity of photosynthesizing leaf tissue (Boote 1981, Peterson et al. 1992).

A potential source of error in damage-simulation studies would be failure to mimic the distribution of injury accurately, either on a single plant or between plants. Shields and Wyman (1984) simulated two insect defoliators in potato by picking leaves from the top leaf down, to simulate defoliation by *Leptinotarsa decemlineata* (Say), and from the lower leaf up, to simulate defoliation be *Peridroma saucia* (Hübner). Generally, potato was more sensitive to top down than bottom up defoliation.

Another error in simulation studies has been not to accurately copy the temporal pattern of damage on a particular plant species. This factor has received little attention. Many damage-simulation studies impose injury in a single day (e.g., Shields and Wyman 1984, Todd and Morgan 1972). However, insects often feed for a number of days, depending on the environmental conditions. Ostlie (1984) found that soybean was less severely damaged by the same total amount of simulated defoliation when it was imposed during a single day as when it was imposed over multiple days (12 days in the study). Ostlie suggests that temporal pattern of injury should mimic as closely as possible the actual pattern of the insect in question. Because injury by early instars accounts for a small percentage of total consumption (usually <15%), this portion is often pooled and imposed the first day of simulated injury (e.g., Hammond and Pedigo 1982, Peterson et al. 1992).

Estimations of insect consumption rates are another potential source of error. Consumption rates can be affected by plant genotype, plant growing conditions and nutritional status, stage of plant development, plant tissue age, and environmental conditions during insect development. Consumption of soybean leaf area was 53.9% less on field-grown leaves than it was in greenhouses, mostly because of differences in specific leaf weight of soybean leaves in the two environments (i.e., greenhouse leaves were thinner than field leaves) (Hammond et al. 1979). Consumption per insect also can be affected by insect-population density. Pedigo et al. (1977) found that *P. scabra* showed a linear response to consumption at densities near the EIL and this may be expected for many insects. If affected, consumption or injury per insect often declines as population density increases (e.g., Bellows et al. 1983).

Statistical Description of the Damage/Loss Relationship

Accurate statistical description of the damage/loss relationship is a critical step in the development of EILs. The relationship between injury intensity and plant yield or utility is described by Pedigo et al. (1986) for a variety of insect/plant systems and is discussed elsewhere in this book (Chapter 3). However, three generalized responses are most often encountered (Poston et al. 1983): tolerant responses, where some injury is tolerated before yield declines linearly with increasing injury; susceptive responses, where yield declines linearly with increasing injury; and hypersusceptive responses, where yield loss is greatest at low levels of injury, and incremental losses become smaller as injury increases. Sometimes, low levels of injury may enhance yield somewhat in tolerant plants, producing an overcompensatory response (Bardner and Fletcher 1974, Pedigo et al. 1986).

Figure 9.1 shows that there usually is an approximately linear relationship between injury or pest numbers and yield in the midranges of injury, on all three response curves. The primary difference in response curves is the relationship between injury and yield at low levels of injury. Because the range of injury needed for EIL determinations often occurs at these lower levels of injury, accurate identification of the true damage-loss response at low injury levels is crucial for accurate EIL calculations. Insect-damage/yield-loss relationships most commonly are described using least-squares regression with a simple linear model, providing a single value that can easily be used in the EIL equation (Poston et al. 1983). Most likely, many of the linear responses reported in the literature probably are actually tolerant or hypersusceptive, but the range of injury studied or variability in the yield-loss relationship do not permit expression of the tolerant or hypersusceptive response (Poston et al. 1983). Nonlinear models that better reflect biological reality of crop response to pest injury probably should be used in most instances, even when significant statistical improvement cannot be demonstrated (Madden 1983). Quadratic models have frequently been used to describe nonlinear damage-response curves, but these models often do not describe well the yield-loss relationship at low levels of injury. Exponential, logistic, cumulative Weibull distribution, and

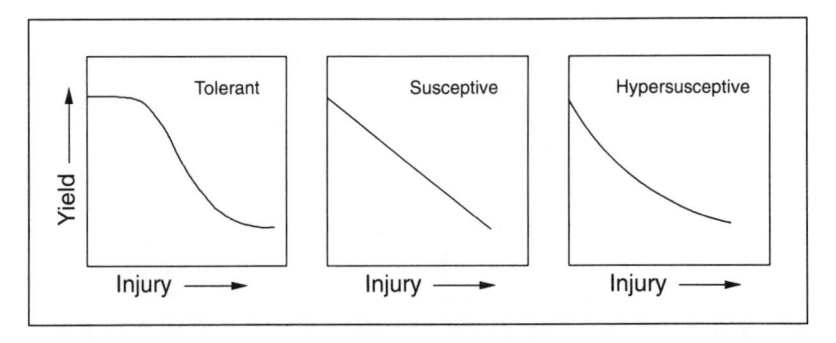

Figure 9.1. Three generalized crop-yield responses to insect injury (redrawn from Poston et al. 1983)

other functions may better describe hypersusceptive and tolerant response curves. Madden et al. (1981) proposed a form of the Weibull function that effectively describes tolerant or hypersusceptive responses of crops to injury by plant pathogens; spline functions, too, are useful for tolerant responses. Spline functions divide the response curve into two sections and provide an estimate of the point of inflection in the curve (i.e., the damage boundary) and separate functions for each section (e.g., Wilson et al. 1991).

Market Value

Commodity or market value, V in the EIL equation, estimates expected economic returns for a given commodity. Typically, most crops have off-farm markets that determine commodity prices and these can be used in EIL calculations. Generally, EILs are calculated for a range of expected commodity prices that are based on historical fluctuations in prices (Table 9.3). EIL values decline as commodity prices increase. Selection of a specific price (i.e., the EIL) for use in a particular situation depends on seasonal fluctuations, future outlook of market prices, personal feelings about future trends, and the financial situation of the producer (Pedigo et al. 1986). If pest damage potentially affects commodity quality, thereby causing a change in commodity grade or standard, the commodity price used to calculate EIL should reflect this change (e.g., Buntin et al. 1992).

Selection of commodity prices becomes more difficult for commodities

Table 9.3. Economic injury levels (EILs) for the alfalfa snout beetle in alfalfa for various combinations of control costs and hay value (Bechinski and Hescock 1990).

Hay value[a]	EIL $(beetles/m^2)$ Control costs $(\$/ha)$[b]		
($/metric ton)	$12.50	$37.50	$62.50
50	16.6	19.6	21.6
72	16.0	18.4	20.1
94	15.5	17.7	19.2
116	15.2	17.2	18.6

[a] Hay values are approximately equal to $45, $65, $85, and $105 per ton.

[b] Control costs include cost of insecticide (carbofuran) and application costs. They are approximately equal to $5, $15, and $25 per acre.

with only a limited off-farm market, or no market, such as forage and pasture crops. One approach is to use replacement-feed cost analysis (Craven and Hasbargen 1979); this estimates the value of forage lost by determining the worth of substitute or replacement feeds, such as corn and soybean meal, that have established commodity prices (e.g., Buntin and Pedigo 1985, Hutchins and Pedigo 1990). Hay also can be used in this analysis if the hay substituted is of similar nutritional quality to the damaged hay (e.g., Buntin and Raymer 1989). Bellows et al. (1983) took a different approach. They set the value of forage produced from rangeland as the expected forage production (yield) per unit area divided by the potential gross return from a cow-calf operation minus the cost of leasing the rangeland. This provided an estimate of forage value that could be used in the calculation of EILs for the range caterpillar, *Hemileuca oliviae* Cockerell, in rangeland.

Management Costs

Probably the simplest component of the EIL to estimate is C, the cost of management activity. C usually consists of the cost of the pesticide plus application cost, but it can also include sampling (scouting) costs. Usually EILs are generated for a range of expected management cost alternatives, with the EIL increasing as management costs increase (Table 9.3). Because of the difficulty of calculating possible environmental and social costs (i.e., externalities), these costs often are not included in the estimate of management costs. However, Higley and Wintersteen (1992) developed a procedure using contingent valuation to assess environmental costs of pesticides that can be directly incorporated into C in the EIL equation (see Chapter 14). EILs for most pesticides usually increase substantially when environmental costs derived by this procedure are included (Pedigo and Higley 1992).

Proportionate Control

Proportionate control, K, has been defined as the expected proportion or degree of control (i.e., population reduction) resulting from a particular management activity (Norton 1976). K usually is a major emphasis in any discussion of pest control by agricultural economists (Mumford and Norton 1984), but typically it is not directly considered by specialists in the pest disciplines (Pedigo et al. 1986). However, recognizing that not all management activities necessarily reduce pest numbers but may simply repel or prevent pest attack, Pedigo and Higley (1992) defined K as the proportion of total pest injury averted by the timely application of a management tactic. Biological control agents often cause a reduction or cessation of feeding well before mortality occurs.

Many entomologists often develop EILs for a "generic" insecticide and assume a constant level of proportionate control, often 100% population reduction (e.g., Yencho et al. 1986) or injury prevention. Yet most reports in any volume of *Insecticide and Acaracide Tests,* published by the Entomological Society of America, clearly show that insecticides vary in effectiveness and level of realized control for most pests. This discrepancy may occur because growers typically are averse to risk and consider a pesticide or management activity unacceptable unless proportionate control is high (>90%) and pest numbers are reduced below the damage boundary. However, if the level of control is less than desired for all available alternatives, then EIL increases as K declines. This is true for the European corn borer, *Ostrinia nubilalis* (Hübner): a single application of any of the most effective insecticides provides only about a 67% reduction (i.e., $K = 0.67$) in the number of first-generation borers in corn (Showers et al. 1989). K also can affect the choice between two alternatives, if the difference in cost of two alternative management activities is large and proportionate control varies substantially between the alternatives (i.e., a much cheaper but less effective alternative may be acceptable).

Calculation of Economic Thresholds from the EIL

The operational decision rule for pest management is the ET. The ET is generated directly from the EIL and incorporates potential changes in pest density caused by the lag time between a management decision and implementation of control. The ET almost always is lower than the EIL, and ideally it should incorporate knowledge about expected pest-population growth and injury rates derived from studies of pest-population dynamics (Pedigo et al. 1989). Weinzierl et al. (1987) proposed an ET that was 75% of the EIL for the western spotted cucumber beetle in snap beans because population studies showed that beetle density never changed more than 25% over a 3- to 4-day period, which is sufficient time to implement a control activity. In many instances, population-dynamics data are lacking, and ET typically is set as a fixed percentage of the EIL. Percentage reductions usually are conservative, with values about 75% being common (Pedigo et al. 1989). Keerthisinghe (1984) discusses generating an ET from the lower confidence interval derived from the slope of the damage/yield-loss relationship. The use of a confidence interval minimizes the error of classifying a population that needs control as noneconomic with a given level of probability. However, when sampling indicates that a population has reached the EIL, because of random sampling variation the true population density actually is below the EIL in half of the cases under study. Consequently, if a grower is averse to using pesticides, it could be argued that the ET should actually be equal to, or possibly higher than, the EIL to insure that pesticides are used only when economically justified. The ET concept is discussed in more detail in Chapter 4.

The EIL and ET concepts have been successfully and widely applied to generate management guidelines for insect pests in many cropping systems. The single greatest challenge to entomologists is describing the relationship between pest injury or numbers and loss of crop utility. A variety of methodologies have been used to define this relationship: each approach has advantages and disadvantages, with characteristics of a specific insect/crop combination and the objectives of the researcher often determining the selection of methodology. Generally, methods that reveal information about the mechanism of injury and crop response to injury provide more insight into insect/crop interactions than procedures that simply relate some measure of pest abundance to crop productivity. Approaches that examine the nature of crop response to injury will permit the development of improved EILs and ETs.

References

Archer, T. L., and E. D. Bynum Jr. 1992. Economic injury level for Russian wheat aphid (Homoptera: Aphididae) on dryland wheat. J. Econ. Entomol. 85:987–992.

Ba-Angood, S. A., and R. K. Stewart. 1980. Economic thresholds and economic injury levels of cereal aphids on barley in southwestern Quebec. Can. Entomol. 112:759–764.

Bardner, R., and K. E. Fletcher. 1974. Insect infestations and their effects on the growth and yield of field crops: A review. Bull. Entomol. Res. 64:141–160.

Bechinski, E. J., and R. Hescock. 1990. Bioeconomics of the alfalfa snout beetle (Coleoptera: Curculionidae). J. Econ. Entomol. 83:1612–1620.

Bellows, T. S. Jr., J. C. Owens, and E. W. Huddleston. 1983. Model for simulating consumption and economic injury level for the range caterpillar (Lepidoptera: Saturniidae). J. Econ. Entomol. 76:1231–1238.

Boote, K. J. 1981. Concepts for modeling crop response to pest damage. ASAE Paper 81-4007. St. Joseph MI, Am. Soc. Agric. Eng.

Bode, W. M., and D. D. Calvin. 1990. Yield-loss relationships and economic injury levels for European corn borer (Lepidoptera: Pyralidae) populations infesting Pennsylvania field corn. J. Econ. Entomol. 83:1595–1603.

Breen, J. P., and G. L. Teetes. 1990. Economic injury levels for yellow sugarcane aphid (Homoptera: Aphididae) on seedling sorghum. J. Econ. Entomol. 83: 1008–1014.

Brook, K. D., A. B. Hearn, and C. F. Kelly. 1992. Response of cotton, *Gossypium hirsutum* L., to damage by insect pests in Australia: Manual simulation of damage. J. Econ. Entomol. 85:1368–1377.

Buntin, G. D., S. L. Ott, and J. W. Johnson. 1992. Integration of plant resistance, insecticides, and planting date for management of the Hessian fly (Diptera: Cecidomyiidae) in winter wheat. J. Econ. Entomol. 85:530–538.

Buntin, G. D., and L. P. Pedigo. 1985. Development of economic injury levels for last-stage variegated cutworm (Lepidoptera: Noctuidae) larvae in alfalfa stubble. J. Econ. Entomol. 78:1341–1346.

Buntin, G. D., and P. L. Raymer. 1989. Hessian fly (Diptera: Cecidomyiidae) damage and forage production of winter wheat. J. Econ. Entomol. 82:301–306.

Cancelado, R. E., and E. B. Radcliffe. 1979. Action thresholds for potato leafhopper on potatoes in Minnesota. J. Econ. Entomol. 72:566–569.

Craven, R. H., and P. R. Hasbargen. 1979. How much is a feed worth? Univ. of Minnesota Agric. Ext. Serv. Fact Sheet FM 560.

Davis, P. M., and L. P. Pedigo. 1990. Yield response of corn stands to stalk borer (Lepidoptera: Noctuidae) injury imposed during early development. J. Econ. Entomol. 83:1582–1586.

Durant, J. A. 1991. Effect of treatment regime on control of *Heliothis virescens* and *Helicoverpa zea* (Lepidoptera: Noctuidae) on cotton. J. Econ. Entomol. 84:1577–1584.

Dutcher, J. D., and J. N. All. 1979. Damage impact of larval feeding by the grape root borer in a commercial concord grape vineyard. J. Econ. Entomol. 72:159–161.

Ferree, D. C. 1979. Influence of pesticides on photosynthesis of crop plants, p. 331–341. *In* R. Marcelle, H. Clijsters, and M. Van Poucke (ed.) Photosynthesis and plant development. W. Junk, The Hague.

Funderburk, J. E., and L. P. Pedigo. 1983. Effects of actual and simulated seedcorn maggot (Diptera: Anthomyiidae) damage on soybean growth and yield. J. Econ. Entomol. 12:323–330.

Hall D. G. IV, and G. L. Teetes. 1982. Yield loss-density relationships of four species of panicle-feeding bugs in sorghum. Environ. Entomol. 11:738–741.

Hammond, R. B., and L. P. Pedigo. 1982. Determination of yield-loss relationships for two soybean defoliators by using simulated insect-defoliation techniques. J. Econ. Entomol. 75:102–107.

Hammond, R. B., L. P. Pedigo, and F. L. Poston. 1979. Green cloverworm leaf consumption on greenhouse and field soybean leaves and development of a leaf-consumption model. J. Econ. Entomol. 72:714–717.

Hartstack, A. W. Jr., R. L. Ridgeway, and S. L. Jones. 1978. Damage to cotton by the bollworm and tobacco budworm. J. Econ. Entomol. 71:239–243.

Higley, L. G., and W. K. Wintersteen. 1992. A novel approach to environmental risk assessment of pesticides as a basis for incorporating environmental costs into economic injury levels. Am. Entomologist 38:34–39.

Hintz, T. R., M. C. Wilson, and E. J. Armbrust. 1976. Impact of alfalfa weevil larval feeding on the quality and yield of first cutting alfalfa. J. Econ. Entomol. 69:749–754.

Hopkins, A. R., R. F. Moore, and W. James. 1982. Economic injury levels for *Heliothis* spp. larvae on cotton plants in the four-true-leaf to pinhead-square stage. J. Econ. Entomol. 75:328–332.

Hutchins, S. H., and L. P. Pedigo. 1989. Potato leafhopper-induced injury on growth and development of alfalfa. Crop Sci. 29:1005–1011.

Hutchins, S. H., and L. P. Pedigo. 1990. Phenological disruption and economic consequences of injury to alfalfa induced by potato leafhopper (Homoptera: Cicadellidae). J. Econ. Entomol. 83:1587–1594.

Jamjanya, T., and S. S. Quisenberry. 1988. Impact of fall armyworm (Lepidoptera: Noctuidae) feeding on quality and yield of coastal bermudagrass. J. Econ. Entomol. 81:922–926.

Jones, V. P., N. C. Toscano, M. W. Johnson, S. C. Welter, and R. R. Youngman. 1986. Pesticide effects on plant physiology: Integration into a pest management program. Bull. Entomol. Soc. Am. 32:103–109.

Jones, V. P., R. R. Youngman, and M. P. Parrella. 1983. Effect of selected acaricides on photosynthesis rates of lemon and orange leaves in California. J. Econ. Entomol. 76:1178–1180.

Keerthisinghe, C. I. 1984. Fiducial inference in economic thresholds. Prot. Ecol. 6:85–90.

Kirby, R. D., and J. E. Slosser. 1984. Composite economic threshold for three lepidopterous pests of cabbage. J. Econ. Entomol. 77:725–733.

LaPre, L. F., F. V. Sances, N. C. Toscano, E. R. Oatman, V. Voth, and M. W. Johnson. 1982. The effects of acaricides on the physiology, growth, and yield of strawberries. J. Econ. Entomol. 75:616–619.

Madden, L. V. 1983. Measuring and modeling crop losses at the field level. Phytopathol. 73:1591–1596.

Madden, L. V., S. P. Pennypacker, C. E. Antle, and C. H. Kingsolver. 1981. A loss model for crops. Phytopathol. 71:685–689.

Mailloux, G., and N. J. Bostanian. 1988. Economic injury level model for tarnished plant bug, *Lygus lineolaris* (Palisot de Beauvois) (Hemiptera: Miridae), in strawberry fields. Environ. Entomol. 17:581–586.

Michels, G. L. Jr., and C. C. Burkhardt. 1981. Economic threshold levels of the Mexican bean beetle on pinto beans in Wyoming. J. Econ. Entomol. 74:5–6.

Morrill, W. L., and G. L. Greene. 1974. Survival of fall armyworm larvae and yields of field corn after artificial infestations. J. Econ. Entomol. 67:119–123.

Mumford, J. D., and G. A. Norton. 1984. Economics of decision making in pest management. Annu. Rev Entomol. 29:157–174.

Norton, G. A. 1976. Analysis of decision making in crop protection. Agro-Ecosystems 3:27–44.

Ostlie, K. R. 1984. Soybean transpiration, vegetative morphology, and yield components following simulated and actual insect injury. Ph.D. Diss., Iowa State University, Ames.

Ostlie, K. R., and L. P. Pedigo. 1984. Water loss from soybeans after simulated and actual defoliation. Environ. Entomol. 13:1675–1680.

Parman, V. R., and M. C. Wilson. 1982. Alfalfa crop responses to feeding by the meadow spittlebug (Homoptera: Cercopidae). J. Econ. Entomol. 75:481–486.

Pedigo, L. P., R. B. Hammond, and F. L. Poston. 1977. Effects of green cloverworm larval intensity on consumption of soybean leaf tissue. J. Econ. Entomol. 70: 159–162.

Pedigo, L. P., and L. G. Higley. 1992. A new perspective of the economic-injury level concept and environmental quality. Am. Entomologist 38:12–21.

Pedigo, L. P., L. G. Higley, and P. M. Davis. 1989. Concepts and advances in economic thresholds for soybean entomology, p. 1487–1493. *In* A. J. Pascale (ed.) Proc. World Soybean Res. Conf. IV, vol. 3. Buenos Aires, Argentina.

Pedigo, L. P., S. H. Hutchins, and L. G. Higley. 1986. Economic injury levels in theory and practice. Annu. Rev. Entomol. 31:341–368.

Peterson, R. K. D., S. D. Danielson, and L. G. Higley. 1992. Photosynthetic responses of alfalfa to actual and simulated alfalfa weevil (Coleoptera: Curculionidae) injury. Environ. Entomol. 21:501–507.

Peterson, R. K. D., S. D. Danielson, and L. G. Higley. 1993. Yield response of alfalfa to simulated alfalfa weevil injury and development of economic injury levels. Agron. J. 85:595–601.

Poston, F. L., L. P. Pedigo, R. B. Pearce, and R. B. Hammond. 1976. Effects of artificial and insect defoliation on soybean net photosynthesis. J. Econ. Entomol. 69:109–112.

Poston, F. L., L. P. Pedigo, and S. M. Welch. 1983. Economic injury levels: Reality and practicality. Bull. Entomol. Soc. Am. 29:49–53.

Rogers, C. E. 1976. Economic injury level for *Contarinia texana* on Guar. J. Econ. Entomol. 69:693–696.

Shelton, A. M., C. H. Hoy, and P. B. Baker. 1990. Response of cabbage head weight to simulated Lepidoptera defoliation. Entomol. Exp. Appl. 54:181–187.

Shields, E. J., and J. A. Wymam. 1984. Effect of defoliation at specific growth stages on potato yield. J. Econ. Entomol. 77:1194–1199.

Showers, W. B., L. V. Kaster, and P. G. Mulder. 1983. Corn seedling growth stage and black cutworm (Lepidoptera: Noctuidae) damage. Environ. Entomol. 12: 421–424.

Showers, W. B., R. E. Sechriest, F. T. Turpin, Z B Mayo, and G. Szatmari-Goodman. 1979. Simulated black cutworm damage to seedling corn. J. Econ. Entomol. 72:432–436.

Showers, W. B., J. F. Witkowski, C. E. Mason, D. D. Calvin, R. A. Higgins, and

G. P. Dively. 1989. European corn borer development and management. North Central Reg. Ext. Publ. 327.

Smith, C. M. 1989. Chapter 7. *In* Plant resistance to insects. Wiley, New York.

Spike, B. P., and J. J. Tollefson. 1989. Relationship of plant phenology to corn yield loss resulting from western corn rootworm (Coleoptera: Chrysomelidae) larval injury, nitrogen deficiency, and high plant density. J. Econ. Entomol. 82:226–231.

Stern, V. M., R. F. Smith, R. van den Bosch, and K. S. Hagen. 1959. The integrated control concept. Hilgardia 29:81–101.

Stewart, J. G., K. B. McRae, and M. K. Sears. 1990. Response of two cultivars of cauliflower to simulated insect defoliation. J. Econ. Entomol. 83:1499–1505.

Stewart, J. G., and M. K. Sears. 1988. Economic threshold for three species of lepidopterous larvae attacking cauliflower grown in southern Ontario. J. Econ. Entomol. 81:1726–1731.

Sutter, G. R., and T. F. Branson. 1980. A procedure for artificially infesting field plots with corn rootworm eggs. J. Econ. Entomol. 73:135–137.

Tammes, P. M. L. 1961. Studies of yield losses. II. Injury as a limiting factor of yield. Tijdschr. Plantenziekten 67:257–263.

Todd, J. W., and L. W. Morgan. 1972. Effects of defoliation on yield and seed weight of soybeans. J. Econ. Entomol. 65:567–570.

Toscano, N. C., F. V. Sances, M. W. Johnson, and L. F. LaPre. 1982. Effect of various pesticides on lettuce physiology and yield. J. Econ. Entomol. 75:738–741.

Trumble, J. T., W. Carson, H. Nakakihara, and V. Voth. 1988. Impact of pesticides for tomato fruitworm (Lepidoptera: Noctuidae) suppression on photosynthesis, yield and nontarget arthropods in strawberries. J. Econ. Entomol. 81:608–614.

Vea, E. V., and C. J. Eckenrode. 1976. Seed maggot injury on surviving bean seedlings influences yield. J. Econ. Entomol. 69:545–547.

Walker, G. P., A. L. Voulgaropoulos, and P. A. Phillips. 1992. Effect of citrus bud mite (Acari: Eriophyidae) on lemon yields. J. Econ. Entomol. 85:1318–1329.

Weinzierl, R. A., R. E. Berry, and G. C. Fisher. 1987. Sweep-net sampling for western spotted cucumber beetle (Coleoptera: Chrysomelidae) in snap beans: Spatial distribution, economic injury level, and sequential sampling plans. J. Econ Entomol. 80:1278–1283.

Welter, S. C. 1991. Responses of tomato to simulated and real herbivory by tobacco hornworm (Lepidoptera: Sphingidae). Environ. Entomol. 20:1537–1541.

Wilson, L. T., P. J. Trichilo, and D. Gonzalez. 1991. Spider mite (Acari: Tetranychidae) infestation rate and initiation: Effect on cotton yield. J. Econ. Entomol. 84:593–600.

Wilson, M. C., R. E. Treece, and R. E. Shade. 1969. Impact of cereal leaf beetle larvae on yields of oats. J. Econ. Entomol. 62:699–702.

Yencho, G. C., L. W. Getzin, and G. E. Long. 1986. Economic injury level, action threshold, and a yield-loss model for the pea aphid, *Acyrthosiphon pisum* (Homoptera: Aphididae), on green peas, *Pisum sativum*. J. Econ. Entomol. 79:1681–1687.

Zadoks, J. C. 1985. On the conceptual basis of crop loss assessment: The threshold theory. Annu. Rev. Phytopath. 23:455–473.

Part 3: New Developments

Robert K. D. Peterson

10
The Status of Economic-Decision-Level Development

It seems especially fitting to review the status of economic decision levels in this, the first book dedicated to economic thresholds for pest management. It has been more than 35 years since the publication of the seminal paper by Stern et al. (1959); although the economic threshold is not the only decision rule for pest management, their concept has become the most widely accepted decision criterion in pest management (Pedigo et al. 1986). Indeed, the economic-decision-level concept provides a practical underpinning for the theory of modern pest management.

In reviewing economic decision levels, a number of questions arise. What theoretical advances in the EIL and the ET concept have occurred since 1959? How have EILs and ETs been calculated? For which insect orders, families, and species have EILs been developed, and why? How have EILs been determined for plant pathogens and weed pests? What is the future of economic decision levels for pest management?

In this chapter, the status of EILs and ETs is reviewed and the questions posed above are addressed. Both conceptual and empirical approaches for economic-decision-level development are addressed, and there is also an analysis of trends in each major area.

To accomplish this, thresholds published in refereed journal articles, book chapters, and conference proceedings from 1959 through 1993 are emphasized. Most articles were obtained from key-word searches in Biological Abstracts (BioSciences Information Service of Biological Abstracts, Philadelphia, Pa.), AGRICOLA (CD-ROM database, National Agricultural Library, Silver Platter Information, Inc., Boston, Mass.), and the Bibliography of Agriculture (National Agricultural Library, Oryx Publishers, Phoenix, Ariz.). In addition, literature-cited sections for economic-decision-level papers were checked for references to other EIL and ET publications. The sample of 218 publications on economic thresholds is not exhaustive; it is likely there is a bias toward English-language publications, but the sample is practically complete and representative of the research and trends in this area.

Thresholds presented in extension publications, without a scientifically published counterpart, are not discussed. Many thresholds presented in extension publications are not based on empirical studies; therefore, they typically are based on the experience of one or more extension specialists. This type of threshold is called a nominal threshold (Poston et al. 1983). Because nominal thresholds are not based on empirical data, they often are too conservative and do not represent an objective decision criterion. Additionally, many thresholds used in extension recommendations may be based on some form of calculated threshold, but the yield-loss data are difficult (if not impossible) to obtain. For these reasons, these types of thresholds were not used in the analysis.

Other papers that will not be discussed are those that do not explicitly address EILs. Many published papers present yield-loss assessments, and from many of these EILs can be calculated (e.g., Berberet et al. 1981), but that did not qualify them for this analysis.

Development of Economic Decision Levels

Although the original concept of the EIL was established in 1959, it was not until 13 years later that a formula for calculating EILs was produced (Stone and Pedigo 1972). Development of calculated EILs increased dramatically after 1972, primarily because of the accessibility and generality of the models of Stone and Pedigo (1972), Norton (1976), and Pedigo et al. (1986). To date, there are more than 200 published articles on EILs and ETs. Exponential growth in EIL development occurred in the 1980s, as indicated by the literature (Fig. 10.1), and continues into the 1990s. In 1990 alone, 21 EIL papers were published.

Although the total number of scientific articles on EILs and ETs increased exponentially in the 1980s, the increase in the number of conceptual papers from 1959 to the present has been more linear (about one per

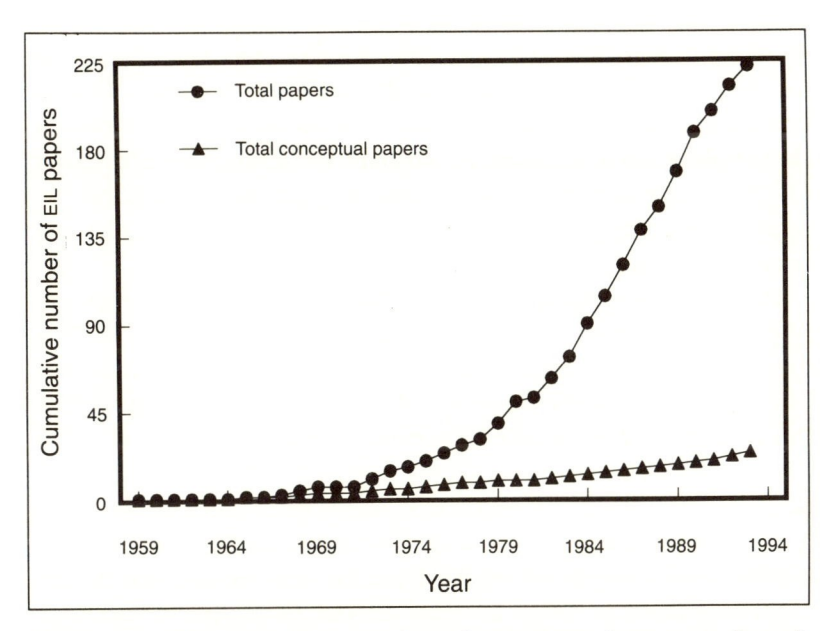

Figure 10.1. The cumulative number of EIL papers from 1959 through 1993. Each marker (triangle and circle) represents a specific year.

year) (Fig. 10.1). In this categorization, "conceptual" papers are theoretical contributions to the formulation or derivation of economic-decision levels; for example, Higley and Wintersteen (1992) calculated EILs for European corn borer, *Ostrinia nubilalis,* and, because the thresholds they calculate include a parameter for environmental risk, this paper is counted as conceptual.

Modern approaches to pest management rest on the philosophy that some level of pests or injury from pests is tolerable. Because it defines what is tolerable, the EIL is the keystone of modern pest management (Pedigo 1989). In light of this, we should expect publications addressing EILs and ETs, both empirically and conceptually, to be numerous, and perhaps dominant, in the literature of integrated pest management; however, since the paper by Stern et al. in 1959, there have been fewer than 250 published articles on economic thresholds. This number of journal articles, proceedings, and chapters may seem substantial, but in reality it is not. Comparisons with other areas related to integrated pest management are striking. During a five-year period, from 1988 through 1992, in the *Journal of Economic Entomology* alone there were 187 articles on insecticide resistance and resistance management; during the same period, there were 316 articles on biological control in the *Journal of Economic Entomology* and *Environmental Entomology*. These five-year totals for insecticide resistance and biological control rival or exceed the greater than thirty-year total of all articles on ETs and EILs.

These figures make it clear that the conceptual importance of the economic-decision level has not been reflected in the scientific literature. Although there have been several important conceptual advances in the area, the development of calculated and comprehensive EILs for many pests is lagging. If ETs form the keystone of modern pest management, the structure at present is rather unsteady.

General Trends

Most of the research on EILs has been associated with insects. More than 80% of all scientific literature concerning ETs (1959–1993) involves insect pests (Fig. 10.2). Scientific articles presenting EILs for mites constitute about 10%; papers presenting EILs for weeds and pathogens (including nematodes) make up the balance.

EIL studies have been biased toward arthropods for several reasons. First, the ET and the EIL concept was initially defined by entomologists. Second, economic decision levels were first recommended for pesticide resistance management and conservation of natural enemies. Resistance management and biological control in the 1960s were issues that primarily affected entomologists. Finally, insect and mite pests are more amenable to curative management techniques than weeds or pathogens. Plant diseases primarily are managed through preventive techniques, such as resistant cultivars and crop rotations. Weeds traditionally have been managed by cultivation and preemergence herbicides. Only relatively recently have postemergence herbicides been available. The availability of these curative herbicides has hastened the development of thresholds for weed pests.

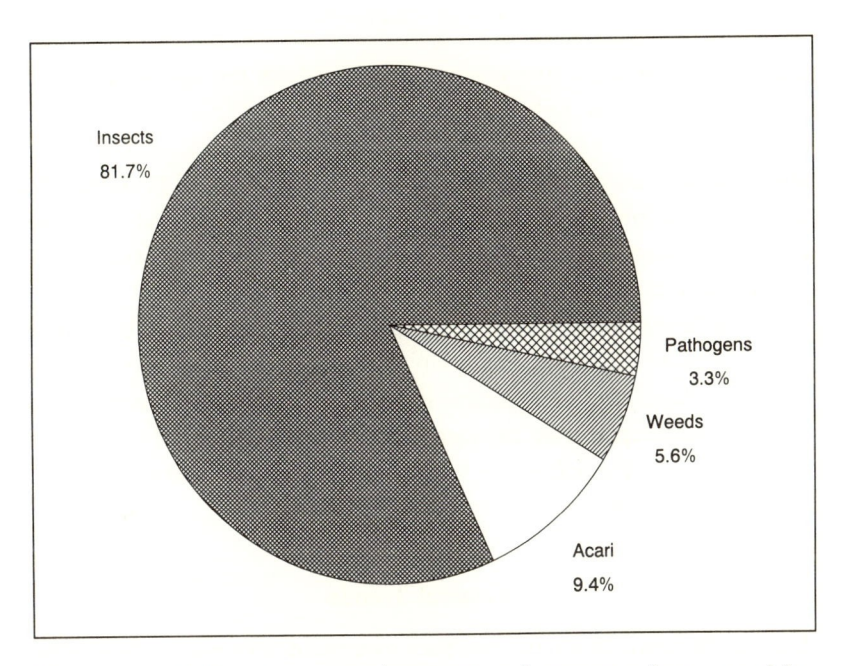

Figure 10.2. Publication comparisons: EIL and ET papers (1959–1993) by pest type.

Thresholds for Arthropods

Among arthropods, almost 50% of EILs are for lepidopteran pests. Approximately 17% are associated with homopterans and 14% with coleopterans (Fig. 10.3). The Lepidoptera, Coleoptera, and Homoptera contain the most important agronomic pest species in the world. Additionally, the types of injury produced by pests in these orders are more amenable to approaches for determining EILs. Together, these factors most likely explain why the majority of ETs are associated with species in these orders.

If insects and mites are grouped according to plant-injury type, 50% of EILs have been calculated for defoliators (including skeletonizers, miners, and stand reducers); about 29% are associated with assimilate sappers, 11% with mesophyll feeders (selective leaf and fruit feeders), and 10% with turgor reducers (root and stem feeders) (Fig. 10.4).

Most EILs have been calculated for defoliators because that injury is one of the most prevalent and important insect-injury types in many crop plants. In addition, the injury is easier to quantify than other injury types,

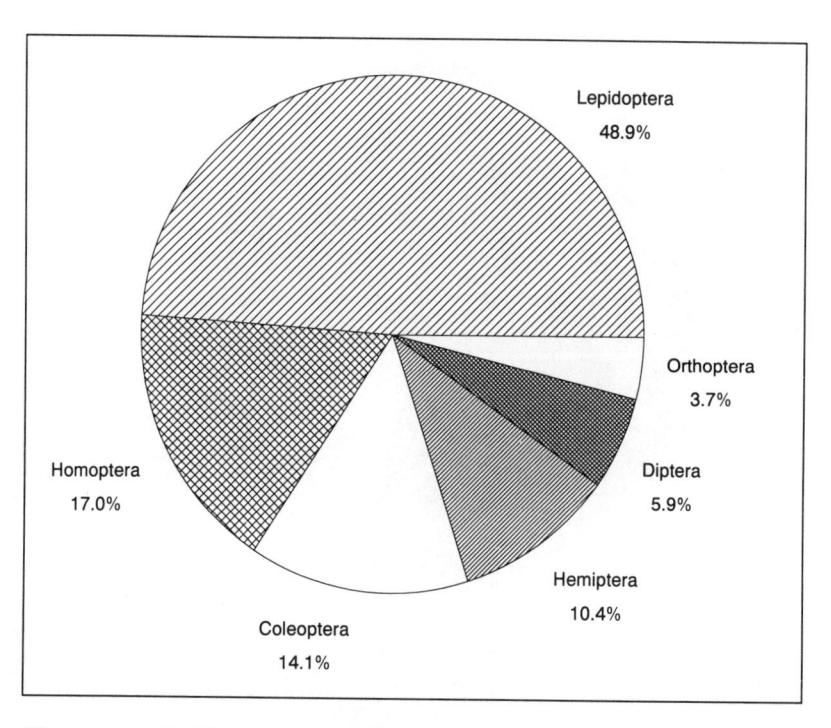

Figure 10.3. Publication comparisons: EIL and ET papers (1959–1993) by insect order.

and much defoliation injury can be simulated by hand removal of leaves (Hammond and Pedigo 1982, Ostlie and Pedigo 1984, Peterson et al. 1992, and Chapter 9).

Economic-decision levels have been determined for insect species in 30 families and for mite or tick species in 3 families (Table 10.1). Within the Lepidoptera, EILs have been developed most frequently for species in the family Noctuidae. There are three reasons for this trend. First, many pest species within this family are among the most deleterious leaf-mass consumers and turgor reducers of cultivated crops. Second, leaf-mass consumption injury and damage is relatively easy to quantify, compared with other injury types. Consequently, injury per insect and damage per unit injury (the *I* and *D* components of the EIL) are easier to determine for these pests. Third, most noctuid species are amenable to curative management techniques (usually in the form of insecticide treatments).

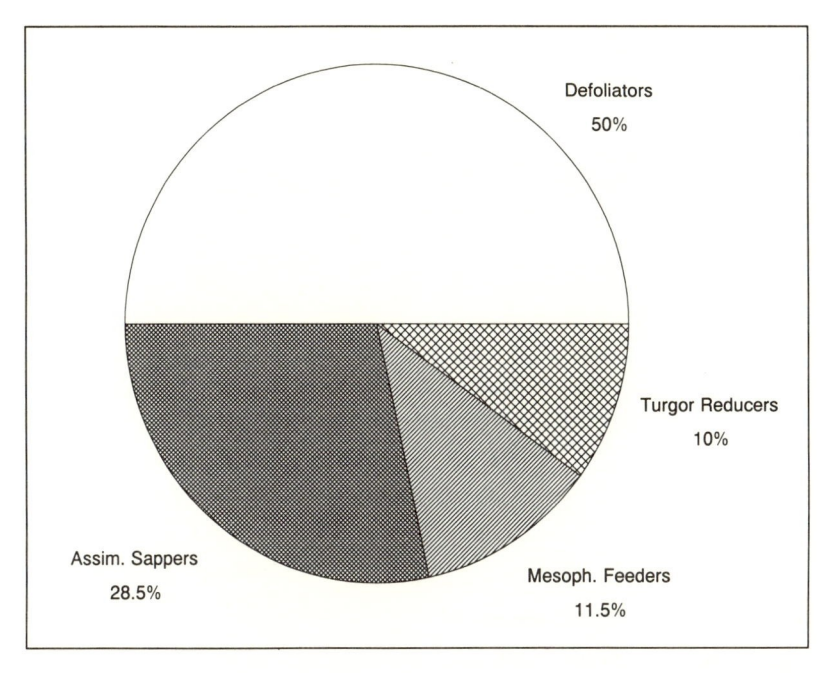

Figure 10.4. Publication comparisons: EIL and ET papers (1959–1993) by arthropod-injury type.

Economic injury levels have been calculated for more than 100 species of insects and weeds (Table 10.1). More EILs have been developed for the cabbage looper, *Trichoplusia ni,* than for any other insect species. Several EIL studies also have been conducted for the twospotted spider mite, *Tetranychus urticae,* potato leafhopper, *Empoasca fabae,* corn earworm, *Helicoverpa zea,* green cloverworm, *Plathypena scabra,* fall armyworm, *Spodoptera frugiperda,* diamondback moth, *Plutella xylostella,* and cabbage butterfly, *Pieris rapae.* EILs have been calculated for the twospotted spider mite on five different crops: apple, common bean, cotton, grape, and strawberry.

Economic injury levels have been determined for 43 commodities, including most major crops (Table 10.1). More EILs have been developed for soybean than for any other commodity; this is because of the numerous species of leaf-mass consumers that injure soybean, especially in the Southern states. A substantial amount of research has been conducted on soy-

Table 10.1. Taxa, species, and commodities with published economic-decision levels. The superscripted numbers refer to literature in the bibliography.

Taxon	Species	Commodity/Host
Insecta		
Acrididae	many species	Rangeland[52, 137, 202]
Aphididae	*Acyrthosiphon kondoi*	Alfalfa[189]
	Acyrthosiphon pisum	Alfalfa[48] Pea[115, 217]
	Diruaphis noxia	Wheat[6, 66]
	Macrosiphum avenae	Barley[8]
	Metapolophium dirhodum	Wheat[98]
	Myzus persicae	Potato[32, 62]
	Rhopalosiphum maidis	Barley[8, 187]
	Rhopalosiphum padi	Barley[8, 187]
	Schizaphis graminum	Wheat[29]
	Sipha flava	Sorghum[26]
	Sitobion avenae	Wheat[98]
Blattellidae	*Blatella germanica*	Human[220]
Cecidomyiidae	*Contarinia sorghicola*	Sorghum[75]
	Contarinia texana	Guar[164]
Cercopidae	*Aeneolamia varia saccharina*	Sugarcane[133]
Chrysomelidae	*Baliosus nervosus*	Soybean[27]
	Ceratoma trifurcata	Soybean[93, 129]
	Diabrotica undecimpunctata	Common bean[210]
	Leptinotarsa decemlineata	Potato[174]
	Odontota horni	Soybean[27]
	Oulema oryzae	Rice[108, 112, 207]
	Pagria signata	Cowpea[155]
Cicadellidae	*Empoasca fabae*	Alfalfa[49, 94] Potato[33, 97, 205] Soybean[135]
Coccinellidae	*Epilachna varivestis*	Common bean[120]
Coreidae	*Clavigralla tomentosicollis*	Cowpea[95]
	Leptoglossus phyllopus	Sorghum[74]
Culicidae	many species	Cattle[184]
Curculionidae	*Anthonomus eugenii*	Bell pepper[160]
	Anthonomus grandis	Cotton[70]

Table 10.1 continued

Taxon	Species	Commodity/Host
	Hypera brunneipennis	Alfalfa[105]
	Hypera postica	Alfalfa[105, 106, 152]
	Hypera variabilis	Alfalfa[68]
	Lissorhoptrus oryzophilius	Rice[122, 203, 207]
	Listronotus oregonensis	Carrot[191]
	Ostiorhynchus ligustici	Alfalfa[15]
	Smicronyx fulvus	Sunflower[141]
Gelechiidae	*Keiferia lycopersicella*	Tomato[151]
	Pectinophora gossypiella	Cotton[101]
Lygaeidae	*Blissus leucopterus leucopterus*	Sorghum[213]
Lyonetiidae	*Perileucoptera coffeella*	Coffee[204]
Membracidae	*Spissistilus festinus*	Soybean[183]
Miridae	*Calocoris angustatus*	Sorghum[176]
	Camplyomma verbasci	Apple[199]
	Lygocoris communis	Apple[118]
	Lygus hesperus	Maize[71]
	Lygus lineolaris	Apple[119] Common bean[193] Strawberry[114, 169]
Muscidae	*Haemotobia irritans*	Cattle[78]
	Stomoxys calcitrans	Cattle[31]
Noctuidae	*Agrotis gladiaria*	Maize[195]
	Agrotis ipsilon	Maize[195]
	Amathes c-nigrum	Grape[216]
	Anticarsia gemmatalis	Soybean[93]
	Earias vittella	Cotton[101]
	Euxoa ochrogaster	Grape[216]
	Helicoverpa zea	Cotton[9, 70, 85, 161, 162] Maize[40] Soybean[218] Tomato[84]
	Heliothis armigera	Cotton[101]
	Heliothis virescens	Cotton[85, 161, 162] Tobacco[109]
	Loxagrotis albicosta	Maize[4]
	Mamestra configurata	Canola[25]
	Papaipema nebris	Maize[51]

(continued)

Table 10.1 continued

Taxon	Species	Commodity/Host
	Peridroma saucia	Alfalfa[28] Peppermint[22, 45] Potato[180]
	Plathypena scabra	Soybean[80, 92, 142, 143, 145, 149, 194]
	Plusia nigrisigna	Cowpea[155]
	Pseudoplusia includens	Soybean[92]
	Spodoptera exigua	Cabbage[35] Soybean[92]
	Spodoptera frugiperda	Coastal bermudagrass[96, 117] Maize[56, 77, 89] Sorghum[117]
	Spodoptera littoralis	Cotton[86]
	Spodoptera litura	Bell pepper[127] Cowpea[155] Eggplant[127] Soybean[167]
	Trichoplusia ni	Cabbage[35, 37, 67, 87, 103, 123, 172, 173, 179, 181, 215] Cauliflower[192]
Otitidae	Tetanops myopaeformis	Sugarbeet[16]
Pentatomidae	Chlorodchroa ligata	Sorghum[74]
	Nezara viridula	Sorghum[74] Tomato[20]
	Oebalus pugnax	Sorghum[74]
Phoridae	Megaselia halterata	Mushroom[163]
Pieridae	Pieris rapae	Cabbage[37, 87, 103, 123, 172, 173, 177, 179, 181] Cauliflower[192]
Plutellidae	Plutella xylostella	Cabbage[35, 37, 87, 103, 123, 173, 177, 179, 181] Cauliflower[192]
Psychidae	Thyridopteryx ephemeraeformis	American arborvitae[156, 157]
Psyllidae	Psylla pyricola	Pear[1, 30]
Pyralidae	Chilo partellus	Maize[168, 175]
	Chilo suppressalis	Rice[2, 111, 207]
	Cnaphalocrocis medinalis	Rice[14, 121, 207]
	Diatraea saccharalis	Sorghum[63]
	Elasmopalpus lignosellus	Peanut[113] Sorghum[38]
	Ostrinia nubilalis	Maize[24]

Table 10.1 continued

Taxon	Species	Commodity/Host
Saturniidae	*Anisota senatoria*	Oak[43, 44]
	Hemileuca oliviae	Blue gamma grass[19]
Sphingidae	*Manduca quinquemaculata*	Tobacco[110]
	Manduca sexta	Tobacco[110]
Tortricidae	*Platynota idaeusalis*	Apple[104]
Acari		
Eriophyidae	*Aceria sheldoni*	Lemon[206]
	Calepitrimerus vitis	Grape[83]
Ixodidae	*Amblyomma americanum*	Cattle[10, 11, 55] Human[125]
Tarsonemidae	*Polyphagotarsonemus latus*	Lime[150]
Tetranychidae	*Colomerus vitis*	Grape[83]
	Eotetranychus willamettei	Grape[211, 212]
	Mononychellus tanajoa	Cassava[58]
	Oligonychus pratensis	Maize[5]
	Panonychus citri	Lemon[76]
	Panonychus ulmi	Apple[18, 88]
	Tetranychus urticae	Apple[88] Cotton[213] Grape[83] Common bean[17] Strawberry[159] *Euonymus alatus*[166]
Weeds		
	Abutilon theophrasti	Soybean[12, 13, 209]
	Alopecurus myosuroides	Wheat[54]
	Amaranthus tuberculatus	Soybean[13]
	Avena fatua	Wheat[47]
	Datura stramonium	Soybean[208, 209]
	Elytrigia repens	Canola[134]
	Helianthus annuus	Soybean[12, 13]
	Sorghum halepense	Soybean[164]
	Xanthium strumarium	Soybean[209]
Pathogens		
	Heterodera glycines	Soybean[33]
	Hoplolaimus columbus	Cotton[128] Soybean[128]
	Puccinia sorghi	Maize[53]

bean responses to the green cloverworm, *Plathypena scabra*. Many EILS also have been calculated for pests of cabbage, corn, cotton, alfalfa, and sorghum.

In a few instances, economic-decision levels have been determined for cattle pests such as the lone star tick, *Amblyomma americanum,* horn fly, *Haematobia irritans,* stable fly, *Musca autumnalis,* and mosquito species. Development of thresholds for livestock pests has proven problematic because of difficulties in quantifying injury. Pests injure livestock by imbibing blood and disrupting normal activity patterns. Additionally, many livestock pests that vector disease do not produce a "quantitative relationship between damage and injury and, therefore, are not amenable to calculation of EILS" (Pedigo et al. 1986). Recently, however, several studies have examined average daily gains of cattle when exposed to various levels of flies (e.g., Campbell et al. 1987, Catangui et al. 1993), and, from these studies, EILS can be calculated.

But—and here we meet the term *nuisance threshold*—many calculated thresholds for livestock pests, especially flies, considered in terms of their annoying effects on humans, may be too high. In these cases, the use of aesthetic thresholds, which are lower than ETs, may be indicated. Few attempts have been made to determine thresholds for pests of humans. Mount and Dunn (1983) proposed thresholds for the lone star tick, *Amblyomma americanum,* for humans in recreational areas. Additionally, thresholds for German cockroaches, *Blatella germanica,* have been calculated (Zungoli and Robinson 1984). Zungoli and Robinson determined aesthetic-injury levels for German cockroaches because an association between these pests and human health has not been well established; therefore, their thresholds were based on public tolerance of the pest infestation, using past infestation levels. Mount and Dunn (1983) indicated that a threshold of about 0.65 attacks by lone star ticks per human per day be allowed before management action is taken. The thresholds for lone star tick and German cockroach essentially are nuisance thresholds and are not EILS in the conventional sense.

Economic injury levels for pests that affect human health are virtually impossible to calculate because it is impracticable to assess market value for human life (Pedigo et al. 1986). This is why there are no EILS for disease vectors. Economic thresholds would require the acceptance of some level of disease and mortality before management action is taken.

Existing thresholds for disease vectors usually are based on insect numbers likely to produce epidemics.

Thresholds for Weeds

Development of economic-decision levels for weeds is still in its infancy; indeed, most EIL research in this area has been conducted only within the past 10 years. The commercial availability of postemergence herbicides has made it possible to manage some weeds after assessment of economic damage. In recent years, EILs have been determined for weed-induced stress in canola, soybean, sugarbeet, and wheat. Some weed species for which EILs have been calculated include *Avena fatua, Datura stramonium, Helianthus annuus, Abutilon theophrasti, Amaranthus tuberculatus,* and *Xanthium strumarium.*

The quantitative relationship between weed stress and yield loss currently is an active area of research (Chapter 7). Crop-loss functions are being determined for many crop species in diverse agronomic production systems. Typically, more than one weed species is present in a field or agroecosystem. This has necessitated research on multiple-species interactions and crop-loss functions. The EIL model used for weeds is similar to that used by entomologists. One important modification has been to include a variable to reflect potential impact of weed seedbanks on subsequent crops (Cousens et al. 1986 and Chapter 7).

Thresholds for Plant Diseases

Economic injury levels and ETs for plant diseases have not been developed as widely as they have for insects and weeds. There are several reasons for this. Pathogen populations are often difficult and impractical to quantify; therefore, yield-loss relationships are difficult to determine and, without this information, it is impossible to calculate an EIL. Additionally, few curative strategies exist for plant-disease problems; most management strategies involve preventive approaches.

For many plant-disease situations, thresholds indicate only if an epidemic is likely to occur within the current growing season or during the next season. If epidemics consistently cause substantial yield losses, thresholds based on yield-loss relationships and pathogen numbers may not be necessary. Thresholds are often determined by counting disease loci, spores, or vectoring insects. Epidemics typically are very sensitive to environmental conditions. Therefore, thresholds can be derived by rating disease severity or monitoring the humidity and temperature of the plant canopy. For many fungal diseases, figures giving total rainfall or the duration of rain events may be used as a threshold. Several of these criteria have been incorporated into computer models that are used to predict the likelihood of an epidemic. If a threshold is reached in one or several of the criteria, management actions may be recommended (Chapter 8).

Considerable research has been directed to establishing yield-loss relationships for plant-parasitic nematodes, but, by and large, practical ETs are not available (Barker and Noe 1987). Research on thresholds development should be the priority in nematology (Barker and Noe 1987).

Novel Approaches

Recent advances in the EIL concept have come not from drastically altering or redefining the EIL model of Pedigo et al. (1986) but rather from capitalizing on the inherent flexibility of its components. Advances have been achieved by adding or modifying the parameters of the model to meet specific needs. This has occurred in four areas: economic thresholds, aesthetic injury levels, environmental EILs, and multiple-species EILs.

Economic Thresholds

Research on ETs has lagged behind research on EILs. In the past decade, however, important conceptual advances in the ET concept have occurred. Most ETs presented in the literature and in extension publications are actually EILs—a misnomer that has caused much confusion among growers and researchers. For example, many agricultural economists and weed scientists typically use the term *economic threshold* to describe the break-even point (e.g., Headley 1972, Palti and Ausher 1986, and Chapter 7). However, the ET, as defined by Stern et al. (1959), is the population density at which control measures should be initiated to prevent an increasing pest population from reaching the EIL. Pedigo et al. (1986) refined the ET as "the injury equivalency of a pest population corresponding to the latest possible date for which a control tactic could be implemented to prevent increasing injury from causing economic damage."

The ET is a direct function of the EIL, but it has been much more difficult to calculate than the EIL. Although the ET is expressed in terms of pest density, it is actually a *time* for management action. As such, its value is influenced by variability associated with time delays for management action and implicit risks associated with predicting when a pest population will exceed the EIL (Pedigo et al. 1986).

Pedigo et al. (1986) and Ostlie and Pedigo (1987) proposed converting insect-population estimates into insect-injury equivalents. An insect-injury equivalent is the "potential total injury potential of one pest, if it were to survive through all injurious stages" (Ostlie and Pedigo 1987). Injury equivalency can be expressed as

$$EI = \sum_{i=1}^{n} e_i \cdot x_i \quad [1]$$

where e is the equivalency coefficient at stage i, x is the number of insects (damaging stages only) per sample in stage, and n is the total number of damaging stages for a pest. Therefore, injury equivalents are determined from estimates of pest-population age structure, pest density, and injury potential. Because many ETs erroneously assume 100% survival of injurious life stages, Ostlie and Pedigo (1987) incorporated survivorship estimates of the green cloverworm, *Plathypena scabra,* into injury equivalents to calculate ETs more accurately. This is an important theoretical advance, because many insect pests experience dramatic mortality before completing their injurious stages.

Similarly, Sterling (1984) suggested using "inaction levels" to account for the action of natural enemies in reducing pest populations. An "inaction threshold" is the density of natural enemies necessary to maintain pests below the economic-injury level. Therefore, no curative action would be taken if natural enemy populations were at or above the inaction level.

An important advance in ET research has been the identification of categories of thresholds. Pedigo et al. (1989) differentiated between subjective ETs and objective ETs. Subjective ETs are not based on calculated EILs; they are based on grower, scientist, or consultant experience and are often static, not changing with changes in market value and control cost (Pedigo

1989). Therefore, they are nominal thresholds (Poston et al. 1983). Subjective ETs are still the most commonly used thresholds and, despite potential inaccuracies, are more progressive than using no ET at all.

Pedigo et al.'s other ETs objective ETs, are based on calculated EILs and, therefore, change with the changing variables in the EIL equation. Pedigo et al. (1989) defined three types of objective ET: fixed, descriptive, and dichotomous. Fixed ETs are usually set at some percentage of the EIL and change as the EIL changes. This is the most commonly used objective ET. For example, an EIL for the alfalfa weevil, *Hypera postica,* may be 3 larvae per stem. If the ET is set at 80% of the EIL, then a curative treatment may be needed when there are 2.4 larvae per stem.

Descriptive ETs are based on estimates of population growth. Typically, sampling procedures are used to estimate the growth potential of the population. If the population is expected to exceed the EIL, action may be taken (Wilson 1985, Ostlie and Pedigo 1987). Peterson et al. (1993) suggested that the ET may be equivalent to the EIL when crop stage-specific EILs are known. For example, they determined EILs for third and fourth instar alfalfa weevils, *H. postica,* for the early bud stage of alfalfa development. Sampling larvae to assess losses at the early bud stage would occur before alfalfa has reached that stage (presumably during the late vegetative stage). Therefore, sampling would involve counting first and second instars per stem. At this point, using the EIL (calculated for third and fourth instars at the early bud stage) as the ET would provide an indication that injury by succeeding third and fourth instars would likely produce economic damage.

Dichotomous ETs rely on statistical procedures to classify a pest population as economic or noneconomic. The time-sequential sampling method is most commonly used (Pedigo and van Schaik 1984).

Categorization represents a significant achievement in the theoretical development of the ET, given the inaccurate presentation of the ET in a portion of the literature and the difficulty in calculating it. Improvements in theory, sampling, and the understanding of pest-population biology will improve the utility of the ET (Pedigo et al. 1989).

Aesthetic Injury Levels

Aesthetic injury levels are not EILs. To determine tolerances to pests based not on the economic value of the commodity but on the aesthetic value of the object—value attributed by the owner or the general public—researchers have used novel approaches: aesthetic attributes, such as form, color, texture, beauty, and general appearance, are not readily definable in economic parameters (Chapter 12). Resources such as lawns, ornamental plants, homes, businesses, and public buildings are examples of sites managed with aesthetic injury levels (AILs) (Chapter 12).

AILs primarily have followed the general EIL model of Pedigo et al. (1986). The cost of control, C, and the market value, V, are relatively easy to estimate, but the injury per pest, I, and the damage per pest, D, are more difficult to ascertain. Raupp et al. (1988) and Coffelt and Schultz (1990, 1993) used amounts of defoliation to estimate the injury per insect for conifers (Raupp et al.) and oak (Coffelt and Schultz). Leaf discoloration was used to estimate injury per spider mite (Sadof and Alexander 1993).

The damage per pest coefficient has been calculated by using the contingent-valuation method to establish damage in economic terms per unit of pest injury. Research has revealed that public tolerance of pest injury is quite low (<10% disfigurement) (Chapter 12). The AIL allows for some tolerance of pests; in other words, it is more accurate than treating with pesticides solely because of the presence of a pest.

Environmental EILs

The potential impacts of pest-management practices (especially pesticide use) on environmental quality has sparked interest in using the EIL concept to address environmental concerns. Environmentally-based EILs have been developed (Higley and Wintersteen 1992, Pedigo and Higley 1992, Higley and Pedigo 1993).

All the variables in the EIL equation reflect potential management decisions that could make pest-management activities more environmentally sustainable (Higley and Pedigo 1993). Pedigo and Higley (1992) discussed the practicality of manipulating the EIL variables (C, V, I, D, and K) to make decision making more environmentally responsive. They identified three variables (C, D, and K) that may be especially amenable to such manipulation. Crop market value (V) cannot be manipulated easily to satisfy environmental concerns because of the supply-and-demand nature of the free-market system; therefore, a safer, but more expensive, management practice may not be reflected in higher market prices. Injury per insect (I)

is difficult to manipulate because it entails changing the attributes of entire populations.

The proportion of total pest injury averted through the application of a management tactic (K) can contribute to environmental sustainability through research to determine the minimum pesticide rate to achieve a K value approximately equal to 1.0. Although the damage per pest (D) is "central to the EIL concept," the relationship between injury and damage is still poorly understood for most pest/host interactions (Higley and Pedigo 1993). One approach to improving environmental sustainability is to develop better understandings of injury and resultant damage which will result in more accurate EILs. Additionally, it may be possible to reduce D by enhancing the host's ability to compensate for injury.

Although not always recognized, a central tenet of integrated pest management is that some amount of pest injury is tolerated. Tolerance, as a resistance mechanism, is very appealing, because it represents a sustainable tactic: tolerant hosts do not select for pest resistance; moreover, EILs for tolerant hosts would be substantially higher than for susceptible hosts. Therefore, reduced pesticide inputs would result. As an integrated pest management tactic, tolerance may be the consummate sustainable strategy (Pedigo and Higley 1992).

Addressing environmental concerns through assigning realistic management costs (C) has been the subject of a considerable conceptual advance for EILs. The variable C traditionally has reflected the cost of the management tactic (e.g., the cost of a pesticide and application). Using contingent valuation methods, Higley and Wintersteen (1992) have estimated environmental costs of several pesticides. By assigning environmental costs to pesticides, EILs can be developed that incorporate a level of environmental protection (Chapter 14).

Multiple-species EILs

The establishment of multiple-species EILs is an important, albeit elusive, goal for pest management. These thresholds would represent a substantial advance in the development of pest management because they would allow us to manage a complex of biotic stressors instead of managing merely single species. Most thresholds developed for weeds are for multiple species because of the similarities in crop response to different species.

Conceptual approaches in multiple-species EILs for insects primarily have used the EIL equation developed by Pedigo et al. (1986) and the distinction made in that study between injury and damage. Characterizing the physiological impact of pest-induced injury on crops or livestock may provide a means to integrate injury from different biotic stressors (Pedigo et al. 1986, Peterson et al. 1992, Higley et al. 1993). An important implication of this approach is that if different pest species produce injuries resulting in similar physiological responses, the species can be placed into injury guilds. These guilds are unlike traditional injury guilds based on taxonomic status of the pest or the physical appearance of the injury; they are strictly based on physiological response and, therefore, are more useful when characterizing damage functions. Recent work suggests that there may be similarities in plant physiological responses associated with each insect injury guild (Welter 1989, Welter 1991, Higley 1992, Peterson et al. 1992, Higley et al. 1993, Peterson and Higley 1993). This is especially evident among leaf-mass consumers (Hutchins et al. 1988, Hutchins and Funderburk 1991, Peterson et al. 1992, and L. G. Higley, unpublished data). Unfortunately, there have not been enough experiments to determine if there are homogeneities of response with other injury types.

Hutchins et al. (1988) used the injury guild and injury equivalency approach to develop multiple-species EILs for defoliating insects on soybean. Because several insect defoliators of soybean produce similar responses, a single damage function was determined for all species (damage functions were based on injury equivalencies). To calculate multiple-species EILs based on their method, several assumptions must be met. Pest species must "1) produce a similar type of injury; 2) produce injury within the same phenological time-frame of the host; 3) produce injury of a similar intensity; and 4) affect the same plant part" (Hutchins and Funderburk 1991).

Few researchers have attempted to construct decision-making criteria for pests in two or more injury guilds. Blackshaw (1986) used a profit-maximization approach to determine the best decision for managing insects and weeds in barley. Hutchins et al. (1988) used a linear programming model to calculate net returns per hectare for optimal control strategies for combinations of three soybean pests: potato leafhopper, *Empoasca fabae,* velvetleaf, *Abutilon theophrasti,* and brown spot, *Septoria glycines.*

Dynamic Factors

Most research effort recently has been devoted to determining calculated EILs (*sensu* Stone and Pedigo 1972). However, several researchers have addressed the importance of understanding the role of dynamic factors (apart from market value and control cost) that affect the EIL, such as host phenology and pest-population dynamics (e.g., Auld and Tisdell 1987, Ostlie and Pedigo 1987, Evans and Stansly 1990, Hoy et al. 1990, Peterson et al. 1993, and Ring et al. 1993).

Recognizing that plant response is dependent on the timing of pest injury, Peterson et al. (1993) determined crop-growth-stage-specific EILs for the alfalfa weevil by simulating larval defoliation during the early bud stage. A yield-loss function was developed for the early bud stage and EILs were calculated. Evans and Stansly (1990) and Ring et al. (1993) found that yield-loss functions were influenced by plant growth stage in corn and cotton, respectively. Consequently, they calculated EILs for different growth stages and development intervals during a season.

These stage-specific responses may vary dramatically. For example, Berberet et al. (1981) documented an $I \times D$ value of 188.1 kg yield loss per hectare per alfalfa weevil larva per stem on vegetative stages of alfalfa. However, Peterson et al. (1993) documented an $I \times D$ value of 62.3 on the early bud stage. Consequently, EILs for larvae on early-bud alfalfa are approximately three times greater than EILs for larvae on vegetative-stage alfalfa.

Allowing for Uncertainty. Several researchers have incorporated the concept of uncertainty into EILs and ETs. Uncertainty is the rule when considering the parameters used to calculate the EIL. Economically, the market value of the commodity is not always known when management action is required. Indeed, the market value component of the EIL is often based on expected returns (Pedigo et al. 1986). Biologically, the pest density is not known precisely. Additionally, interactions of the crop with the environment may not be predictable. Finally, the reduction in injury (such as the efficacy of the pesticide or other management tactic) may not be known.

Plant (1986) introduced a random variable into an EIL equation (he terms it an ET equation) to represent uncertainty in the proportion of the pest population killed by a pesticide (the K parameter). He found that uncertainty may have a substantial effect on the calculated EIL. However, levels of uncertainty may be extremely difficult to estimate; hence, the technique may not be of any practical use. Auld and Tisdell (1987) used criteria such as maximization of net gain, minimax gain, and mean-risk analysis to determine whether to control or not to control weed pests when potential weed densities and the form of the crop-loss function were uncertain.

The Future of Economic Decision Levels

If the goal of integrated pest management is to reduce economic losses from pests, reduce environmental and human health hazards, and enhance agricultural sustainability, then economic-decision levels clearly play the key role. Although the economic-decision level is the most important aspect of pest-management programs, research to establish EILs and ETS has not kept pace with other research areas, such as pesticide resistance, biological control, and pest sampling. Indeed, many sampling programs for insect pests are remarkably sophisticated, at least in theory. Yet the ETS used in these sampling programs are often based on nothing more than intuition about damaging levels (i.e., nominal thresholds).

Calculated EILs exist for many of the major insect pests of the United States and Canada. However, when considered in toto, the literature reveals that calculated EILs exist for relatively few pests, and fewer still have anything more than simple EILs and ETs.

Because soybean has more EIL citations than any other commodity, reviewing the status of EILs in soybean helps illustrate where we stand regarding threshold development. There are at least 39 arthropod pests of soybean in the United States (Higley and Boethel 1994); of these, eight species are identified as the most damaging. The literature search (Table 10.1) indicates five of these have calculated EILs. These species are bean leaf beetle, *Ceratoma trifurcata,* velvetbean caterpillar, *Anticarsia gemmatalis,* corn earworm, *Helicoverpa zea,* green cloverworm, *Plathypena scabra,* and soybean looper, *Pseudoplusia includens.* Calculated EILs are being developed for southern green stink bug, *Nezara viridula,* and green stink bug, *Acrostrenum hilare,* although these are not yet in the literature (Lindgren, personal communication). Of the most damaging pests, only Mexican bean beetle lacks calculated EILs.

As stated above, there are more EILs for soybean than for any other commodity. Yet, not all perennially or occasionally important soybean pests have calculated EILs. For example, the twospotted spider mite, *Tetranychus urticae,* and the Mexican bean beetle, *Epilachna varivestis,* do not have calculated EILs for soybean. It is difficult to develop EILs for these pests because of the difficulty in determining relationships between injury and yield response, although EILs exist for some of these pests on other crops.

The literature review (Table 10.1) indicates nine calculated EILs for soybean insects. Because some papers report yield-loss data necessary for EILs, but not EILs themselves, the summary in Table 10.1 may slightly underreport calculated EILs. Extension recommendations list EILs for many additional soybean pests; most of these probably are nominal, rather than calculated, thresholds.

Two important observations follow from a consideration of thresholds with soybean insects. First, almost all major pests have calculated EILs, and some of these EILs are quite sophisticated. For example, three different types of EIL exist for bean leaf beetle to reflect seedling injury, leaf feeding, and pod feeding. Additionally, thresholds for different soybean stages exist within each of these categories. Similarly, EILs and ETs for other major soybean pests include considerations of soybean water status and the impact of natural enemies. Also, recent efforts have been directed at developing thresholds for injury guilds, which are likely to expand the number of pest species covered by calculated EILs.

Second, although EIL development in soybean clearly exceeds that of any other single commodity, either in number of species with calculated EILs or sophistication of the EILs themselves, only about one-third of the 39 potential pest species have calculated EILs. Not surprisingly, EIL development has focused on major pests and has lagged for species where determining yield-loss relationships is difficult.

Clearly, calculated EILs are lacking for many pests in many commodities. Even in soybean, which as noted has calculated EILs of considerable sophistication for almost all major pests, EIL development has yet to cover most potential insect pests. In other commodities, EIL development is not nearly as advanced as in soybean. Certainly, we cannot expect to have calculated EILs for every insect pest on every commodity, but it seems questionable whether or not we have achieved even a minimally acceptable level of EIL development in most commodities.

The lack of attention to the importance of economic-decision levels within the framework of integrated pest management is of concern. A reflection of the neglect is evident by the fact that this book is the first to focus on EILs—more than 35 years after the inception of the concept. How is it possible that a concept central to modern pest management has been so neglected and (as noted earlier) misunderstood for so many years?

The difficulty in establishing thresholds is well known. To manage pests effectively within an integrated pest management framework, we must understand the quantitative relationship between injury and damage (or when economic damage occurs). Plant responses to biotic stressors vary depending on the plant part injured, the timing of injury, the intensity of injury, the injury type, and interactions with the environment. Researchers must consider these factors when characterizing plant responses, determining injury per pest, and calculating EILs (Pedigo et al. 1986, Peterson et al. 1993, and Chapter 9).

For many years, entomologists primarily conducted the research to develop EILs for insect pests of crop plants. Many entomologists have only limited training in plant biology and physiology. Therefore, characterizing the relationship between injury and crop yield (one of the most important components of the EIL) often is difficult. Crop-growth-stage-specific EILs have not been developed for many crops, even though one pest species may injure the plant during several of its growth stages. Understanding host physiology is crucial to the development of multiple-growth-stage EILs, which are necessary for the development of comprehensive thresholds.

This is not to imply that only complicated thresholds are of value. Indeed, overly sophisticated thresholds and threshold calculations may not be amenable to practical application. They may be more accurate than simple calculated thresholds, but they may be so difficult to use that they are not accepted by growers, farm managers, or consultants. Sophistication must be balanced with accessibility.

There continues to be much confusion over the terms *economic injury level* and *economic threshold*. Several references are made to "static EILs" and "dynamic EILs." Static EILs are nonexistent and *dynamic EILs* is redundant. EILs, by their very definition, are dynamic. The variables that determine EILs (market value, management cost, injury per production unit, and damage per production unit) change throughout the growing season; therefore, the EILs also change. Several authors have used the term *multi-*

dimensional economic threshold or *dynamic action threshold* to incorporate biological criteria into the definition of an economic threshold (e.g., Gholson 1987, Nordh et al. 1988). For example, Nordh et al. (1988) write: "Because factors such as the pest population's age distribution, the crop maturity and vigor, and weather all influence the pest-density/yield-loss relationship, pest control decisions should be based upon a multidimensional ET." These authors fail to recognize that the EIL concept and the EIL equation, as presented by Pedigo et al. (1986), *is* biologically based. The factors mentioned by Nordh et al. (1988) can be included in the I and D components of the equation. Therefore, there is no need to distinguish between EILs and multidimensional EILs.

An important goal for integrated pest management is the ability to make management decisions for many pests. Progress toward comprehensive EILs continues to be slow. This most likely is a reflection of the large amount of time and research required to characterize damage functions for the host and injury per pest. Additionally, many pests produce injury types resulting in different host physiological responses. If this is so, it may be impossible to calculate EILs for multiple pests in different injury guilds (Pedigo et al. 1986); i.e., it may be impossible to extend the EIL concept to these situations.

Progress toward understanding host responses to injury guilds has occurred in recent years (e.g., Hutchins et al. 1988, Hutchins and Funderburk 1991, Peterson et al. 1992, Higley et al. 1993). The development of EILs for multiple pests within injury guilds would allow us to extend the usefulness of the EIL and ET concept and address several of the goals of integrated pest management and sustainable agriculture.

The paucity of EILs for pest management is perhaps indicative of a larger problem: the lack of an articulated paradigm for integrated pest management. Although there are dozens of different definitions for integrated pest management, there is no single treatise containing its philosophy and central tenets (Higley and Pedigo 1993). Continued progress in EIL development will occur only with concomitant progress toward a more conceptual approach to integrated pest management.

More than twenty years after the first calculated EILs were derived, Hutchins and Gehring (1993) stressed the need for a renewed emphasis on the research and development of economic thresholds in light of growing concerns about pesticide misuse. Higley and Pedigo (1993) suggested that "continued work within and upon the EIL concept" has been and will continue to be important for "conserving environmental quality and sustaining agricultural production." The EIL concept is conceptually strong and can accommodate changes in market value, control costs, and damage functions. In its traditional form, it essentially has incorporated aesthetic criteria, multiple pests, and environmental risk to expand its applicability. Reliance on calculated and comprehensive thresholds will enhance our ability to manage plant and livestock stress, resulting in agricultural and urban systems with minimal destabilizing effects from pests.

Acknowledgment

I thank M. Peterson for her help with database development. The database file (Microsoft Access version 1.1) is available on request from the author.

References

Auld, B. A., and C. A. Tisdell. 1987. Economic thresholds and response to uncertainty in weed control. Agric. Syst. 25:219–227.

Barker, K. R., and J. P. Noe. 1987. Establishing and using threshold population levels, p. 75–81. *In* J. A. Veech and D. W. Dickson (ed.) Vistas on nematology: A commemoration of the twenty-fifth anniversary of the Society of Nematologists. Soc. of Nematologists, Hyattsville MD.

Berberet, R. C., R. D. Morrison, and K. M. Senst. 1981. Impact of the alfalfa weevil, *Hypera postica* (Gyllenhal) (Coleoptera: Curculionidae), on forage production in nonirrigated alfalfa in the southern plains. J. Kans. Entomol. Soc. 54:312–318.

Blackshaw, R. P. 1986. Resolving economic decisions for the simultaneous control of two pests, diseases or weeds. Crop Prot. 5:93–99.

Campbell, J. B., I. L. Berry, D. J. Boxler, R. L. Davis, D. C. Clanton, and G. H. Deutscher. 1987. Effects of stable flies (Diptera: Muscidae) on weight gain and feed efficiency of feedlot cattle. J. Econ. Entomol. 80:117–119.

Catangui, M. A., J. B. Campbell, G. D. Thomas, and D. J. Boxler. 1993. Average daily gains of Brahman-Crossbred and English x Exotic feeder heifers exposed to low, medium, and high levels of stable flies (Diptera: Muscidae). J. Econ. Entomol. 86:1144–1150.

Coffelt, M. A., and P. B. Schultz. 1990. Development of an aesthetic injury level to decrease pesticide use against orangestriped oakworm (Lepidoptera: Saturniidae) in an urban pest management project. J. Econ. Entomol. 83:2044–2049.

Coffelt, M. A., and P. B. Schultz. 1993. Quantification of an aesthetic injury level and threshold for an urban pest management program against orangestriped oakworm (Lepidoptera: Saturniidae). J. Econ. Entomol. 86:1512–1515.

Cousens, R., C. J. Doyle, B. J. Wilson, and G. W. Cussans. 1986. Modelling the economics of controlling *Avena fatua* in winter wheat. Pestic. Sci. 17:1–12.

Evans, D. C., and P. A. Stansly, 1990. Weekly economic injury levels for fall armyworm (Lepidoptera: Noctuidae) infestation of corn in lowland Ecuador. J. Econ. Entomol. 83:2452–2454.

Gholson, L. E. 1987. Adaptation of current threshold techniques for different farm techniques. Plant Dis. 71:462–465.

Hammond, R. B., and L. P. Pedigo. 1982. Determination of yield loss relationships for two soybean defoliators by using simulated insect defoliation techniques. J. Econ. Entomol. 75:102–107.

Headley, J. C. 1972. Defining the economic threshold, p. 100–108. *In* Pest control strategies for the future. Nat. Acad. Sci., Washington DC.

Higley, L. G. 1992. New understandings of soybean defoliation and their implications for pest management, p. 56–65. *In* L. G. Copping, M. G. Green, and R. T. Rees (ed.) Pest management in soybean. Elsevier Science Publishers, London UK.

Higley, L. G., and D. J. Boethel (ed.) 1994. ESA handbook of soybean insects. Entomol. Soc. Am., Lanham MD.

Higley, L. G., J. A. Browde, and P. M. Higley. 1993. Moving towards new understandings of biotic stress and stress interactions, p. 749–754. *In* D. R. Buxton, R. Shibles, R. A. Forsberg, B. L. Blad, K. H. Asay, G. M. Paulson, and R. F. Wilson (ed.) International crop science I. Crop Sci. Soc. Am., Madison WI.

Higley, L. G., and L. P. Pedigo, 1993. Economic injury level concepts and their use in sustaining environmental quality. Agric. Ecosystems Environ. 46:233–243.

Higley, L. G., and W. K. Wintersteen. 1992. A novel approach to environmental risk assessment of pesticides as a basis for incorporating environmental costs into economic injury levels. Am. Entomologist 38:34–39.

Hoy, C. W., C. E. McCulloch, A. J. Sawyer, A. M. Shelton, and C. A. Shoemaker. 1990. Effect of intraplant insect movement on economic thresholds. Environ. Entomol. 19:1578–1596.

Hutchins, S. H., and J. E. Funderburk. 1991. Injury guilds: A practical approach for managing pest losses to soybean. Agri. Zool. Rev. 4:1–21.

Hutchins, S. H., and P. J. Gehring. 1993. Perspective on the value, regulation, and objective utilization of pest control technology. Am. Entomologist 39:12–15.

Hutchins, S. H., L. G. Higley, and L. P. Pedigo. 1988. Injury equivalency as a basis for developing multiple-species economic injury levels. J. Econ. Entomol. 81:1–8.

Lindgren, S. Personal communication. Department of Entomology, Louisiana State University.

Mount, G. A., and J. E. Dunn. 1983. Economic thresholds for lone star ticks (Acari: Ixodidae) in recreational areas based on a relationship between CO_2 and human subject sampling. J. Econ. Entomol. 76:327–329.

Nordh, M. B., L. R. Zavaleta, and W. G. Ruesink. 1988. Estimating multidimensional economic injury levels with simulation models. Agric. Syst. 26:19–33.

Norton, G. A. 1976. Analysis of decision making in crop protection. Agro-Ecosystems 3:27–44.

Ostlie, K. R., and L. P. Pedigo. 1984. Water loss from soybeans after simulated and actual insect defoliation. Environ. Entomol. 13:1675–1680.

Ostlie, K. R., and L. P. Pedigo. 1987. Incorporating pest survivorship into economic thresholds. Bull. Entomol. Soc. Am. 33:98–102.

Palti, J., and R. Ausher. 1986. Crop value, economic damage thresholds, and treatment thresholds. Crop protection monographs: Advisory work in crop pest and disease management. Springer-Verlag, New York.

Pedigo, L. P. 1989. Entomology and pest management. Macmillan, New York.

Pedigo, L. P., and L. G. Higley. 1992. A new perspective of the economic-injury level concept and environmental quality. Am. Entomologist 38:12–21.

Pedigo, L. P., L. G. Higley, and P. M. Davis. 1989. Concepts and advances in economic thresholds for soybean entomology, p. 1487–1493. In A. J. Pascale (ed.) Proc. World Soybean Res. Conf. IV, Vol. 3, Asociación Argentina de la Soja, Buenos Aires, Argentina.

Pedigo, L. P., S. H. Hutchins, and L. G. Higley. 1986. Economic injury levels in theory and practice. Annu. Rev. Entomol. 31:341–368.

Pedigo, L. P., and J. W. van Schaik. 1984. Time-sequential sampling: A new use of the sequential probability ratio test for pest management decisions. Bull. Entomol. Soc. Am., 30:32–36.

Peterson, R. K. D., S. D. Danielson, and L. G. Higley. 1992. Photosynthetic responses of alfalfa to actual and simulated alfalfa weevil (Coleoptera: Curculionidae) injury. Environ. Entomol. 21:501–507.

Peterson, R. K. D., S. D. Danielson, and L. G. Higley. 1993. Yield responses of alfalfa to simulated alfalfa weevil injury and development of economic injury levels. Agron. J. 85:595–601.

Peterson, R. K. D., and L. G. Higley 1993. Arthropod injury and plant gas exchange: Current understandings and approaches for synthesis. Trends Agric. Sci. 1:93–100.

Plant, R. E. 1986. Uncertainty and the economic threshold. J. Econ. Entomol. 79:1–6.

Poston, F. L., L. P. Pedigo, and S. M. Welch. 1983. Economic injury levels: Reality and practicality. Bull. Entomol. Soc. Am. 29:49–53.

Raupp, M. J., J. A. Davidson, C. S. Koehler, C. S. Sadof, and K. Reichelderfer.

1988. Decision-making considerations for aesthetic damage caused by pests. Bull. Entomol. Soc. Am. 34:27–32.
Ring, D. R., J. H. Benedict, J. A. Landivar, and B. R. Eddleman. 1993. Economic injury levels and development and application of response surfaces relating insect injury, normalized yield, and plant physiological age. Environ. Entomol. 22:273–282.
Sadof, C. S., and C. M. Alexander. 1993. Limitations of cost-benefit-based aesthetic injury levels for managing twospotted spider mites (Acari: Tetranychidae). J. Econ. Entomol. 86:1516–1521.
Smith, R. F. 1969. The importance of economic injury levels in the development of integrated pest control programs. Qualitas Plant. Matero. Veg. 17:81–92.
Sterling, W. 1984. Action and inaction levels in pest management. Tex. Agric. Exp. Stn. Bull. 1480.
Stern, V. M. 1965. Significance of the economic threshold in integrated pest control. Proc. FAO Symp. Integ. Pest Cont. 2:41–56.
Stern, V. M., R. F. Smith, R. van den Bosch, and K. S. Hagen. 1959. The integrated control concept. Hilgardia 29:81–101.
Stone, J. D., and L. P. Pedigo. 1972. Development and economic-injury level of the green cloverworm on soybean in Iowa. J. Econ. Entomol. 65:197–201.
Sylven, E. 1968. Threshold values in the economics of insect pest control in agriculture. Nat. Swedish Inst. Plant Prot. Contr. 14:65–79.
Welter, S. C. 1989. Arthropod impact on plant gas exchange, p. 135–150. *In* E. A. Bernays (ed.) Insect-plant interactions. Vol. 1. CRC Press, Boca Raton FL.
Welter, S. C. 1991. Responses of tomato to simulated and real herbivory by tobacco hornworm. Environ. Entomol. 20:1537–1541.
Wilson, L. T. 1985. Developing economic thresholds in cotton. *In* R. E. Frisbie and P. L. Adkisson (ed.) Integrated pest management on major agricultural systems. Texas Agric. Exp. Sta. MP-1616, College Station TX.
Zungoli, P. A., and W. H. Robinson. 1984. Feasibility of establishing an aesthetic injury level for German cockroach pest management programs. Environ. Entomol. 13:1453–1458.

John B. Campbell and Gustave D. Thomas

11
Economic Thresholds for Veterinary Pests

In animal agriculture, use of the EIL concept as the primary element in making a control decision for arthropod pests of livestock is relatively new. There are reasons (listed below) for this difference between livestock producers and crop producers in adapting integrated-pest-management (IPM) practices in pest control.

– Livestock are often a secondary commodity, except in commercial feedlots, ranches, dairies, and swine units, whereas crops are often primary in a farm operation.

– Arthropod pests of livestock often are subtle in their effect on the animal; as successful parasites, they do not kill their host.

– Economic thresholds for arthropod vectors of animal disease are difficult to determine.

– Many of the livestock arthropod pests attack animals only as highly mobile adults.

– Control methodologies are either inefficient and expensive, very efficient and inexpensive, or nonexistent.

As just noted, control methodologies for livestock insects can be grouped into three categories: expensive and inefficient, economical and very efficient, and nonexistent or not effective. Examples of livestock insect pests that fit into the control category of expensive and inefficient would be the filth fly pests of confined livestock. Control of these pests requires strict sanitation on a relatively rigid schedule, combined with judicious use of pesticides. This methodology is expensive when viewed on a short-term basis, but it may be economically sound when viewed on a long-term basis.

Reduction of fly breeding requires good drainage, leak-free watering systems, development of mounds for drainage and animal comfort, well-maintained feedbunks, and a good system of animal-waste management.

An example of the second category—an economical and very efficient control methodology for livestock insect pests—would be the use of systemic insecticides for control of cattle grubs. The products available are very effective and, if used on animals that remain on the farm or ranch for a period of three or four years, will practically eradicate cattle grubs for that producer. In addition, these products will reduce lice populations and any blood-feeding Diptera that are present at treatment time. This treatment also is one method recommended for the management of pyrethroid-resistant horn flies.

The newer, biologically produced pesticides (e.g., ivermectin) are so broad in spectrum that they control not only cattle grubs but also blood-feeding ectoparasites and parasitic nematodes. Because of the additional benefits of these low-cost pesticides, cattle are treated for grubs without regard to grub-infestation levels. When insecticides are inexpensive and efficient, treatment decisions usually are not based on economic thresholds of the pest.

An example of the third category—where control methodology does not exist—is in the aquatic Diptera complex. This includes black flies, tabanids, mosquitoes, and *Culicoides*. These pests breed in aquatic habitats that are impracticable to treat. The only method of control is the use of animal sprays, which are expensive, generally short-lived, and ineffective. Ticks and stable flies feeding on range and pasture cattle could be included in this category.

The EILs of livestock insects are often very difficult to determine because of several compounding factors. The weight-gain response of cattle to horn flies, *Haematobia irritans* (L.), may be dependent on factors such as breed of cattle, quantity and quality of forage, and heat stress as much as on numbers of horn flies. In cases of other insect pests, the EIL for animal response depends on how efficient the insect is in transmitting diseases. An EIL for face flies, *Musca autumnalis* DeGeer, on cattle is probably unnecessary; that is, if eye diseases are not considered (see below). Another factor that makes determining EILs for livestock insects difficult to determine may be the mobility of the insect, taken in combination with the inefficiency of current control methodologies. A case in point is the EIL for stable flies on range and pasture cattle. No effective treatments are available for stable flies; therefore, trials comparing weight gains of cattle free of flies with cattle infested with flies cannot be conducted. Other examples of complexity complicating EIL determination include the interaction of stable flies and heat stress on feedlot cattle, and interactions between animal nutrition and cattle-lice infestations: cattle under attack by stable flies tend to bunch, which may increase heat stress; cattle-lice populations tend to decline on cattle being fed a highly nutritious finishing ration.

The rest of this chapter will address EILs for the major arthropod pests of livestock on an individual basis. In addition to this summary, several reviews on losses to livestock by arthropods (Steelman 1976, Drummond et al. 1981, and Drummond 1987) are pertinent to issues involved with EILs.

Arthropod Pests of Cattle

Ticks

Globally, the most serious ectoparasites of livestock are probably ticks. They are serious pests in Africa, Central and South America, Australia, and the southern half of the United States. Lancaster and Meisch (1986) state that, worldwide, various species of ticks infest 800 million cattle and an equal number of sheep. This review indicates that ticks affect livestock in several ways: physiologically by irritation, allergic response, and loss of blood that results in reduced weight gains, weight losses, and, in extreme cases, death; transmission of diseases such as anaplasmosis, epizootic bovine abortion, and babesiosis; paralysis; and damage to carcass, fleece, or hide from feeding and secondary myiasis from flies invading tick bites or wounds.

Despite the magnitude of the tick problem to livestock (cattle in particular), not many studies in the literature relate tick numbers to economic losses. In the review by Drummond (1987) of research conducted by Lancaster et al. (1955), Barnard (1985), Barnard et al. (1986), and Ervin et al. (1987), the injury threshold for lone star ticks, *Amblyoma americanum* (L.), was 15 or more engorging females; the ET was 26 to 38 feeding females. The data of Ervin et al. (1987) indicated that the presence of 40 feeding female ticks reduced weight gains of stocker cattle by 26 kg over a 100-day trial. These results are somewhat different, but that may be because of different trial circumstances. In several of these studies, *Bos taurus* cattle had greater numbers of ticks and were more affected by ticks than were *B. indicus* or *B. indicus* × *B. taurus* crossbred cattle. Stacey et al. (1978) and Williams et al. (1978) obtained similar results with these two types of cattle exposed to Gulf Coast ticks, *Amblyomma maculatum* Koch. In addition to decreased weight-gain performance, cattle infested by Gulf Coast ticks have a condition known as gotch ear, in which the tick feeding destroys the ear cartilage. Gladney et al. (1977) noted that calves with gotch ear often bring $4 per hundredweight less than those with undamaged ears.

Horn Flies

The horn fly, *Haematobia irritans* (L.), has been in the United States since the late 1880s. Morgan and Thomas (1974, 1977) have compiled much of the world literature on the species into an annotated bibliography. Although the horn fly is widespread throughout North America and recently has spread to Central and South America, there is little research data in the literature on the ET of the pest. Many of the published economic studies are comparisons of weaning or yearling weight gains between treated and untreated cattle where numbers of flies on untreated cattle were generally high. Drummond (1987), Drummond et al. (1981), and Campbell and Thomas (1992) have reviewed the recent literature on economic losses to cattle from the horn fly. In studies by Schreiber et al. (1987) and Haufe (1982), the economic population threshold for weaned calves was 200 or more flies per cow and 235 flies per yearling steer. In most weight-gain studies, horn fly populations in a given area are about the same each year (Duren and O'Keefe 1972, Campbell 1976, Kunz et al. 1984, Kinzer et al. 1984, and Quisenberry and Strohbehn 1984); thus, it is difficult to determine the effect of different horn fly populations on cattle weight gains. Most researchers feel that temperature, breed of cattle, cattle management, and quality and quantity of forage production, in addition to numbers of horn flies, interact in determining the economic threshold for horn flies.

The advent of pyrethroid resistance in horn fly populations, where insecticide-impregnated ear tags have been widely used, has been instrumental to some extent in making cattle producers aware of the economic threshold for the pest. One of the recommendations for management of pyrethroid resistance is to treat only when population thresholds reach 200 flies per animal. Cattle producers resist this management practice because it requires a roundup of cattle from summer pastures for ear-tag treatment, which may offset some of the gain from treatment.

Cattle Grubs

There are two species of cattle grubs in the United States: *Hypoderma lineatum* (deVillers), the common grub, and *H. bovis* (L.), the northern grub. Scholl et al. (1988) pointed out disagreements in the literature on the economic thresholds for cattle grubs (Scharff 1950, Cox et al. 1967, Rich 1970, Collins and DeWhirst 1971, Campbell et al. 1973, Smith 1976, and Khan and Kozub 1981).

Several factors can influence the degree of depressed animal performance because of cattle grubs. One of these factors is the length of time before the cattle were weighed, posttreatment. In some of the senior author's studies, it was noted that treated cattle had gained less than nontreated cattle at 28-days posttreatment. Sharff et al. (1962) reported serious effects on calves treated with systemic insecticides. This probably indicates the insecticides caused some stress to cattle for a period of time. However, overall, the treated cattle gained significantly more weight than did the untreated cattle during the 140-day trial (Campbell et al. 1973). In another study, Campbell et al. (1974) showed that the impact of grubs on weight gains of yearling cattle was greater in February, shortly before grubs emerged from the back, than it was 60 days later. This implies, as discussed by Scholl et al. (1988), that compensatory gain may occur after grubs have emerged and the wounds have healed.

There seems to be a consensus among packers that the economic threshold for cattle grubs is about five grubs per animal. At that threshold, they reduce the price of the animal about $5 per hundredweight. However, this number has little meaning to the cattle producer: cattle grubs are internal parasites; thus, the extent of infestation is not known until four or five months after the treatment decision had to be made. The treatment decision is made by the producer on the basis of potential infestation, efficacy, and cost of treatment and the added benefit of the grub treatment in reducing or controlling populations of other parasitic species (cattle lice and/or nematodes). Cattle ranchers generally include grub treatment as a standard herd-health procedure, regardless of the infestation potential.

Cattle Lice

Few economic studies in the literature address cattle lice. Generally, these studies are comparisons of treated and louse-infested cattle, using whatever louse population level was present (Scharff 1962, Ely and Harvey 1969, Gibney et al. 1985). In all of these studies, the impact of louse infestations on cattle weight gains seemed to interact with nutritional levels at which cattle were being fed. If cattle were at lower levels of nutrition (growing vs. finishing), the reduction in weight gains was greater. Nelson (1984) reviewed the interaction between nutrition and parasitology and reported that a high nutrition level may offset the normal reduction in animal performance caused by the parasite. Studies on cattle lice seem to verify this concept.

In the economic studies on lice cited, the authors indicate that a heavy infestation level (10 or more lice per 6.452 cm^2 [1 in^2] [Hoffman et al. 1969]) generally was required to reduce weight gains of cattle in a growth mode. In all of the studies cited, except that of Gibney et al. (1985), the trial cattle were infested with only one louse species. In the Gibney et al. study, cattle were infested with three species, *Bovicola bovis* (L.), *Linognathus vituli* (L.), and *Solenopotes capillatus* (Enderlein), simultaneously.

However, cattle producers must consider not only the possible effect of cattle lice on animal weight-gain performance but also the physiological effects of the blood-feeding species. Several studies implicate cattle lice in lowering erythrocyte numbers and the hemoglobin ratios that causes anemia. Anemic animals are more susceptible to disease, particularly respiratory diseases prevalent in the winter months when lice populations are the highest (Smith and Richards 1955, Shemanchuk et al. 1960, Haufe 1962, Scharff 1962, Collins and DeWhirst 1965, Freer and Gahan 1968, Nickel et al. 1970).

Other factors that inhibit consideration of an ET by producers when making a treatment decision are: (1) treatment for cattle grubs will reduce lice populations; (2) lice may not represent an economic infestation when cattle are processed in the fall but may reach that level as the winter progresses (particularly in severe winters); (3) cattle producers do not want cows to transmit lice to calves when they are born in the spring; and (4) lice-infested cattle are destructive to pasture fences because cattle rub on fences in response to itching caused by the lice.

The Face Fly

The face fly, *Musca autumnalis* DeGeer, was first reported in North America in 1952 (Vockeroth 1953). After its introduction, it spread from the East Coast to the West Coast of southern Canada and the northern United States in less than 20 years. However, for reasons unknown, it is presently not found in economically significant numbers in the Northern Great Plains and Intermountain regions (Peitzmier et al. 1992). Although the face fly has long been present in Europe, Asia, and Africa, it was considered only a minor pest until it was introduced into North America. Morgan et al. (1983) published an annotated bibliography of the face fly that contains literature citations dating to the early nineteenth century. It is possible that cattle in the United States are more susceptible to pinkeye (infectious bovine keratoconjunctivitis) than European cattle. The bacterium, *Moraxella bovis* (Hauduroy), generally is thought to be the causative agent for pinkeye (Hughes et al. 1965, Wilcox 1968, Pugh and Hughes 1975, Baptista 1979). Hall (1984), in a review of the literature on the relationship of the face fly and pinkeye, indicates that face fly feeding is only one of the factors that may predispose cattle to infection by pinkeye. Steve and Lilly (1965) and Burton (1966) determined that *M. bovis* could survive for a few days on inert objects, but the organism disappeared within 48 hours (Glass et al. 1982) from the digestive tract of face flies. Several authors correlated size of face fly populations with the incidence of pinkeye (Cheng 1967, Gerhardt et al. 1982). In studies that examined face flies collected from herds of cattle, Berkebile et al. (1981) found less than 1% of the flies were contaminated with *M. bovis;* Gerhardt et al. (1982) found 8.8% contamination.

Shugart et al. (1979) determined that face fly feeding caused abrasions of the ocular tissue of cattle. Broce and Elzinga (1984) examined the mouth parts of face flies with scanning electron microscopy and found the prestomal teeth were long and had jagged terminal points that accounted for the damage to the conjunctival tissue observed by Shugart et al. (1979).

Brown and Adkins (1972), Arends et al. (1982b), and Shugart et al.

(1979) determined that cattle fed on by face flies prior to exposure to *M. bovis* were more susceptible to pinkeye infection than cattle not previously exposed to face fly feeding. However, most researchers agree that the transmission of the organism by the fly is primarily mechanical.

The economic impact of pinkeye on cattle performance is well known (Thrift and Overfield 1974, Killinger et al. 1977, Ward and Nielson 1979). However, in the absence of pinkeye, the economic impact of face flies on cattle may be slight. Recent studies by Schmidtmann et al. (1981), Arends et al. (1982a), McMillan et al. (1982), and Schmidtmann et al. (1984) indicate the face fly does not suppress weight gains or quantity or quality of milk.

A further complication of the issue of the economic impact of the face fly is the association of the fly with other diseases. *Brucella abortus* Bang was cultured from contaminated flies for 12 hours but not 72 hours, and there was no midgut replication, which infers the possibility of only mechanical transmission of the disease by face flies (Cheville et al. 1989). Johnson et al. (1991) found that Bovine herpesvirus–1 could be detected for 48 hours after face fly feeding, but they concluded that the face fly is unlikely to transmit the disease, either mechanically or biologically.

Chitwood and Stoffolano (1971) reported the presence of the eye worm, *Thelazia* spp., in face flies. Later Geden and Stoffolano (1981) determined the face fly is a biological vector of the worm to livestock. The face fly is also a vector of *Parafilaria bovicola* (Tubangui) in Europe (Bech-Nielsen et al. 1982).

To achieve a satisfactory level of control for face flies, the producer must make a special effort (spray, add dust bags or oilers, or use feed additives) in addition to or instead of the insecticide ear tags. As indicated earlier, horn flies have developed resistance to pyrethroid insecticides used in ear tags, but resistance has not yet been reported for face flies. This decision is usually made only when producers note bunching of cattle from fly annoyance, excessive eye weeping (particularly among calves), disrupted grazing patterns, and an increase in the incidence of pinkeye. It is difficult to say at what population level of face fly these events occur, but probably it is at levels of 10 or more face flies per animal.

Stable Flies and House Flies

The stable fly, *Stomoxys calcitrans* (L.), and the house fly, *Musca domestica* L., are found worldwide, wherever there are domestic livestock. The life cycles of the two species are similar, except the house fly may complete its life cycle in from one to two weeks while the stable fly requires two to three weeks in summer. The two species generally are closely associated, since both species breed in wet, decaying organic matter (including wet manure mixed with soil); however, the house fly can breed in pure manure, allowing it to complete its life cycle in arid climates and in confined beef and dairy units where manure is collected in pits beneath the housing unit.

The EIL for livestock is probably quite different in terms of the impact stable flies or house flies have on animal production. However, two factors negate these differences. One factor is the aesthetic value livestock producers place on having production facilities free of flies, both for appearance and because of the annoyance to livestock and workers. A second factor is the threat of nuisance lawsuits, particularly in areas where urban housing is encroaching on agricultural production areas (Thomas and Skoda 1993). In either case, the fly species is not an issue, because both are considered part of the problem—and, generally, neither the livestock producer nor the urban dweller may be able to differentiate between them.

The economic effect of stable flies and house flies on livestock results from the transmission of diseases or depression of animal production (weight gain, milk, or eggs). There is far more evidence in the literature concerning the role of the house fly in disease transmission than there is for the stable fly (Greenberg 1973). Conversely, the literature implicates the stable fly, more than the house fly, for its direct effect on animal performance (Greenberg 1973, Campbell et al. 1977, Campbell et al. 1987).

The generally accepted ET for stable flies is five flies per front leg of confined feedlot cattle (McNeal and Campbell 1981). Bruce and Decker (1958) correlated numbers of flies with milk and butterfat production and found that an average of one stable fly per dairy cow reduced milk production 0.7% and butterfat production 0.65%. Later Miller et al. (1973) could show no effect of stable flies on milk production. However, the two studies were conducted under completely different conditions (farmer-owned-and-managed dairy herds versus a research herd with controlled environment and optimum condition). Wieman et al. (1992) indicate that 71.5% of the weight-gain depression caused by stable flies is from increased heat stress (bunched cattle) and 28.5% from irritation, energy loss from fighting flies, and physiological effects. Campbell et al. (1987), using pooled data, showed significant losses in feeder-cattle weight gain from as few as 2.58 flies per front leg. However, in 1992, the coolest summer on record in Nebraska, Campbell (unpublished data) did not see a significant weight-gain depression from 12 flies per front leg on feeder cattle.

Very little in the literature pertains to the effects of house flies on cattle performance. Freeborn et al. (1925) reported a 3.3% decrease in milk production as a result of house flies, but the same authors reported no effect three years later (Freeborn et al. 1928). Campbell et al. (1981), in a two-year study, saw no significant difference in weight gains at densities of between 32, 34, and 49 house flies per animal side. However, these authors point out that, at times, the number of house flies around feedbunks is so high that cattle appear reluctant to feed.

The Aquatic Complex

The economic thresholds for cattle afflicted by the blood-feeding aquatic species are largely unknown.

Mosquitoes and Tabanids. Steelman and Schilling (1977) and Steelman (1979) summarized the data concerning the economics of protecting cattle from mosquito attack. In this study, injury thresholds for Hereford, Brahman, and Hereford × Brahman crossbred steers were 27, 50, and 33 blood-fed mosquitoes per 0.09 m^2 resting area near the cattle. Mosquito populations at that magnitude caused weight-gain reductions of 0.045 kg/day/steer.

Perich et al. (1986) reported Hereford heifers protected from six species of tabanids averaged weight gains of 0.9 kg/day more than unprotected heifers, and the protected heifers were 16.9% more feed efficient. The daily tabanid attack rate in these studies varied, but ranged from a few to several hundred. Bruce and Decker (1951) reported reduced cattle weight gains of 0.29 kg/day in a 38-day study, with 0–9 tabanids per cow.

Ceratopogonids and Simuliids. There are no economic studies in the literature on *Culicoides* spp. (Diptera: Ceratopogonidae) except for their role in transmitting disease (primarily blue tongue virus). The numbers required to increase the incidence of disease is unknown.

The simuliids (black flies) are small, blood-feeding hump-backed flies. Lancaster and Meisch (1986) reviewed the literature on the economic effects of black flies on livestock. An outbreak of black flies in Romania caused the death of 16,000 animals in 1923. When feeding, the flies inject a vasoactive histamine that causes a toxic reaction. Death of the animal probably requires great numbers of flies. One report stated an average of 25,000 bites brought about the death of one animal. In some cases, asphyxiation may occur when animals inhale flies. There are no economic studies relating numbers of black flies to economic losses. Severe effects result only when outbreak numbers are present.

Black flies also are pests of poultry, particularly turkeys. Noblet et al. (1975) determined that *Leucocytozoon smithi* Laveran and Lucet, a protozoan blood parasite of turkeys, was transmitted by at least two black fly species. Jones et al. (1972) found that breeder turkeys infested with the disease suffered significantly higher mortality and decreased egg production when compared with turkeys protected from the black flies. Garris and Noblet (1976) reported 19 species of black flies present in the *L. smithi* epizootic area of South Carolina.

Pests of Swine

The major arthropod pest of swine is the mite *Sarcoptes scabiei* (DeGeer), which causes mange. Other pests are the hog louse, *Haematopinus suis* (L.), and the filth flies.

Mange Mites

Economic thresholds for the mange mite have not been established, but mange-infested pigs have reduced growth rates and lowered feed efficiency (Williams 1985). Mites increase in winter months and decline in the summer but may be present year-round. The mites are spread by contact between animals. Mites are usually more of a problem in poorly managed swine units where nutrition is not optimal. In such situations, mite problems will become more severe over time unless control action is taken.

Hog Louse

The hog louse also seems more prevalent in swine production units with poor management. As in the case of mites, there are no economic thresholds for this pest, because the hog louse will increase beyond reasonable thresholds if left unchecked (Williams 1985).

Filth Flies

The filth flies are the house fly and the stable fly. House flies and, to a much lesser extent, stable flies are pests at swine units. House flies will be present in great numbers around swine units unless controlled. Campbell et al. (1984) and Moon et al. (1987) found no economic effects of flies on feeder pigs. The potential for disease transmission by house flies, however, is of concern to swine producers, particularly in confined housing.

Pests of Poultry

DeVaney (1978, 1986) and Axtell and Arends (1990) have reviewed the ecology and management of arthropod pests of poultry. Axtell and Arends (1990) indicate that any factor, including parasites, that increases poultry stress results in decreased feed efficiency. The cost-profit margin is so low in modern poultry production that even minor increases in the cost of production can result in financial losses.

Northern Fowl Mite

Ornithonyssus sylviarum (Canestrini and Fanzago), the northern fowl mite, is considered the major arthropod parasite of chickens (Matthysse et al. 1974). Losses from the northern fowl mite have been reported in caged layer and breeder flocks but not in broiler production (DeVaney 1979). Although the life cycle of the mite is short (5 to 12 days) broilers usually are not in a house long enough for economically significant levels of the fowl mite to develop. Arends et al. (1984) reported that losses due to mites resulted in a decrease in egg production and an increase in feed consumption. An ET is not considered in making a treatment decision for the northern fowl mite, because, if left untreated, populations invariably increase to economic levels.

The Chicken Body Louse

Maracanthus stramineus (Nitzsch), the chicken body louse, is the most common louse species found in poultry facilities. DeVaney (1976) reported that infestations of the body louse cause decreases in hen weights, clutch size, and feed consumption. Axtell and Arends (1990) indicate that economic-loss data from several ectoparasites of poultry, including the body louse, are in conflict in the literature; the discrepancy may be because the genetics of birds currently in production is considerably different than that of birds used in trials in past years.

Chicken Mite and Bedbug

Dermanyssus gallinae DeGeer, the chicken mite, unlike the northern fowl mite and the chicken body louse, does not remain on the bird. Its feeding causes lesions on the breast and lower legs of the birds. The mites are most often a problem in breeder houses. Financial losses due to chicken mites have not been quantified, but anemia, disease transmission, reduced production, and increases in feed consumption have been recorded (Axtell and Arends 1990).

The bedbug, *Cimex lectularius* L., is a pest of poultry and poultry-associated humans worldwide (Axtell and Arends 1990). Bedbugs feed on the blood of poultry or humans mostly at night. Both immatures and adults feed on blood but remain on the host only long enough to engorge. They are a pest primarily of breeder houses. When not feeding, all the life stages are found concealed in cracks and crevices in the breeder house. Financial losses have not been quantified, but Kulash (1947) reported allergic reactions of humans, egg spots from fecal deposits, lower production, and decreased feed consumption resulting from bedbugs.

Filth Flies and Litter Beetles

Other arthropod pests of poultry include the filth-fly complex and habitat pests such as the litter beetles. The flies are threats because of the potential for disease transmission and nuisance lawsuits. In either case, the usefulness of ETs is a moot point. The litter beetles (the lesser mealworm, *Alphitobuis diaperinus* [Panzer] and the hide beetle, *Dermestes maculatus* DeGeer) also are a threat for disease transmission and may serve as reservoir hosts for several internal parasites of poultry. In addition, they can cause considerable structural damage, primarily to building insulation. The biggest economic problem with the litter beetles is the lack of a good method of control.

Pests of Sheep

Sheep, like most animals, suffer losses from ectoparasites specific to their own species as well as losses from general pests such as flies.

Sheep Ked

The most noted of the sheep insects is the sheep ked, *Melophagus ovinus* (L.). This insect is often called the sheep tick, but it is actually a fly with a very specialized life cycle. As with several lice and mite species, the sheep ked population increases in the fall and winter and declines in the spring and summer. Losses from the sheep ked are discussed by Lloyd (1985) and Meyer et al. (1988). Infestations of the sheep ked cause reductions in not only dressed carcass weight and dry weight of fleece but also value of the hide. This latter is because of "cockles," a condition caused by the feeding punctures of the sheep ked. There are no studies in the literature on the economic threshold of sheep ked. Pfadt et al. (1953) found that sheep ked populations on range lambs decline when the lambs are placed in the feedlot and fed a finishing ration. In Wyoming and several surrounding states, an effort has been made to eliminate the sheep ked by treating all sheep at shearing time. In states where eradication is the objective, an ET is not considered.

Sheep Bot Fly

The sheep bot fly, *Oestrus ovis* L., deposits larvae in the nostrils of sheep. The larvae migrate to the head sinuses. When the larval stages are completed, the last-stage larvae migrate back to the nostrils, whence they are expelled. Pupation occurs in loose soil. There may be two generations per year, particularly in the southwestern United States.

The ET for the sheep bot fly larva is unknown. Generally, infested sheep are predisposed to bacterial infection and abscess formation. Reduced weight gains of 4% have been reported (Meyer et al. 1988). Ranchers usually treat when the sheep exhibit behavior associated with nose bot infestation.

Wool Maggots

Several species of blow fly may infest sheep. The most common species are the black blow fly, *Phormia regina* (Meigen), the northern black blow fly, *Protophormia terraenovae* (Robineau-Desvoidy), and the green bottle fly, *Phaenicia sericata* (Meigen). Blow flies deposit eggs (fly strike) in areas of the fleece contaminated with feces, urine, blood, or other body fluids. The larvae feed on skin, causing distress to the sheep and loss of wool. With heavy infestation, sheep may die from heat loss or from stress-related diseases (Broce 1985). Economic thresholds for wool maggots are unknown. Prevention—by clipping dirty wool from the crotch and other areas, and early shearing—is the most common approach to avoiding such maggot problems.

Biting Midges

The biting midges—*Culicoides variipennis* (Coquillet) in particular—are important in livestock economics, primarily because of their role as disease vectors. With sheep, bluetongue, a viral disease, is of particular importance: its mortality rate for infected sheep may reach 50% (Price and Hardy 1954).

For most disease-vector insects, ETs are not known: an ET would depend on vector efficiency. Specifically, in the case of *C. variipennis* the question of an ET does not arise, because there is no efficient method of control. Prevention through grazing management is stressed; where it is possible, sheep are moved away from areas where *Culicoides* breed.

Other Pests

Sheep are hosts for several other pests—species of lice, ticks, and blood-feeding Diptera. There are no ETs for any of these pests. Generally, if the pests are present in substantial numbers, the animals are treated.

This review of the literature on ETs for many of the insects of veterinary importance indicates a need for further research. It seems probable that this need will increase in the future because of increased resistance of pest populations to pesticides, fewer pesticides, public concern over food safety, and environmental constraints on the use of pesticides.

The major restraint to conducting the needed research is the cost: replicated trials with animals, particularly cattle, are very expensive. Entomologists seldom have control of research cattle, and it is usually not feasible to superimpose entomological trials over animal-science or range-science cattle trials in a manner that would meet the statistical requirements of research in both disciplines. The animals are either "treated" or "untreated"; i.e., the level of infestation by the arthropod pest is at one level, and—in any given geographical area—that level may be about the same year after year.

There is another factor that increases the need for research on economic thresholds of livestock insect pests: much of the older data may no longer be valid. Cattle being produced now (crossbred English x Continental) are quite different genetically from the straightbred English herds of two or three decades ago. These modern cattle gain weight more rapidly than the cattle produced earlier, and to some extent they appear to overcome stress from ectoparasites.

In addition, range- and pasture-management practices have improved in many major cattle-production areas; hence, the nutritional level for cattle is higher, which, along with the improved breeding, may offset some parasite-induced weight-gain reductions.

Cattle-management practices also have changed. Many cattle producers now use veterinarian-developed animal-health programs designed for their particular herd. These programs include disease vaccinations and treatment for both ecto- and endoparasites on a calendar-scheduled basis. These programs use treatment on the basis of prevention rather than on an economic-threshold basis. Consequently, neither IPM practices nor ETs are considered in making treatment decisions. In time, because of the limited number of parasiticides being developed, these practices probably will lead to pesticide resistance. The cost of production is increased when unnecessary treatments are used, and this impacts the profit margin of the livestock producer and the cost to consumers.

Insecticide resistance in horn fly populations has induced ranchers to consider the horn fly ET. One of the resistance-management recommendations is to delay treatment until horn flies reach 200 or more per animal. At populations lower than 200, treatment is considered noneconomic. Sometimes, adversity can be viewed in the long term as a benefit; e.g., horn fly resistance is forcing producers to consider IPM practices.

Livestock producers are far more apt to adopt IPM practices where insecticide-control methodologies are expensive or inefficient. An example can be seen with stable fly and house fly control in confined-livestock units: fly breeding-source reduction is more efficient than use of insecticides for fly control. In addition, the existence of regulations (EPA, state, and local) encourages producers to consider developing a good waste-management plan, and such plans generally reduce fly production.

Another area where research is needed is on livestock-insect vectors of disease. Economic-threshold data that consider vector efficiency and incidence of disease pathogens in the vector population are not available. The data cited earlier in this chapter indicated that the percentage of face flies contaminated with *Moraxella bovis* was quite low, but control decisions are made primarily because face flies can transmit pinkeye.

References

Arends, J. J., P. B. Barto, and R. E. Wright. 1982b. Transmission of *Moraxella bovis* in the laboratory by the face fly (Diptera: Muscidae). J. Econ. Entomol. 75:816–818.

Arends, J. J., S. H. Robertson, and C. S. Payne. 1984. Impact of northern fowl mite on broiler breeder flocks in North Carolina. Poult. Sci. 63:1457–1461.

Arends, J. J., R. E. Wright, K. S. Lusby, and R. W. McNew. 1982a. Effect of face flies on weight gains and feed efficiency in beef heifers. J. Econ. Entomol. 75:794–797.

Axtell, R. C., and J. J. Arends. 1990. Ecology and management of arthropod pests of poultry. Annu. Rev. Entomol. 35:101–126.

Baptista, P. J. H. P. 1979. Infectious bovine keratoconjunctivitis: A review. Br. Vet. J. 135:225–242.

Barnard, D. R. 1985. Injury thresholds and production loss functions for the lone star tick, *Amblyomma americanum* (Acari: Ixodidae) on pastured preweaner beef cattle, *Bos taurus*. J. Econ. Entomol. 78:852–855.

Barnard, D. R., R. T. Ervin, and F. M. Epplin. 1986. Production system-based model for defining economic thresholds in preweaner beef cattle, *Bos taurus*, infested with lone star tick, *Amblyomma americanum* (Acari-Ixodidae). J. Econ. Entomol. 79:141–143.

Bech-Nielsen, S., S. Bornstein, D. Christensson, T. Wallgren, G. Zakrisson, and J. Chirico. 1982. *Parafilaria bovicola* (Tubangui) in cattle: Epizootic-vector studies and experimental transmission of *Parafilaria bovicola* to cattle. Am. J. Vet. Res. 43:948–954.

Berkebile, D. R., R. D. Hall, and J. J. Webster. 1981. Field association of female face flies with *Moraxella bovis*, an etiological agent of bovine pinkeye. J. Econ. Entomol. 74:475–477.

Broce, A. B. 1985. Myiasis-producing flies, p. 83–100. *In* R. E. Williams, R. D. Hall, A. B. Broce, and J. Scholl (ed.) Livestock insects. John Wiley, New York.

Broce, A. B., and R. J. Elzinga. 1984. Comparison of prestomal teeth in the face fly *(Musca autumnalis)* and the house fly *(Musca domestica)* (Diptera: Muscidae). J. Med. Entomol. 21:82–85.

Brown, J. F., and T. R. Adkins. 1972. Relationships of feeding activity of face fly *(Musca autumnalis)* to production of keratoconjunctivitis. Am. J. Vet. Res. 33:2552–2555.

Bruce, W. N., and G. C. Decker. 1951. Tabanid control on dairy and beef cattle with synergized pyrethrins. J. Econ. Entomol. 44:154–159.

Bruce, W. N., and G. C. Decker. 1958. The relationship of stable fly abundance to milk production in dairy cattle. J. Econ. Entomol. 51:269–274.

Burton, R. P. 1966. Transmission studies of *Moraxella bovis* (Hauduroy) by *Musca autumnalis* DeGeer. M.S. thesis, Univ. of Wyoming, Laramie WY.

Campbell, J. B. 1976. Effect of horn fly control on cows as expressed by increased weaning weights of calves. J. Econ. Entomol. 69:711–712.

Campbell, J. B., I. L. Berry, D. J. Boxler, R. L. Davis, D. C. Clanton, and G. H. Deutscher. 1987. Effects of stable flies (Diptera: Muscidae) on weight gain and feed efficiency of feedlot cattle. J. Econ. Entomol. 80:117–119.

Campbell, J. B., D. J. Boxler, D. M. Danielson, and M. A. Crenshaw. 1984. Effects of house flies and stable flies on weight gain and feed efficiency of feeder pigs. Southwest. Entomol. 9:273–274.

Campbell, J. B., D. J. Boxler, J. I. Shugart, D. C. Clanton, and R. Crookshank. 1981.

Effects of house flies on weight gains and feed efficiency on yearling heifers on finishing rations. J. Econ. Entomol. 74:94–95.

Campbell, J. B., and G. D. Thomas. 1992. The history, biology, economics, and control of the horn fly, *Haematobia irritans*. 1992. Agri-Practice 13:31–36.

Campbell, J. B., R. G. White, and H. Stokely. 1974. Controlling grubs on cattle. College of Agric., Univ. of Nebraska, Farm, Ranch and Home Quarterly 21:11.

Campbell, J. B., R. G. White, J. E. Wright, R. Crookshank, and D. C. Clanton. 1977. Effects of stable flies on weight gains and feed efficiency of calves on growing and finishing rations. J. Econ. Entomol. 70:592–594.

Campbell, J. B., W. Woods, A. F. Hagen, and E. C. Howe. 1973. Cattle grub insecticide efficacy and effects on weight gain performance on feeder calves in Nebraska. J. Econ. Entomol. 66:429–432.

Cheng, T. H. 1967. Frequency of pinkeye incidence in cattle in relation to face fly abundance. J. Econ. Entomol. 60:598.

Cheville, N. F., D. G. Rogers, W. L. Deyoe, E. S. Krafsur, and J. C. Cheville. 1989. Uptake and excretion of *Brucella abortus* in tissues of the face fly *(Musca autumnalis)*. Am. J. Vet. Res. 50:1302–1306.

Chitwood, M. B., and J. G. Stoffolano. 1971. First report of *Thelazia* spp. (Nematode) in the face fly, *Musca autumnalis*, in North America. J. Parasit. 57:1363–1364.

Collins, R. C., and L. W. DeWhirst. 1965. Some effects of the sucking louse, *Haematopinus eurysternus*, on cattle on unsupplemented range. J. Am. Vet. Med. Assoc. 146:129–132.

Collins, R. C., and L. W. DeWhirst. 1971. The cattle grub problem in Arizona. II. Phenology of common cattle grub infestations and their effects on weight gains of preweaning calves. J. Econ. Entomol. 64:1467.

Cox, D. D., M. T. Mullee, and A. D. Allen. 1967. Cattle grub control with feed additives (coumaphos and fenthion) and pour-on (fenthion and trichlorfon). J. Econ. Entomol. 60:522.

DeVaney, J. A. 1976. Effects of the chicken body louse, *Menocauthus stramineus*, on caged layers. Poult. Sci. 55:430–435.

DeVaney, J. A. 1978. A survey of poultry ectoparasite problems and their research in the United States. Poult. Sci. 57:1217–1220.

DeVaney, J. A. 1979. The effects of the northern fowl mite, *Ornithonyssus sylvarum*, on egg production and body weight of caged white leghorn hens. Poult. Sci. 58:191–194.

DeVaney, J. A. 1986. Ectoparasites. Poult. Sci. 65:649–656.

Drummond, R. O. 1987. Economic aspects of ectoparasites of cattle in North America, p. 9–24. *In* W. H. D. Leaning and J. Guerro (ed.) The economic impact of parasitism in cattle. Proc. MSD Ag. Vet. Symp., XXIII World Vet. Cong., Canada.

Drummond, R. O., G. Lambert, H. E. Smalley Jr., and C. E. Terrill. 1981. Estimated losses of livestock to pests, p. 111–127. *In* D. Pimmentel (ed.) CRC handbook of pest management, Vol. I. CRC Press, Boca Raton FL.

Duren, E., and L. E. O'Keefe. 1972. Horn fly control with dust bags: Effect on weight gain. Animal Nut. and Health 27:3–4.

Ely, D. G., and T. L. Harvey. 1969. Relation of ration to short-nosed cattle louse infestations. J. Econ. Entomol. 62:341–344.

Ervin, R. T., F. M. Epplin, R. L. Byford, and J. A. Hair. 1987. Estimating and economic implications of lone star tick (Acari: Ixodidae) infestation on weight gain of cattle, *Bos taurus* and *B. taurus* × *B. indicus*. J. Econ. Entomol. 80:443–445.

Freeborn, S. B., W. M. Regan, and A. H. Folger. 1925. The relation of flies and fly sprays to milk production. J. Econ. Entomol. 18:779–790.

Freeborn, S. B., W. M. Regan, and A. H. Folger. 1928. The relation of flies and fly sprays to milk production. J. Econ. Entomol. 21:494–501.

Freer, R. E., and P. J. Gahan. 1968. Controlling lice on beef herds—Is it economic? Agric. Gaz., NSW, Australia 79:308–309.

Garris, G. I., and R. Noblet. 1976. Investigations on black flies in Chesterfield County, South Carolina, an area epizootic for *Leucocytozoon smithi* of turkeys. S. Carolina Agric. Exp. Sta. Tech. Bull: 1056.

Geden, C. J., and J. G. Stoffolano. 1981. Geographic range and temporal patterns of parasitization of *Musca autumnalis* (Diptera: Muscidae) by *Thelazia* spp. (Nematode: Spirurata) in Massachusetts, with observations of *Musca domestica* (Diptera: Muscidae) as an unsuitable intermediate host. J. Med. Entomol. 18:449–456.

Gerhardt, R. R., J. W. Allen, W. H. Green, and P. C. Smith. 1982. Role of face flies in an episode of infectious bovine keratoconjunctivitis. J. Am. Vet. Med. Assoc. 180:156–159.

Gibney, V. J., J. B. Campbell, D. J. Boxler, D. C. Clanton, and G. H. Deutscher. 1985. Effects of various infestation levels of cattle lice (Mallophaga: Trichodectidae and Anoplura: Haematopinidae) on feed efficiency and weight gains of beef heifers. J. Econ. Entomol. 78:1304–1307.

Gladney, W. J., M. A. Price, and O. H. Graham. 1977. Field tests for insecticides for control of Gulf Coast tick on cattle. J. Med. Entomol. 13:579–586.

Glass, H. W. Jr., R. R. Gerhardt, and W. H. Green. 1982. Survival of *Moraxella bovis* in the alimentary tract of the face fly. J. Econ. Entomol. 75:545–546.

Greenberg, B. 1973. Flies and disease. Vol. II. Princeton Univ. Press, Princeton NJ.

Hall, R. D. 1984. Relationship of the face fly (Diptera: Muscidae) to pinkeye in cattle: A review and synthesis of the relevant literature. J. Med. Entomol. 1:361–365.

Haufe, W. O. 1962. Control of cattle lice. Can. Dept. Agric. 1006:1–8.

Haufe, W. O. 1982. Growth of range cattle protected from horn flies *(Haematobia irritans)* by ear tags impregnated with fenvalerate. Can. J. Anim. Sci. 62:567–573.

Hoffman, R. A., R. O. Drummond, and O. H. Graham. 1969. Insects affecting livestock and domestic animals, p. 87–89. *In* Survey Methods for Livestock Insects. USDA Agr. Res. Serv. 81–83.

Hughes, D. E., G. W. Pugh Jr., and T. McDonald. 1965. Ultraviolet radiation and *Moraxella bovis* in the etiology of bovine infectious keratoconjunctivitis. Am. J. Vet. Res. 26:1331–1338.

Johnson, G. D., J. B. Campbell, H. C. Minocha, and A. B. Broce. 1991. Ability of *Musca autumnalis* (Diptera: Muscidae) to acquire and transmit bovine herpesvirus-1. J. Med. Entomol. 21:361–365.

Jones, J. E., B. D. Barnett, and J. Solis. 1972. The effect of *Leucocytozoon smithi* infection on reproductive factors of broad breasted white turkey hens. Poult. Sci. 51:1543–1545.

Kahn, M. A., and G. C. Kozub. 1981. Systemic control of cattle grubs (*Hypoderma* spp.) in steers treated with Warbex and weight gains associated with grub control. Can. J. Comp. Med. 45:15–19.

Killinger, A. A., D. Valentine, M. E. Mansfield, G. E. Rickets, G. F. Cmarck, H. H. Newmann, and H. W. Warton. 1977. Economic impact of infectious bovine keratoconjunctivitis in beef calves. J. Econ. Entomol. 72:618–620.

Kinzer, H. G., W. E. Houghton, J. M. Reeves, S. E. Kunz, J. D. Wallace, and N. S. Urquhart. 1984. Influence of horn flies on weight loss in cattle with notes on prevention of loss by insecticide treatment. Southwest. Entomol. 9:212–217.

Kulash, W. M. 1947. DDT for control of bedbugs in poultry houses. Poult. Sci. 26:44–47.

Kunz, S. E., J. A. Miller, P. L. Sims, and D. C. Meyerhoeffer. 1984. Economics of controlling horn flies (Diptera: Muscidae) in range cattle management. J. Econ. Entomol. 77:656–660.

Lancaster, J. L. Jr., C. J. Brown, and R. S. Hones. 1955. Steers gained more when ticks were controlled. Arkansas Farm Res. 4:6.

Lancaster, J. L. Jr., and M. V. Meisch. 1986. Arthropods in livestock and poultry production. Halsted: Div. of John Wiley, New York.

Lloyd, J. E. 1985. Arthropod pests of sheep, p. 253–267. *In* R. E. Williams, R. D. Hall, A. B. Broce, and P. J. Scholl (ed.) Livestock entomology. Wiley, New York.

Matthysse, J. G., C. J. Jones, and A. Purnasiri. 1974. Development of northern fowl mite population on chickens, effects on the host and immunology. Search Agric. 4:1–39.

McMillan, I., J. H. Burton, G. A. Surgeoner. 1982. Bovaid ear tags for fly control in dairy cattle. J. Dairy Sci. Suppl. 1:155 (Abstract).

McNeal, C. D. Jr., and J. B. Campbell. 1981. Insect pest management in Nebraska feedlots and dairies: A pilot integrated pest management project. Univ. of Nebr. Entomol. Dept. Rep. No. 10.

Meyer, H. J., D. D. Kopp, and R. G. Hengen. 1988. Insect pests of sheep in North Dakota. N. Dak. St. Univ. Ext. Publ. EB-50.

Miller, R. W., L. G. Pickens, W. O. Morgan, R. W. Thimijan, and R. L. Wilson. 1973. Effect of stable flies on feed intake and milk production of dairy cows. J. Econ. Entomol. 66:711–713.

Moon, R. D., L. D. Jacobson, and S. G. Cornelius. 1987. Stable flies (Diptera: Muscidae) and productivity of confined nursery pigs. J. Econ. Entomol. 80: 1025–1027.

Morgan, C. E., and G. D. Thomas. 1974. Annotated bibliography of the horn fly, *Haematobia irritans* (L.), including references on the buffalo fly, *H. exigua* (deMeijere) and other species belonging to the genus *Haematobia*. USDA Misc. Publ. 1278.

Morgan, C. E., and G. D. Thomas. 1977. Supplement 1: Annotated bibliography of the horn fly, *Haematobia irritans* (L.), including references on the buffalo fly, *H. exigua* (deMeijere) and other species belonging to the genus *Haematobia*. USDA Misc. Publ. 1278.

Morgan, C. E., G. D. Thomas, and R. D. Hall. 1983. Annotated bibliography of the face fly, *Musca autumnalis* DeGeer. J. Med. Entomol. Suppl. 4.

Nelson, W. A. 1984. Effects of nutrition of animals on their ectoparasites. J. Med. Entomol. 21:621–635.

Nickel, N. E., J. H. Hyland, I. Gjurekovic, and D. Brondke. 1970. The problem of cattle lice. Proc. Ann. NSW (Australia) Vet. Assoc.: 1–5.

Noblet, R., J. B. Kissani, and T. R. Adkins Jr. 1975. *Leucocytozoon smithi:* Incidence of transmission by black flies in South Carolina (Diptera: Simulidae). J. Med. Entomol. 12:111–114.

Peitzmeier, B. A., J. B. Campbell, and G. D. Thomas. 1992. Insect fauna of bovine dung in northeastern Nebraska and their possible effect on the face fly, *Musca autumnalis* (Diptera: Muscidae). J. Kans. Ent. Soc. 65:267–274.

Perich, M. J., R. E. Wright, and K. S. Lusby. 1986. Impact of horse flies (Diptera: Tabanidae) on beef cattle. J. Econ. Entomol. 79:128–131.

Pfadt, R. E., L. H. Paules, and G. R. DeFoliart. 1953. Effect of the sheep ked on weight gains of feeder lambs. J. Econ. Entomol. 46:95.

Price, D. A., and W. T. Hardy. 1954. Isolation of the bluetongue virus from Texas sheep—Culicoides shown to be the vector. J. Am. Vet. Med. Assoc. 124:255–258.

Pugh, G. W., and D. E. Hughes. 1975. Bovine infectious keratoconjunctivitis: Carrier state of *Moraxella bovis* and the development of preventative measures against disease. J. Am. Vet. Med. Assoc. 167:310–313.

Quisenberry, S. S., and D. R. Strohbehn. 1984. Horn fly control on beef cows with permethrin-impregnated ear tags and effect on subsequent calf weight gains. J. Econ. Entomol. 32:463–478.

Rich, G. B. 1970. The economics of systemic insecticide treatment for the reduction of slaughter trim loss caused by cattle grubs, *Hypoderma* spp. Can. J. An. Sci. 50:301–310.

Scharff, D. K. 1950. Cattle grubs—Their biologies, their distribution and experiments in their control. Mont. Agric. Exp. Stn. Bull. 471.

Scharff, D. K. 1962. An investigation of the cattle louse problem. J. Econ. Entomol. 55:684–688.

Scharff, D. K., G. A. M. Sharman, and P. D. Ludwig. 1962. Illness and death in calves induced by treatments with systemic insecticides for the control of cattle grubs. J. Am. Vet. Med. Assoc. 141:582–584.

Schmidtmann, E. T., D. Berkebile, R. W. Miller, and L. W. Douglas. 1984. The face fly (Diptera: Muscidae) effect on Holstein milk production. J. Econ. Entomol. 77:1200–1205.

Schmidtmann, E. T., M. E. Valla, and L. E. Chase. 1981. Effect of face flies on grazing time and weight gain in dairy heifers. J. Econ. Entomol. 74:33–39.

Scholl, P. J., R. Hironaka, and J. Weintraub. 1988. Impact of cattle grub (*Hypoderma* spp.) (Diptera: Oestridae) infestations on performance of beef cattle. J. Econ. Entomol. 81:246–250.

Schreiber, E. T., J. B. Campbell, S. E. Kunz, D. C. Clanton, and D. B. Hudson. 1987. Effects of horn fly (Diptera: Muscidae) control on cows and gastrointestinal worms (Nematode: Trichostrongylidae) treatment for calves on cow and calf weight gains. J. Econ. Entomol. 80:451–454.

Shemanchuk, J. A., W. O. Haufe, and C. O. M. Thompson. 1960. Anemia in range cattle heavily infested with the short-nosed sucking louse, *Haematopinus eurysternus* (Anoplura: Haematopinidae). Can. J. Comp. Med. 24:158–161.

Shugart, J. L., J. B. Campbell, D. B. Hudson, C. M. Hibbs, R. G. White, and D. C. Clanton. 1979. Ability of the face fly to cause damage to eyes of cattle. J. Econ. Entomol. 72:633–635.

Smith, C. L., and R. Richards. 1955. Evaluations of some new insecticides against lice on livestock and poultry. J. Econ. Entomol. 48:566.

Smith, D. L. 1976. Weight gain in calves in response to control of cattle grubs with insecticides. Manitoba Entomol. 10:5–8.

Stacey, B. R., R. E. Williams, R. G. Buckner, and J. A. Hair. 1978. Changes in weight and blood composition of Hereford and Brahman steers in drylot and infested with adult Gulf Coast ticks. J. Econ. Entomol. 71:967–970.

Steelman, C. D. 1976. Effects of external and internal arthropod parasites on domestic livestock production. Annu. Rev. Entomol. 21:155–178.

Steelman, C. D. 1979. Economic thresholds for mosquitoes. Mosq. News 38:324–329.

Steelman, C. D., and P. E. Schilling. 1977. Economics of protecting cattle from mosquito attack. J. Econ. Entomol. 69:499–502.

Steve, P. C., and J. H. Lilly. 1965. Investigations on transmissibility of *Moraxella bovis* by the face fly. J. Econ. Entomol. 58:444-446.

Thomas, G. D., and S. R. Skoda. 1993. Rural flies in the urban environment, N. Cent. Reg. Res. Publ. No. 335 and Univ. of Nebraska Agric. Res. Division Res. Bull. No. 317.

Thrift, F. A., and J. R. Overfield. 1974. Impact of pinkeye (infectious bovine keratoconjunctivitis) on weaning and postweaning performance of Hereford calves. J. Anim. Sci. 38:1179-1184.

Vockeroth, J. 1953. *Musca autumnalis* DeGeer in North America (Diptera: Muscidae). Can. Entomol. 85:422-423.

Ward, J. K., and M. K. Nielson. 1979. Pinkeye (bovine infectious keratoconjunctivitis) in beef cattle. J. Anim. Sci. 49:361-366.

Wieman, G. A., J. B. Campbell, J. A. DeShazer, and I. L. Berry. 1992. Effects of stable flies (Diptera: Muscidae) and heat stress on weight gain and feed efficiency of feeder cattle. J. Econ. Entomol. 75:1835-1842.

Wilcox, G. E. 1968. Infectious bovine keratoconjunctivitis: A review. Vet. Bull. 38:349-360.

Williams, R. E. 1985. Arthropod pests of swine, p. 239-252. *In* R. E. Williams, R. D. Hall, A. B. Broce, and P. J. Scholl (ed.) Livestock entomology. John Wiley, New York.

Williams, R. E., J. A. Hair, and R. W. McNew. 1978. Effects of Gulf Coast ticks on blood composition and weights of pastured Hereford steers. J. Parasitol. 64: 336-342.

Clifford S. Sadof and Michael J. Raupp

12
Aesthetic Thresholds and Their Development

While most agricultural resources are managed for their yield of food or fiber, many are managed for their aesthetic value measured in terms of beauty, general appearance, and the level of satisfaction they provide. These resources include ornamental plants grown in greenhouses, nurseries, landscapes, and in and around structures where people reside and work. The value of these resources can be diminished by the presence or action of pests. Moreover, the value of these resources is often based on the subjective perception of those who utilize or manage them. With these restrictions, it is sometimes difficult to quantify in economic terms decision-making rules useful for managing pests associated with aesthetic injury. In this chapter, we will discuss how decision-making guidelines can be generated for economically vague commodities and resources. Approaches to establishing decision-making rules such as aesthetic injury levels and thresholds will be described and the limitations of these methods discussed.

Aesthetics and Decision Making in Pest Management

History

The economic-injury-level (EIL) concept is based on an economic-decision-making model that balances the cost of control with the benefits derived from averting future damage (Chapter 1). In practical terms, this model is used to calculate "the lowest population density that will cause economic damage" to a crop (Stern et al. 1959). Two key assumptions of this model are that the crop has an economic value that can be measured, and that this value will be reduced as the crop accumulates injury. This model's utility stems from its dependence on economics: agronomic yields measured in terms of bulk or nutritional content can be readily quantified in economic terms. In contrast, the aesthetic value of a beautiful flowering tree or a home relatively free of cockroaches is more difficult to assess in this manner.

Olkowski (1974) recognized this operational difficulty and suggested that the EIL concept was impracticable for systems where value was based largely on aesthetics. He proposed the existence of a parallel decision-making rule called an aesthetic injury level (AIL) that could alleviate problems associated with economically based models. He suggested that this new level could be described as the lowest population density that would cause aesthetic injury. In managing insect pests of urban street trees, he and his colleagues demonstrated the utility of this concept to replace prophylactic applications of pesticide with decision-based treatments (Olkowski et al. 1976, 1978). Despite pragmatic advantages of using decision-making rules not based on economic parameters, it is difficult to discount the merits of an economic approach to managing the pests of an industry of substantial monetary value. Ornamental plants grown in greenhouses and nurseries generated more than $8 billion in gross receipts in 1990 (USDC 1992). Furthermore, a recent evaluation of the urban forest in the United States estimated that it was comprised of more than 61 million street trees with an aggregate value of $18 to $30 billion (Kielbaso 1990). In addition to street trees, Kielbaso (1990) estimated that an additional 600 million trees are grown in yards and parks in the United States. The amount of money spent to manage pests of landscape plants further demonstrates the enormous worth of plants to their owners. Nationwide, about 27 million households used pesticides at least once to control insect pests of lawns, trees, and gardens (NGA 1990). Annual expenditures associated with these treatments exceeded $1 billion.

In response to a need for economically-based decision-making guidelines for ornamental plant systems, methods were developed to quantify objectively the economic loss caused by aesthetic injury in a retail nursery (Raupp et al. 1988). This approach, known as contingent valuation (Higley and Wintersteen 1992), used survey techniques to calculate the cost of losing potential revenue from insect injury. This procedure allowed the calculation of an environmental EIL using the standard EIL model.

Sadof and Alexander (1993) found that the wide variation in the cost of controlling a pest problem limited the utility of the EIL model that balanced the cost of control with the benefit of reduced damage. In a wholesale nursery, where the cost of control was low relative to the value of the plant, a quantitative EIL could be used to produce plants desirable to the vast majority of potential consumers. However, in a landscape setting where the pesticide costs for an individual plant were significantly greater than the nursery and the value of the control approached its purchase price, the adherence to an EIL as a guideline for pesticide applications resulted in injury deemed undesirable to many homeowners (Sadof and Alexander 1993). In this case, Sadof and Alexander suggested using an AIL model *sensu* Olkowski (1974) that produces thresholds based on the expected homeowner dissatisfaction associated with a known level of infestation.

Choosing the Appropriate Decision-Making Model

The appropriateness of an EIL or an AIL depends more on the management objective of the grower or system manager than on the species of plant being managed. In theory, an EIL would be an appropriate decision-making rule if wholesale or retail nurserymen were concerned only with the number, weight, or volume of plants produced. However, in these production systems, aesthetic considerations contribute to the value of the plant. Once a plant is installed in a landscape and managed for its aesthetic qualities alone, an EIL is more difficult to define and may be inappropriate as a decision-making rule. In the latter case, an AIL is the more appropriate model because of the subjective value of the plant and the damage caused by pests. In the former case, where the benefit of a plant to the producer is determined by a combination of quantitative factors such as size and qualitative factors such as beauty, we propose the use of a Hybrid EIL model as the decision-making rule. The relevance of each of these models to AILs is discussed below.

The EIL Model

The management objective will be to optimize one or more attributes of productivity such as yield expressed in terms of weight, volume, or nutritional value. These attributes are usually readily definable in terms of economic parameters such as bushels per acre. Examples include most agronomic crops such as field corn, soybeans, or small grains. In these cases, EIL, *sensu* Stern et al. (1959), will be the predominant decision-making rule. This model will rarely be used in systems where aesthetic considerations are of primary importance.

The AIL Model

The management objective will be to optimize one or more aesthetic attributes of quality expressed in terms of form, color, texture, beauty, or general appearance. These attributes are often qualitative and not readily definable in economic parameters. Examples of commodities and resources managed with these objectives include ornamental plants in landscapes or interiorscapes, lawns, and structures such as domiciles, food establishments, places of business, and public buildings (e.g., schools). In these cases, the AIL, *sensu* Olkowski (1974), is likely to be the appropriate decision-making rule.

The Hybrid EIL Model

The management objective will be to optimize production measured in quantitative units such as number, weight, or volume. However, in this case, subjective attributes such as form, texture, color, or quality will influence price. Examples include fresh market vegetables, organically grown crops, some fruits, and nursery and greenhouse crops, including sod. The decision-making rule will be a complex EIL based on objective economic parameters but modified by subjective ratings of valuation.

Approaches to Establishing AILs and Hybrid EILs for Ornamentals

Several studies have attempted to estimate AILs, Hybrid EILs, or components thereof, including value, injury, or damage functions (Table 12.1). AILs and Hybrid EILs attempt to define densities of pests that produce unacceptable aesthetic damage. Because of their qualitative nature, it is not surprising that a variety of methods have been used to establish subjective decision-making rules. We will discuss several of these methods.

Table 12.1. Quantitative investigations of AILS, Hybrid EILS, and their component model elements.

Pest	System	Valuation Approach		Parameters Calculated	Author
Argyresthis cupressella	*Juniperus* spp. *Thuja* spp. *Chamaecyparis* spp.	Method: Audience: Medium:	expert ranking entomologists live plants	I	Koehler and Moore 1983
Cotinus nitida	Turf	Method: Audience: Medium:	expert ranking golf course superintendents 35 mm slides	I,D	Salvaggio 1989
Blatella germanica	Structural	Method: Audience: Medium:	pest density - tolerance apartment dwellers survey	D	Zungoli and Robinson 1984
Dendroctonus frontalis	Forest	Method: Audience: Medium:	intervally scaled preference professionals and laymen 35 mm slides	D	Buyhoff and Leuschner 1978
Dendroctonus ponderosae Choristoneura occidentalis	Forest	Method: Audience: Medium:	intervally scaled preference laymen 35 mm slides	D	Buyhoff et al. 1982
Cecidomyia spp.	*Pinus radiata*	Method: Audience: Medium:	market survey retail customers choose-and-cut plantation	D	Paine et al. 1990
Liriomyza trifolii	Chrysanthemum	Method:	contingent valuation	D	Larew et al. 1984

(continued)

Table 12.1 Continued

Pest	System	Valuation Approach	Parameters Calculated	Author
		Audience: greenhouse growers		
		Medium: illustrations and live plants		
Thyriodopteryx ephemaraeformis	*Thuja occidentalis*	Method: contingent valuation Audience: retail nursery customers Medium: photographic chart	V,C,I,D,AIL Hybrid EIL	Raupp et al. 1988
Tetranychcus urticae	*Euonymus alatus*	Method: contingent valuation Audience: retail nursery customers nursery industry professionals Medium: photographic chart residence history	V,C,I,D,AIL Hybrid EIL	Sadof and Alexander 1993, Sadof 1993
Anisota senatoria	*Quecus* spp.	Method: contingent valuation treatment request records Audience: home owners Medium: photographic chart residence history	V,C,I,D,AIL	Coffelt and Shultz 1990, 1993
Macrosiphum liriodendri	*Liriodendron tulipifera*	Method: treatment request records Audience: city dwellers Medium: residence history	I,D D	Dreistadt and Dahlsten 1988
Xanthogaleruca luteola	*Ulmus procera*	Method: expert ranking Audience: entomologists Medium: live plants	I	Dahlsten et al. 1993

Expert Estimations

Expert Approximations. In this method, discipline experts draw on their knowledge of pest biology and the injury potential of a pest to estimate a pest density at which injury becomes intolerable or intervention should be considered. This method is purely subjective. The expert simply uses his or her experience in the field to guess the number of insects that cause enough injury to create what he or she believes to be a problem in a given management situation. Nielsen (1989) provided expert approximations for several pests of woody landscape plants. He suggested that these be used to decide when chemical intervention was necessary.

Expert Rankings. This method is only partly subjective. Experts use their experience to guess the aesthetic injury caused by objectively determined densities of pests. Quantitative studies are conducted to determine the relationship of pest density to these subjective aesthetic rating schemes. Koehler and Moore (1983) used this technique to determine the susceptibility of various conifers to injury caused by the cypress tip miner, *Argyresthia cupressella,* on conifers. Uninfested plants were given an "unsightliness" rank of 1, whereas plants with the greatest disfiguration were given a rank of 5. After infesting plants with tip miners, Koehler and Moore regressed unsightliness on tip miner density to establish the relationship between injury rating and pest density. Salvaggio (1989) used a similar technique to relate the injury caused by the green June beetle, *Cotinus nitida* (L.), to its abundance in potted turf and turf plots on golf courses. Dahlstein et al. (1993) developed action thresholds for elm leaf beetle, *Xanthogaleruca luteola,* by using a predictive degree day model to establish the relationship between egg-mass density and an expert ranking of leaf injury.

This approach can be extremely useful to landscape managers wishing to use a rating scheme to determine levels of injury tolerable to their clients. By knowing the relationship between pest density and injury rank, they could manage the resource below threshold pest density. A problem with this method is that it relies on the ability of a manager to consult with an individual client about pest tolerance. In some instances, establishing a tolerable injury level will be difficult at best. For example, when managing cockroaches in dwellings, few pest control operators would feel comfortable directly asking their clients the number of cockroaches that they would be willing to tolerate. Surveys such as those by Wood et al. (1981) and Zungoli and Robinson (1984) on the tolerance of public-housing residents to cockroaches may take the place of personal interviews by pest managers. Such surveys could substantiate decision-making rules when used in conjunction with expert rankings.

Survey Approaches

Several survey approaches have been used to take objective measure of public attitudes toward pests and the injury and damage they cause. Of these approaches, pest density—tolerance surveys are the most useful in generating AILs or Hybrid EILs because they actually establish the relationship between pest density and the injury or damage that results. Surveys that establish only the critical level of tolerable injury must be coupled with studies of the relationship between pest density and resource injury to generate a useful decision-making rule.

Pest-Density-Tolerance Surveys. This is the most direct approach toward determining an AIL. Respondents are simply asked the number of pests that they would be willing to tolerate, and the threshold is set accordingly. This method is most appropriate when the public is responding to the pest, such as to cockroaches in a dwelling. Wood et al. (1981) and Zungoli and Robinson (1984) conducted pest-density-tolerance surveys for cockroaches in public housing units. Raupp et al. (1988) used a graphical interpretation of these results to establish an AIL for cockroaches in apartments based on the response of the population median of residents surveyed. Although this approach has been used to determine an AIL for this pest, this decision-making rule may have limited utility in multiple-family dwellings. In these instances the critical density of cockroaches in a single domicile may depend on the population dynamics and dispersal of the population in the entire housing unit (Wood et al. 1981, Zungoli and Robinson 1984). Zungoli and Robinson (1984) also suggested that AILs would be difficult to estimate for cockroaches because the tolerance for cockroaches would likely decrease as managers reduced the overall pest population.

Intervally-Scaled-Preference Surveys. Foresters have applied intervally-scaled-preference surveys to establish the relationship between injury caused by pests and injury expressed as a loss of visual preference of "psychological disutility" (Buhyoff and Leuschner 1978, Buhyoff et al. 1982). This approach used photographs (35 mm slides) of landscapes with varied levels of pest injury. Survey subjects were shown all possible pairings of slides and were told to indicate the landscape they liked least. These paired comparisons were used to establish relationships between visual preference values and the proportion of a landscape injured by pests. They found that a 10% level of forest injury by southern pine beetle was sufficient to cause most respondents to report a visual aversion to a landscape (Buhyoff and Leuschner 1978).

This approach is especially useful to managers of recreational areas because it enables them to determine management thresholds in a variety of complex situations. The modifying effects of other landscape features were later demonstrated by Buhyoff et al. (1982) who found that damage caused by spruce budworm and mountain pine beetle was ignored when vistas contained a large number of scenic high peaks.

Market Surveys. Market surveys provide a method to determine the economic impact of pest injury on retail sales. This method involves estimating the marketability of plants with known amounts of pest injury and determining the relationship between these parameters. This method is not commonly used because most businesses involved in retail sales are reluctant to sell plants known to be injured. This method was used to determine if a pine resin midge was economically significant to Christmas tree growers (Paine et al. 1990). Paine et al. were unable to establish any significant relationship between pest injury and tree sales.

The problems associated with using this method stem from the complexity involved in a customer's decision to buy a plant in a retail nursery. Many factors, including local supply, are likely to influence the decision to accept a particular plant. The effect of pest injury on the marketability of a plant may be obscured or outweighed by other factors.

Contingent Valuation. This method for determining the relationship between pest injury and aesthetic or economic damage avoids some of the problems associated with market studies by restricting the factors associated with customer choice. This process is called contingent valuation because the respondents must decide among a limited range of purchase choices and extrapolation from this trend is contingent on the representational nature of the group of plants presented in the survey (Cummings et al. 1986). For this approach to be valid, the survey group must be representative of the management situation. For example, "homeowners" should be surveyed when determining AILs for landscapes, whereas "retail customers" should be used when determining hybrid EILs for retail nurseries (Larew et al. 1984, Raupp et al. 1988, Sadof and Alexander 1993, Sadof 1993).

In cases where this approach has been used, charts, plant displays, or series of photographs are prepared that encompass the range of pest injury, including an uninjured plant and a severely injured plant. Survey respondents are allowed to examine the range of injury and formulate opinions regarding the acceptability of each injury level. When determining an EIL for a retail nursery customer, survey respondents are asked how much they are willing to pay for a plant with a defined amount of defoliation (Raupp et al. 1988); when establishing an AIL for plants in the landscape, survey respondents are asked whether they find the plants unacceptable (Coffelt and Shultz 1990). An interesting generality emerged from studies that determined that relationship between levels of injury and the decline in acceptability. In all four studies that employed this methodology, <10% injury was sufficient to cause the majority of the respondents to consider a plant unacceptable (Larew et al. 1984, Raupp et al. 1988, Coffelt and Schultz 1990, Sadof and Alexander 1993).

Records of Requests for Treatment. When developing a decision-making rule for a landscape, one empirical approach involves determining the level of pest injury that is needed to trigger a complaint and a request for treatment. This approach has been used to determine tolerance to honeydew produced by tulip tree aphid and defoliation caused by orangestriped oakworms on street trees (Dreistadt and Dahlstein 1988, Coffelt and Schultz 1990). Dreistadt and Dahlstein monitored the production of honeydew, correlated honeydew production with aphid abundance, and matched complaint records with timed honeydew counts to establish a treatment threshold.

Coffelt and Schultz demonstrated that treatment requests generated a different management threshold from that determined by contingent valuation. They found that homeowners accustomed to caterpillar treatments administered on demand were reluctant to forgo sprays on trees with acceptable levels of defoliation. Coffelt and Schultz effectively argued that many residents were more concerned about frass falling from large, wandering caterpillars than tree appearance. Clearly, complex factors, including historical management practices, may influence decision-making rules based on complaint records.

The Hybrid EIL

The Aesthetic Component

For many ornamental plant systems such as nurseries, sod farms, and greenhouses, the management objective of the grower is clearly economic. However, the value of the commodity produced and the reduction in this value caused by pests will have a strong aesthetic component. In these circumstances, we suggest the use of a Hybrid EIL as a decision-making rule. The Hybrid EIL model for ornamental plants has an economic basis and uses the general form of the EIL model proposed by Pedigo et al. (1986) as follows: $EIL = C/VIDK$, where C = the cost of control (e.g., $/plant), V = economic value and undamaged good ($/plant), I = proportion of injury per unit of pest density (discoloration or defoliation/pest/plant), D = proportion of consumers perceiving damage per unit of pest injury ($ lost/unit of injury/plant), and K = effectiveness of control.

The key difference with the Hybrid model is that the aesthetic quality of the plant is the primary consideration when calculating the value *(V)* and damage *(D)* coefficients.

Cost of Control. The cost of control per plant is the sum of the costs for materials and their delivery to the target area. The managed system must be clearly defined in order to determine the cost of control. Large variations in costs have been reported for the ornamental plants because of the economies of scale. For example, Sadof and Alexander (1993) found the cost to apply a treatment to a 1-acre planting of burning bush euonymus of $.025 per plant. The same spray delivered to a 10-plant hedge in the landscape was $3.00. A spot treatment to an individual plant was $30.

Determining Plant Value. Difficulties can arise when objectively quantifying the economic value of a commodity sold for its aesthetic quality. The approach to establishing this value will depend on the management objective. When plants are designated for sale, market surveys can be conducted to determine the average price that a customer is willing to pay for a plant (Raupp et al. 1988, Sadof and Alexander 1993). When plants are already in the landscape, valuation becomes a complex function of replacement costs and aesthetic value. Standard procedures for estimating this value have been adopted by the landscape industry and are available for assigning values to plants (Neely 1988).

Injury and Damage. For a given pest, I is determined by measuring the relationship between pest density and the injury that it causes. The slope of the line that describes this relationship is the proportional rate of change in injury per unit of pest density, I. For example defoliation was measured when calculating I for defoliators of oak (Coffelt and Schultz 1990, 1993) and a conifer (Raupp et al. 1988). Similarly, leaf discoloration was used to assess the injury caused by sucking arthropods (Sadof and Alexander 1993). For some pests, such as honeydew producers, plant appearance is less of a concern than the nuisance caused by the accumulation of excrement on objects beneath infested trees. In this instance, because people respond to the honeydew rather than direct or indirect injury to the plant, complaints about the honeydew can be determined directly from the relationship between pest density and the accumulation of honeydew on moisture-sensitive cards (Dreistadt and Dahlsten 1988).

The damage coefficient, D, can be determined by using the contingent valuation method objectively to determine the relation between damage measured in economic terms, such as value lost per unit of pest injury. The slope of the line describing this relationship is the damage coefficient for the EIL equation. In the few instances where the form of this relationship has been established, the shape has been asymptotic (Buyhoff and Leuschner 1978, Raupp et al. 1988, Coffelt and Schultz 1990, Sadof and Alexander 1993). Damage increases rapidly at low rates of injury and peaks well below the 100% injury level. The portion of the curve prior to the inflection point can often be approximated with a straight line, whose slope can be used as a simple estimate for D in the EIL calculation.

Effectiveness of Control. The studies we have discussed in relation to aesthetic injury have ignored the determination of effectiveness of control, K. The factors that contribute to control are complex. In light of limited information concerning K, these studies simply assume 100% efficacy of control and assign K a value of 1.

CASE HISTORY

Determining a Hybrid EIL, and an AIL for twospotted spider mite on burning bush *(Euonymus alatus 'compacta')*

We will show how a Hybrid EIL can be calculated for the twospotted spider mite on burning bush managed under three different regimes (Sadof and Alexander 1993). We will then compare the aesthetic impact of each EIL using an AIL curve that describes the relationship between spider-mite densities and the proportion of retail customers that would consider the plant damaged.

Description of Management Regimes

In these three cases, the management objectives are to produce a plant for sale or to maintain plants so they do not appear damaged to property owners.

Case 1. A commercial nursery has a 1-acre block of burning bush designated for retail sale.

Case 2. A plant-care firm has a burning bush hedge of 10 plants at a building foundation on a property in its care.

Case 3. A plant-care firm has a single burning bush plant at a building foundation on a property in its care.

Control Costs (C). The costs per plant for these three cases include the cost of materials and delivery as described above. They are respectively estimated to be $0.025, $3.00, and $30.00.

Plant Value (V). The value of an undamaged plant for the commercial nursery, Case 1, was determined from a survey of retail nursery customers to be $25.16 for an undamaged plant. For the two landscape cases, this only represents the cost of the plant itself. The value of the plant in the landscape (Neely 1988) is determined as follows: $V = (P+R)\,CL$, where P = retail value, R = labor costs of removal and replacement, C = correction factor for plant condition, and L = the correction factor for location. For

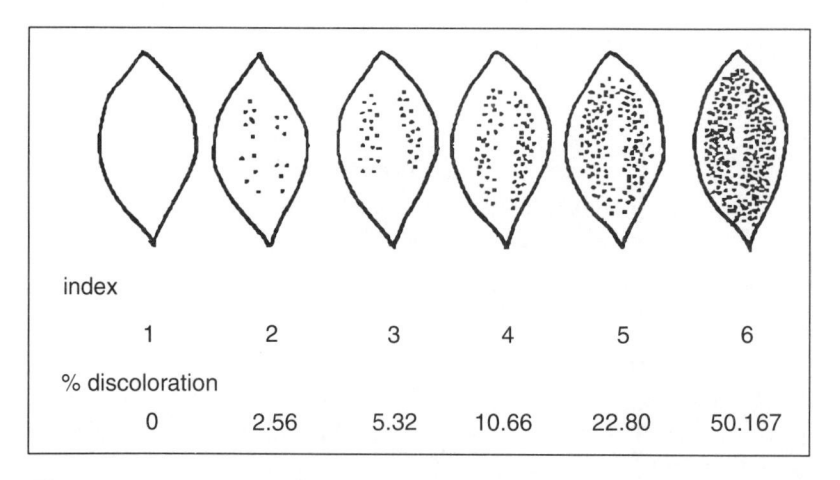

Figure 12.1. Leaf-discoloration index rank and percentage leaf discoloration caused by *T. urticae* on *E. alatus*. Reprinted from Sadof and Alexander (1993); copyright by the Entomological Society of America.

this situation where the P = $25.16, R = $75.48, or three times the retail value, C = 1 for an undamaged plant, and L = 0.9, the average correction factor for a plant in a foundation planting, V = $90.58 per plant.

Injury Coefficient (I). This was determined from the relationship between the cumulative density of spider mites during a season and the average leaf discoloration of plants. For illustrative purposes we will only discuss the relationship determined for a single study year. The details of the sampling method and the validity of this example are discussed in Sadof and Alexander (1993). Mites were sampled using a beating technique at two-week intervals from the earliest date of detection to the end of the growing season to determine their cumulative density. The discoloration caused from mite feeding was estimated from leaf samples that were scored according to Sadof and Alexander's rating scheme (Fig. 12.1). The injury coefficient is the slope of the line that describes the relationship between pest density and the injury ranking (Fig. 12.2).

Damage Coefficient (D). The damage coefficient was determined using the contingent-valuation technique. Surveys were taken in the same retail nurseries as those in which the retail value was determined. The damage

[Graph showing leaf discoloration index vs cumulative mites per beat, with regression line: y = 1.684 + 0.00066 (± 0.00011)x]

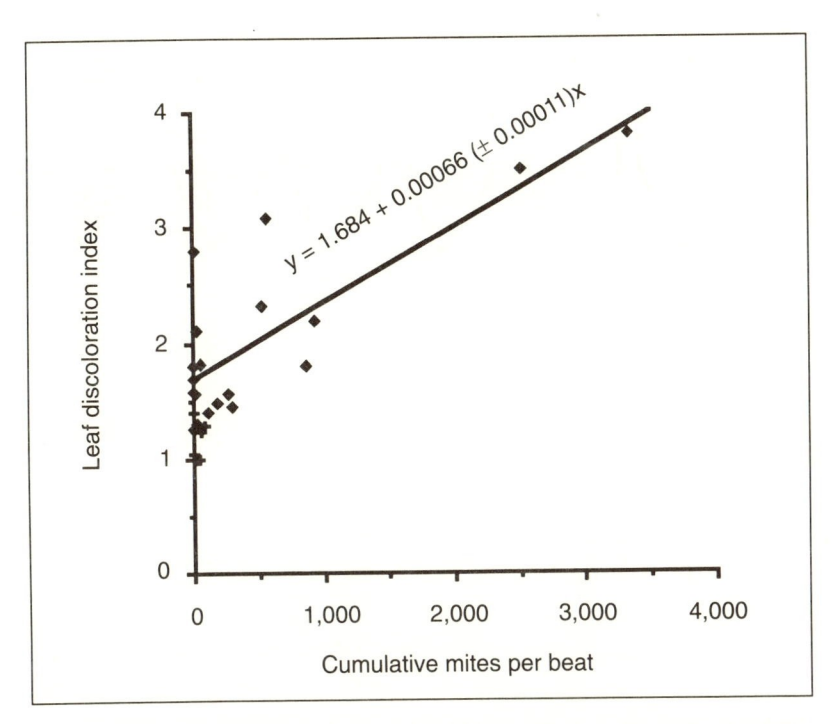

Figure 12.2. Relationship between cumulative density of mobile *T. urticae* falling from a beaten branch onto a 17 × 22 cm sheet of paper and average discoloration ranking of 25 leaves. F = 32.35; df = 1, 18; P<0.007; R^2 = 0.643. Reprinted from Sadof and Alexander (1993); copyright by the Entomological Society of America.

coefficient, D, was the slope of the line that described the relationship between damage recognition and injury. Although the relation was somewhat sigmoid, the linear approximation gave a reasonably good fit (Fig. 12.3). It is important to note that 50% of the survey respondents considered a plant with a discoloration of 2.81 (<5% discoloration) to be damaged.

Effectiveness of Control (K). Two levels of control effectiveness, 100% (K = 1) and 50% (K = 0.5), are used to examine the importance of control effectiveness in the model. A 50% level of control is probably a more realistic approximation for the activity of many materials as used in the landscape.

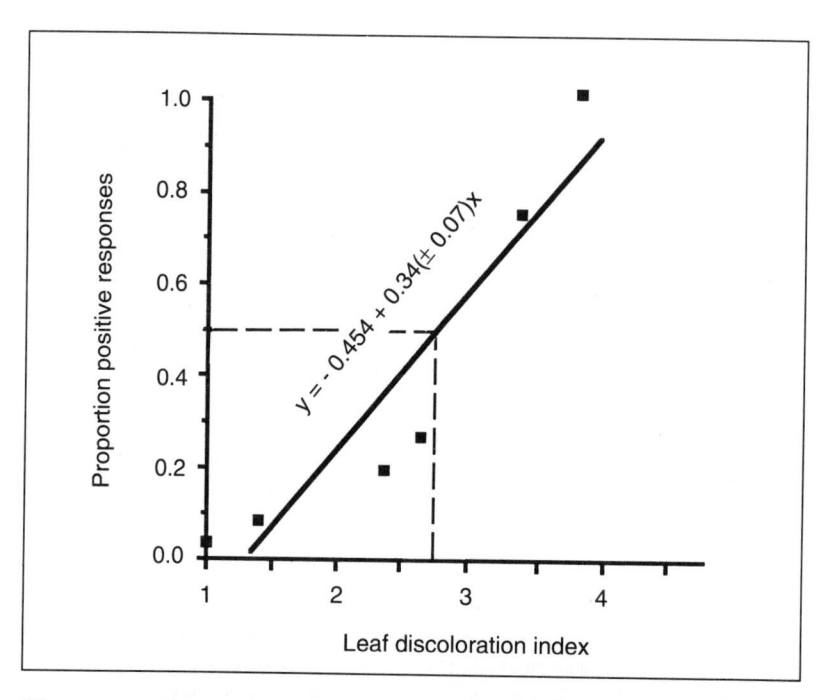

Figure 12.3. Proportion of 134 retail nursery customers responding "Yes" when asked, "Is this plant damaged?" and shown a plant with known leaf discoloration indices. Actual responses are represented by solid squares. Solid line with equation describes linear regression (F = 26.01; df = 1, 4; P<0.007; R^2 = 0.867). Dashed lines predict point in the linear regression where 50% of nursery customers would be dissatisfied with the level of leaf discoloration. Modified from Sadof and Alexander (1993); copyright by the Entomological Society of America.

Comparing Hybrid EILs. The EILs for each management case are listed in Table 12.2. As expected, the EIL for a pesticide application on a plant is greater in the managed landscape than that in the nursery because the cost of control is closer to the replacement value of an individual plant. When control effectiveness is reduced to 50%, the Hybrid EIL is doubled.

Table 12.2. Hybrid economic-injury levels for *T. urticae* on *E. alatus* 'compacta' for three management regimes. Data from Sadof and Alexander (1993).

Case No. (see text)	Cumulative Mite Density[a] (mobile mites per beat)		Expected Customer[b] Dissatisfaction (%)		C:V Ratio[c]
	$K = 1$[d]	$K = 0.5$	$K = 1$	$K = 0.5$	
1. Nursery production	4.4	8.9	−5.9	−5.7	0.001
2. Landscape maintenance 10-plant spray	147.6[e]	295.2	−2.4	1.1	0.033
3. Landscape maintenance single-plant spray	1,475.9[e]	2,951.9	29.44	64.87	0.330

[a] Total number of mobile *T. urticae* to fall on a 17 × 22 cm sheet of paper during a season. [b] Estimated from regressions in Figure 12.4. [c] Ratio of control costs to plant value. [d] K = effectiveness of control. [e] Plant value based on ISA method described in text.

Checking for Aesthetic Impact

For the Hybrid EIL model to be effective, plants maintained at this level should be aesthetically pleasing. This objective can be tested by determining the aesthetic impact from previously described regressions for pest density and injury (Fig. 12.2), as well as injury and damage (Fig. 12.3). Using the substitution method, these two regression equations can be solved to describe the relation between pest density and customer dissatisfaction (Fig. 12.4).

When the aesthetic impact of each EIL is determined, it is clear that the Hybrid EIL model produces aesthetically acceptable plants for nursery producers and landscapers managing blocks of plants even when the control effectiveness is only 50% (Table 12.2). It is equally clear that the Hybrid EIL model is not an acceptable decision-making tool for landscapers managing individual plants. An AIL, *sensu* Olkowski, may be the better approach in this situation. Landscape managers simply cannot stay in business when large numbers of clients consider plants in their care to be damaged.

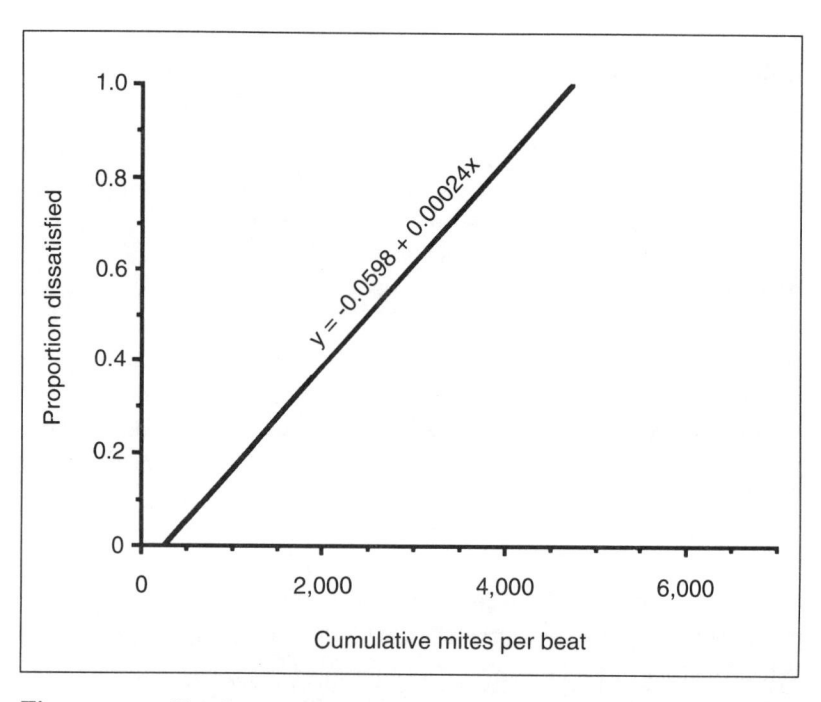

Figure 12.4. Simulated relationship between predicted retail customer dissatisfaction and cumulative *T. urticae* density, where y = expected proportion of clients who would consider a plant to be damaged.

This aesthetic analysis illustrates the relationship between control effectiveness *(K)* and the pest density that is calculated from the standard EIL equation (Pedigo et al. 1986). As control effectiveness decreases, the critical pest density increases. This has the rather curious effect of increasing allowable consumer dissatisfaction when controls are less effective. The implications of this property of *K* on decision making are beyond the scope of this chapter and need to be addressed elsewhere.

Impact of Cost-to-Value Ratio on Utility of Hybrid EILs

The ratio of control costs to plant value (CV ratio) is crucial to determining the utility of a Hybrid EIL because it is inherently derived from a cost/benefit model. As costs of control approach the value of the plant, the Hybrid EIL increases because there is less benefit to the grower or home owner who spends money on control. Consequently, more damage is allowed to offset the reduced gain from control actions. Thus, if control costs equal half the value of a plant, the model allows a plant to lose half of its value before it reaches the EIL. When expressed in terms of dissatisfied retail customers, half should be dissatisfied with the plant's appearance when $K = 1$ and cost equals half the plant value.

This theoretical 1:1 relationship between the CV ratio and the percentage of dissatisfied retail customers was tested with the data collected for spider mites injuring euonymus. Here we calculated the Hybrid EIL for a landscape plant when CV ratios were sequentially halved, starting from where control costs equaled plant value to where these costs constituted 1/2048 of the plant value. We used the relationship between pest density and retail-customer dissatisfaction to simulate the estimated proportion of dissatisfied customers for each Hybrid EIL when $K = 1$ (Fig. 12.5). The resulting model was very close to what was predicted from the theoretical relationship. The slope of this linear model was only slightly greater than 1 and predicted a proportional customer dissatisfaction of 0.475 when the CV ratio was 0.5. This small deviation in slope was caused by negative values in proportion of dissatisfied customers when the cost of control was equal to or less than 5.6% of the plant value. This originated from damage and injury relationships that indicate complete satisfaction of all customers with plants having very low spider mite densities (Fig. 12.4), or leaf discoloration (Fig. 12.3).

Implications of Aesthetics Research

Considerations for Future Perception of Injury and Pest Presence

The studies we have reviewed indicate that the public has a relatively low tolerance to insect injury or pest presence. Generally, most survey respondents believe that plants with less than 10% disfigurement are damaged. Similarly, small numbers of cockroaches had a similar effect on public opinion. The nature of this response should be examined in a wide variety of systems to assess its generality.

These low tolerances may partly be explained as an entomophobic response to perceived filth or uncleanliness associated with the presence of insects. This was clearly the situation when the public responded to cock-

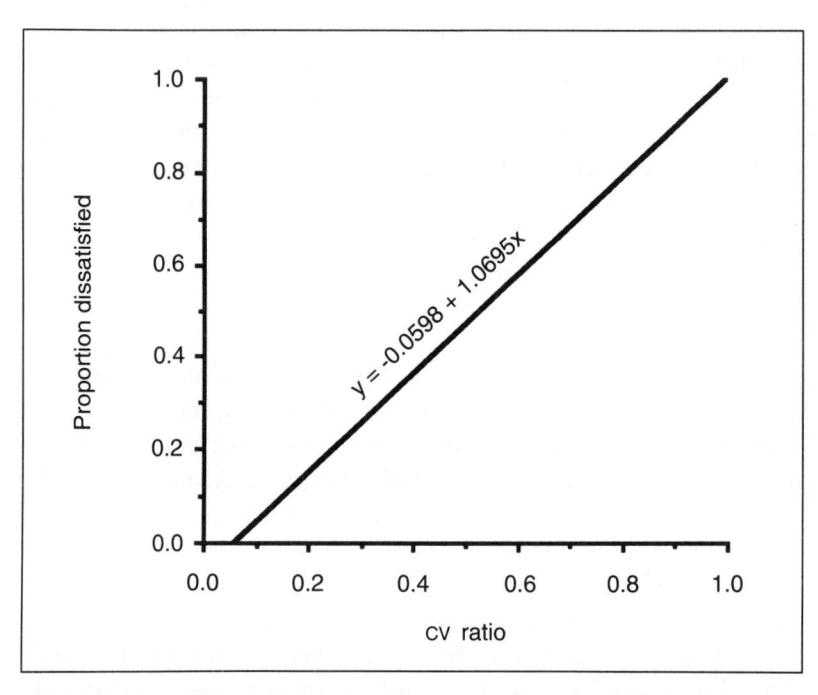

Figure 12.5. Simulated relationship between predicted retail-customer dissatisfaction with plants managed at the Hybrid EIL and the ratio of control costs to the replacement value of the plant (CV ratio), where y = expected proportion of clients who would consider a plant to be damaged.

roach presence or to fecal droppings from trees with orangestriped oakworms (Zungoli et al. 1984, Coffelt and Schultz 1990). Entomophobia also affected respondents' attitudes in defoliation studies (Buyhoff and Lueshner 1978, Buyhoff et al. 1982). Survey audiences who had been told they were looking at forests damaged by southern pine beetle had a lower threshold for injury than an uninformed group. Entomophobia, however, cannot explain the entire response of people to insect injury on plants. The tolerance for two groups of foresters examining forests injured by southern pine beetle was still less than 10% (Buyhoff and Leuschner 1978). The appearance of plants in the landscape was the primary determinant of their aesthetic quality.

Recognition of this low tolerance is likely to help researchers focus on aesthetically important levels of plant injury. Studies should accurately determine the number of pests that are likely to cause 10% disfigurement. We do not, however, suggest abandoning future research on the perception of injury. Opinion research is still needed to be sure that the injury caused by the pest is aesthetically relevant (Paine et al. 1990). Furthermore, this research is needed to calculate the damage constant used in determining a Hybrid EIL.

Cost/Benefit Approach to Decision Making

The cost/benefit basis of the EIL equation has important implications regarding decision-making rules for aesthetically valuable crops. Both plant value and the cost of control vary widely in the landscape industry. We described earlier how this decision-making model could produce unrealistic decision-making rules in a situation when the cost of control approached the replacement value of a plant. Thus, we suggested the need to use an AIL model as an alternative approach to evaluate the aesthetic impact of a particular injury level.

When plant value is high in relation to control costs, the EIL will approach zero (Table 12.2). These low levels make scouting for pests economically unfeasible compared with "calendar" spraying of pesticides. This apparent conundrum stems from the focus of previous studies on pesticides as the principal method of control. Thus, the control costs are assumed to be contained in a single application that is 100% effective ($K = 1$). For many pests, especially those like spider mites used in the example, multiple applications of pesticides can be needed because the application is often less than 100% effective. Even when a more realistic assumption of 50% effectiveness ($K = 0.5$) is used, the EIL threshold still remains low (Table 12.2).

One unfortunate result of zero thresholds for high value crops is that it provides economic justification for routine applications of pesticides in the absence of scouting. When pesticides are the predominant management tactic, this model is likely to provide justification for increased pesticide use. However, when biological control can be effective and the cost of the biological control agent approaches that of a traditional pesticide, this model can be used to justify inundative releases of natural enemies in high-value crops, and woody ornamentals support this interpretation of the model (van Lenteren and Woets 1988, Heinz and Parella 1990, Bellows et al. 1992).

In summary, the calculation of Hybrid EILs for aesthetic systems is possible by using objective methods to quantify the economic significance of changes in the aesthetic value of plants. Methods such as contingent valuation analysis make it feasible to determine a damage coefficient that can be used in the calculation of a Hybrid EIL. Most of the research suggests that the public has a low tolerance (<10% disfigurement) for pest injury. Application of the Hybrid EIL model to high-value crops may provide economic justification for routine application of either pesticides or biological controls to manage pest populations. The utility of the Hybrid EIL model is limited to situations where the unit cost of control is low in relation to the unit crop value. When the ratio of cost to plant value is high, plants managed at the Hybrid EIL may be aesthetically unacceptable. In these situations, the more appropriate decision-making rule would be an AIL that is based on the relation between pest density and aesthetic acceptability.

Acknowledgment

This is publication 13994 of the Indiana Agricultural Experiment Station and scientific article number A6602, contribution number 8816, of the Maryland Agricultural Experiment Station. We greatly acknowledge K. Nagarajan and R. Foster of Purdue University for their helpful comments on an earlier version of this manuscript.

References

Bellows, T. S., T. D. Paine, J. R. Gould, L. G. Bezack, and J. C. Bull. 1992. Biological control of ash whitefly: A success in progress. Calif. Agric. 46:24–28.

Buhyoff, G. J., and W. A. Leuschner. 1978. Estimating psychological disutility from damaged forest stands. Forest Sci. 24:424–432.

Buyhoff, G. J., J. D. Wellman, and T. C. Daniel. 1982. Predicting scenic quality for mountain pine beetle and western spruce budworm-damaged forest vistas. Forest. Sci. 28:827–838.

Coffelt, M. A., and P. B. Schultz. 1990. Development of an aesthetic injury level to decrease pesticide use against orangestriped oakworm (Lepidoptera: Saturniidae) in an urban pest management project. J. Econ. Entomol. 83:2044–2049.

Coffelt, M. A., and P. B. Schultz. 1993. Quantification of an aesthetic injury level and threshold for an urban pest management program against orangestriped oakworm (Lepidoptera: Saturniidae). J. Econ. Entomol. 86:1512–1515.

Cummings, R. G., S. Brookshire, and W. D. Schulze. 1986. Valuing environmental goods: An assessment of the contingent valuation method. Rowman and Allanheld, Totowa NJ.

Dahlsten, D. L., S. M. Tait, D. L. Rowney, and B. J. Gingg. 1993. A monitoring system and development of ecologically sound treatments for elm leaf beetle. J. Arboric. 19:181–186.

Dreistadt, S. H., and D. L. Dahlsten. 1988. Tulip tree aphid honeydew management. J. Arboric. 14:209–214.

Heinz, K. M., and M. P. Parella. 1990. Biological pest control of insect pests on greenhouse marigolds. Environ. Entomol. 19:825–835.

Higley, L. G., and W. K. Wintersteen. 1992. A novel approach to environmental risk assessment of pesticides as a basis for incorporating environmental costs into economic injury levels. Amer. Entomologist 38:34–39.

Kielbaso, J. J. 1990. Trends and issues in city forests. J. Arboric. 16:69–75.

Koehler, C. S., and W. S. Moore. 1983. Resistance of several members of the Cupressaceae to the cypress tip miner, *Argyresthia cupressella*. J. Environ. Hort. 1:87–88.

Larew, H. G., J. Knodel-Montz, and S. L. Poe. 1984. Leaf miner damage: How much is too much? Greenhouse Manager 3(8):53–55.

National Gardening Association. 1990. National Gardening Survey. Burlington VT.

Neely, D. 1988. Valuation of landscape trees, shrubs, and other plants. International Society for Arboriculture, Urbana IL.

Nielsen, D. G. 1989. Integrated pest management in arboriculture: From theory to practice. J. Arboric. 15:25–30.

Olkowski, W. 1974. A model ecosystem management program. Proc. Tall Timbers Conf. Ecol. Anim. Cont. Habit. Manag. 5:103–117.

Olkowski, W., H. Olkowski, R. van den Bosch, and R. Hom. 1976. Ecosystem management: A framework for urban pest control. Bioscience 26:384–389.

Olkowski, W., H. Olkowski, T. Drlik, N. Heidler, M. Minter, R. Zuparko, L. Laub, and L. Orthel. 1978. Pest control strategies: Urban integrated pest management, p. 215–233. *In* E. H. Smith, D. Pimentel (ed.) Pest control strategies. Academic Press, New York.

Paine, T. D., C. S. Koehler, and M. K. Malinoski. 1990. Pine resin midges (Diptera: Cecidomyiidae) in Monterey pine plantations: Control decisions and perception of aesthetic injury. J. Econ. Entomol. 83:485–488.

Pedigo, L. P., S. H. Hutchins, and L. G. Higley. 1986. Economic injury levels in theory and practice. Annu. Rev. Entomol. 31:341–368.

Raupp, M. J., J. A. Davidson, C. S. Koehler, C. S. Sadof, and K. Reichelderfer. 1988. Decision-making considerations for aesthetic damage caused by pests. Bull. Entomol. Soc. Amer. 34:27–32.

Sadof, C. S. 1993. Keeping up appearances. Amer. Nurseryman 177:83–85.

Sadof, C. S., and C. Alexander. 1993. Limitations of cost-benefit-based aesthetic injury levels for managing twospotted spider mites. J. Econ. Entomol. 86:1516–1521.

Salvaggio, R. S. 1989. Aesthetic injury level and threshold for managing green June beetle larvae *(Coleoptera: Scarabaeidae)* on golf course fairway turfgrass. M.S. thesis, University of Maryland, College Park.

Stern, V. M., R. F. Smith, R. van den Bosch, and K. S. Hagen. 1959. The integrated control concept. Hilgardia 29:81–101.

United States Department of Commerce. 1992. Statistical Abstract of the United States. Washington DC.

van Lenteren, J. C., and J. Woets. 1988. Biological and integrated pest control in greenhouses. Annu. Rev. Entomol. 33:239–269.

Wood, F. E., W. H. Robinson, S. K. Kraft, and P. A. Zungoli. 1981. Survey of attitudes and knowledge of public housing residents towards cockroaches. Bull. Entomol. Soc. Amer. 27:9–13.

Zungoli, P. A., and W. H. Robinson. 1984. Feasibility of establishing an aesthetic injury level for German cockroach pest management programs. Environ. Entomol. 13:1453–1458.

Stephen C. Welter

13
Thresholds for Interseasonal Management

Many of our current pest-management concepts and programs consider interseasonal consequences of pest-management decisions. The notion that cropping cycles are temporally and biologically integrated has long been implicitly recognized in many cultural-control practices including crop rotation, sanitation of potential overwintering sites, and strategies to avoid disrupting predator/prey interactions. Similarly, more recent concepts, such as insecticide-resistance management or management of biotypes resistant to new crop cultivars, also recognize the temporal linkage of insect populations. In contrast, scientists have not, in general, considered long-term effects of herbivory when determining economic-injury levels (EILs).

In this chapter, I will discuss the need to consider such effects on host plants, the unique problems that perennial crops present in developing EILs, and potential directions for future research. I will not review interseasonal thresholds involving weeds (Chapter 7), although interseasonal effects, particularly regarding the weed seedbank, can be an issue in thresholds for weeds. Whereas this review is focused exclusively on insects, many of the principles apply to other pestiferous organisms or other long-term effects such as abiotic factors. I will focus primarily on perennial cropping systems, but many of the ideas are equally applicable to systems that repeatedly plant the same annual crop plant in the same field or region. Functionally, growers have created perennial crop systems using annual crop plants.

Perhaps the initial omission of long-term effects in our thinking about EIL stems from the origin of the terms. Stern et al. (1959) presented the original concepts in reference to two cropping systems, cotton and alfalfa. In each case, the cost of the immediate control practice was balanced against the losses already incurred. Although long-term effects were discussed, decisions were based on past damage, associated pest densities, and current control costs, and not on projected costs and projected benefits. The strength of the original ideas lay in their attempts to move IPM (expressed as integrated control) away from prophylactic treatments that were sometimes used to prevent infestations of imaginary occurrence or of trivial economic consequence. While the goals of Stern et al. (1959) remain as valid today as they were 30 years ago, refinement of the EIL concept for perennial systems is necessary. Consideration of long-term effects will require estimating probabilities of infestations, projections for control, and estimates of future losses and control costs. As such, inclusion of long-term effects may move EILs away from sole application to cost-justifying immediate losses and toward more prophylactic and potentially uncertain decisions; however, if interseasonal models are correctly parameterized, optimal solutions should yield, on average, maximum profits with minimal pesticide use.

Differences in EILs for Annuals and Perennials

Discussions of arthropod effects on perennial systems need to reflect innate differences between pest management of annual systems and that of perennial systems. Perennial plants often have reserve systems that serve both as buffers between cropping cycles and functionally, as pest-management memories (for review of plant storage, see Chapin et al. 1990). The old saying, "the son must pay for the sins of the father," may be true for perennial-crop farmers. As will be discussed, the effects of arthropod injury may continue to manifest themselves for years after the initial damage was done. Therefore, the notion of long-term management of pest-population levels incorporates managing total crop productivity and profits over extended time periods. Taking the long term into account, pest-management personnel, in essence, will have to consider the history of a site, expected future use patterns and returns, and expected effects of various management tactics. Each of these concepts introduces factors that are either unknown, difficult to determine, or uncertain. As discussed below, uncertainty will directly and indirectly affect the determination of long-term EILs.

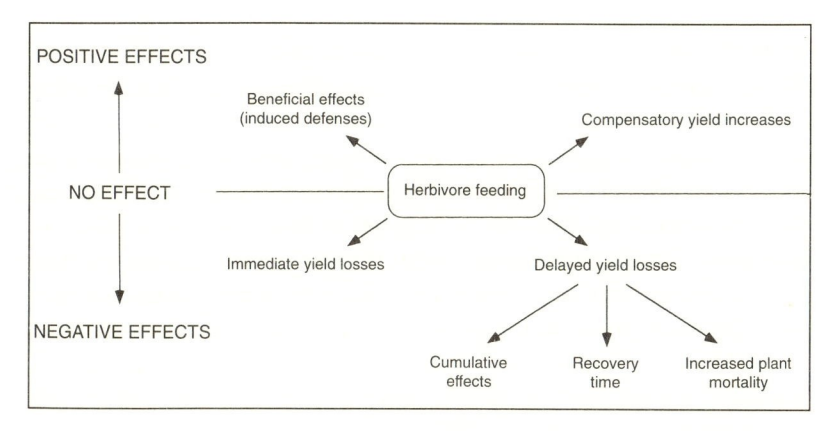

Figure 13.1. Effects, both immediate and interseasonal, of herbivory on perennial crops.

Furthermore, pest-management options often are more limited for perennial cropping systems. By definition, annual crop rotation is impossible for perennial plants. Resistant crop cultivars may require longer to produce, since many perennial crops (e.g., fruit-tree crops) require four to five years to produce their first commercially acceptable crop for evaluation; thus, the investment for plant breeders in each potential cultivar, in both time and money, can be considerable. Because many insect pests will live permanently within perennial crop systems, the consequences of pest-management decisions often span several years, affecting pest or natural-enemy population levels at the start of each growing cycle. Combined, these factors have led to speculation among some entomologists that perennial crop growers may have a different attitude toward long-term management than that of growers of annual crops.

Arthropod Effects on Productivity of Perennial Crop Plants

Three ways that economic losses can be manifested in perennial systems are shown in Figure 13.1: there can be immediate losses within the current season, delayed effects on subsequent yields, and cumulative effects over time.

Immediate effects are defined for this review as damage by arthropods expressed within the same season as the injury. This includes direct damage to the marketable portion of the crop and indirect damage that results in crop loss even if a within-season lag between injury and damage exists. *Delayed effects* are defined as damage that is manifested one season or more after the injury. Damage may or may not increase in severity over time. *Cumulative effects* include damage that becomes progressively greater as the number of damage cycles increases. As such, injury for a single season may not generate damage that is economically or statistically measurable, whereas the same injury level over successive seasons may produce significant levels of damage. These factors are themselves modified by a multitude of factors that have been discussed by Pedigo et al. (1986); however, some factors may be of unique importance to perennial plants, for example:

historical levels of injury and their impact on reserve systems

chronic effects of repeated low levels of damage

interseasonal lags in expression of damage

differential rates of recovery from herbivory

increased uncertainty in projecting benefits and losses

Long-term effects on EILs include long-term benefits from control strategies and long-term negative effects on crop productivity or longevity. Consideration of either possibility is conceptually similar in the distribution of the effects over time. Development of interseasonal thresholds requires an ability to predict or understand factors such as future pest-population dynamics, pest damage by yield-reduction functions over time, and the duration of effects of control tactics. In general, uncertainty increases for predictions that reach farther into the future; thus, interseasonal thresholds by definition have a greater level of uncertainty than thresholds for annual crops, which are based exclusively on current losses and costs. Various authors have discussed the role of uncertainty as it pertains to economic injury levels or pesticide use patterns (Norgaard 1976, Feder 1979, Plant 1986). Increasing uncertainty produces lower EILs either because of decreased expected mortality from pesticides (Plant 1986) or from the perception of risk by risk-averse growers (Norgaard 1976, Feder 1979).

In addition, incorporation of future losses into the EIL must be based in part on projected value of the product and likelihood of loss through insect injury. As such, control practices targeting the elimination of future losses will be employed unnecessarily in some years. In essence, prevention of future losses may represent a return to the prophylactic treatments that Stern et al. (1959) were trying to avoid, but on average prevention should deliver economically and biologically optimal solutions.

In a discussion of including future benefits in an EIL model, Torell et al. (1989) examine the example of interseasonal control of rangeland insects. Assuming probabilities for infestation levels and occurrence can be developed with accuracy, then economic solutions for these models are possible to optimize economic return. A variety of key factors were identified, including the value of forage, treatment costs, pest-population dynamics between years, the duration of the treatment effect, and treatment efficacy. Torell et al. (1989) concluded that the uncertainty of estimating many necessary parameters severely limited the accuracy of determining interseasonal EILS. We thus see that the grower sits on the horns of uncertainty: on the one side, the models suggest that long-term effects from insects will have strong effects on estimating EILS; on the other side, our ability to incorporate the data and concepts effectively is severely constrained by fluctuations.

To illustrate the effects of herbivory on perennial crops, I will use citations for arthropod effects primarily from three perennial cropping systems: fruit and nut tree crops, perennial legumes, and grapes. More specific crop-oriented references for arthropod effects can be obtained for these and other crops; the citations in this chapter illustrate specific issues and do not attempt to provide a litany of all possible references for long-term effects. Recent reviews of arthropod effects on plants include Pedigo et al. 1986, Crawley 1989, and Trumble et al. 1993.

Immediate Effects

Determining direct losses in productivity within the current year for perennial systems generally is no more difficult to determine than it is in annual crops. Quantifying losses to direct pests such as citrus bud mite on lemons (Walker et al. 1992) or to mirid bugs (Wieres et al. 1985, Uyemoto et al. 1986, Michaud et al. 1989, Purcell and Welter 1991) is estimated by standard sampling techniques; e.g., estimation of percentage of infested fruit at harvest, assuming no preharvest losses. Development of thresholds for direct pests is outlined in Pedigo et al. (1986), using various market and control costs. Similarly, losses caused by indirect pests can be determined and measured (Westigard et al. 1966, Hoyt et al. 1979, Beers et al. 1987). Dramatic reductions in yield up to 48% have been associated with mite damage in apples (Baker 1983), and these shifts have been associated with a number of factors, including number of fruit (Chapman et al. 1952, Hardman et al. 1985) and both fruit size and number of fruit (Asquith and Hull 1979).

Even the estimation of the costs of direct damage may involve interseasonal considerations. For example, estimation of navel orangeworm damage at harvest of almonds is made easily by direct determination of the percentage of infested almonds, but what have usually not been considered are the costs of leaving residual populations for the next year. Estimates of overwintering populations of navel orangeworm in almonds have been correlated with infestation levels the following spring (Engle and Barnes 1983). Elimination of almonds remaining on the trees after harvest eliminates the overwintering site for navel orangeworm. The cost of the program could be justified solely on the grounds of the recapture and sale of almonds that failed to shake loose in the initial harvesting; however, the value of the program, in total, includes the minimization of infestation levels the following year and the potential elimination of broad-spectrum insecticide use. In the southern San Joaquin Valley of California, the latter often results in outbreaks of secondary pests. In essence, the value of the approach is not determined by the cost of management equal to current losses, but the cost of the control strategy compared with prevention of immediate and future infestations, prevention of secondary pest outbreaks, and the minimizing of selection for insecticide resistance in navel orangeworm. The positive effects escalate.

Delayed Effects

The cost of arthropod damage is often borne within the current season, but it also may be delayed for multiple seasons; thus, the delay between injury and damage to the crop often uncouples the perceived linkage between pest infestation and reductions in yield. Delayed effects have been fairly well studied in fruit tree crops, perennial legumes, and grapes.

Fruit Tree Crops. The focus of most studies of fruit tree crops has been on the effects of indirect folivores on yield or growth as it relates to yield. Many different authors have shown an effect of damage on pome fruit (Chapman et al. 1952, Lienk et al. 1956, Dunstan and Stevenson 1961, Westigard et al. 1966, Light and Ludlam 1972, Zwick et al. 1976, Hoyt et al. 1979, Boivin and Stewart 1982, Ames et al. 1984, Hardman et al. 1985, Wieres et al. 1985, Hamilton et al. 1986, Michaud et al. 1989, Pfeiffer et al. 1989, Varn and Pfeiffer 1989, Beers et al. 1990, Beers and Hull 1990, Hull and Beers 1990, Baufeld and Freier 1991, Hluchy and Pospisil 1992), on stone fruit trees (Bailey 1979, McClernan and Marini 1986), and on nut crops (Tedders and Wood 1985, Barnes and Andrews 1978, Barnes and Moffitt 1978, Bolkan et al. 1984, Welter et al. 1984, Uyemoto et al. 1986, Purcell and Welter, 1991). Where Westigard et al. (1966) showed immediate yield losses caused by spider mite feeding injury, McNab and Jerie (1993) demonstrated dramatic reductions in yield, up to 66%, during the following year. These reductions resulted from changes in the numbers of fruit set per flower cluster rather than a reduction in the number of flowers produced the following season. Although a reduction in number of fruit occurred, increases in mean fruit weight were only slight and did not offset the reduction in fruit number. The effects of fruit set and yield disappeared by the second season after the injury. Unfortunately, no yield data were taken by McNab and Jerie (1993) during the first year of infestation. A pattern of no effect the first year followed by a subsequent reduction in the following year was observed by Welter et al. (1984) for spider mite feeding on almonds.

In one of the few studies to look at the timing of injury for perennial crops, Beers and Hull (1990) also looked at delayed effects on flowering and fruit set: early-season damage had little effect, midseason damage had the strongest reductions in mean fruit weight, fruit set, and crop load, and late-season damage affected only return bloom. Timing of damage relative to plant phenology and potential effects on plant growth or yield also were discussed in Welter et al. (1984).

Several authors have discussed the possibilities of fruit load as a mitigating factor that determines the relative effects of indirect damage (Zwick et al. 1976, Forshey and Elfving 1977, Hoyt et al. 1979, Ames et al. 1984, Beers et al. 1987, Hare et al. 1990). The results have varied between studies, but for apples the typical interactions of crop load and mite effects have not been significant. What is unclear is the role of tree reserves in determining mite effects. Given that reserves can affect fruit development (Weinbaum et al. 1984), the variability in results may reflect innate differences in susceptibility between species, the history of the field or orchard, and the relative carrying capacity of the fruit. Measurements of absolute fruit load are relatively meaningless unless contrasted with available reserves, a variable not measured in most entomological studies. Tedders and Wood (1985) did demonstrate strong effects of aphids on pecan productivity; and earlier, Wood and Tedders (1982) correlated aphid infestations with depressions in carbohydrate reserves.

The phenomenon of a delay in the expression of damage until the following season presents logistical problems for researchers and difficulties for extension personnel implementing EILS. Often orchards will return the following year without apparent damage from the previous injury. The lack of controls in nonexperimental situations prevents growers from differentiating between lag effects from arthropod injury and interyear variability in yield caused by other factors. Extension personnel can thus have problems convincing growers of potential economic savings or in preventing growers from overemphasizing the effects from the previous year's injury.

Grapes. Delayed effects of herbivory also have been shown for another perennial crop—grapes. Studies on the effects of mite feeding injury restricted to a single year failed in some cases to show significant effects from foliar damage by the Pacific mite (Laing et al. 1972, Kinn et al. 1974). In contrast, Flaherty and Huffaker (1970) found that the Pacific spider mite did affect grape quality and productivity in the first year of injury, whereas no effect was correlated with damage by another mite species, the Willamette mite, on a vigorous grape cultivar. In another study on Willamette mite effects, working in a different growing region and with a different cultivar, McNally and Farnham (1985) demonstrated losses in fruit quality during the first year of damage. Subsequent studies using the same cultivar showed a strong effect of Willamette mite on grape quality for the first year but no effect on yield. However, although significant losses in yield were detected the following year, the cumulative effects over two years of damage did not show an increase (Welter et al. 1989a, 1991). These general trends were repeated in studies by Karban and English-Loeb (1990). Here, the effects of the mite species were hypothesized to depend on the particular cultivar or plant architectural structure, as well as on potential behavioral differences between the two mite species.

Studies restricted to a single year on the effects of foliar damage by natural infestations of Japanese beetle failed to show any significant effect on fruit quality, yield, or shoot growth, whereas controlled infestations affected grape quality, but not yield (Boucher and Pfeiffer 1989). Significant depressions in grape quality during the first year of injury are generally consistent with the studies looking at mite effects; however, we might also predict significant effects on yield the subsequent season.

In contrast, Jubb et al. (1983) looked at the effects of multiple years of infestation by the leafhopper, *Erythroneura comes* Say, on grape yields and quality. No significant depressions in fruit production or quality were detected, despite three years of repeated damage; however, only five single-vine replicates were used for each of three treatments. In these circumstances, the lack of significant differences may have resulted from either no herbivore effect or the lower number of replicates used in the trial, compared with other studies.

Legumes. A variety of herbivores, including defoliating and sucking insects, suppress the immediate growth of legumes after injury and reduce subsequent within-season harvests. Many different taxa reduce legume growth and quality, including weevil species (Berberet et al. 1981, Norris et al. 1984, Shaw et al. 1986, Wilson and Quisenberry 1986, Alverson et al. 1991, Latheef et al. 1992), noctuid larvae (Buntin and Pedigo 1985a, 1985b), aphids (Cuperus et al. 1982, Summers and Coviello 1984), mirids—but only for dry weight and stem growth—(Jensen et al. 1991), and leafhoppers (Johnson 1936, Poos and Johnson 1936, Wilson et al. 1955, Kouskolekas and Decker 1968, Kindler et al. 1973, Faris et al. 1981, Cuperus et al. 1983, Hower and Flinn 1986, Shaw et al. 1986, Hower 1989, Wilson et al. 1989). Delayed effects on the subsequent year's growth and yield were demonstrated early in alfalfa pest-management research (Wilson et al. 1955) and continue to be demonstrated (Alverson et al. 1991). Some papers have shown that the effects of single species or low levels of damage could disappear within the same season if further injury was prevented with insecticides; or they could have serious delayed effects within the same season if damage levels were sufficiently severe (Summers and Coviello 1984, Wilson and Quisenberry, 1986).

It is far from clear why authors fail to consider effects beyond the first year of infestation despite similarities of results between taxa and the previous literature—similarities that clearly demonstrate the potential for long-term effects. Although the effects appear to be the most severe during the same year as infestation, the effects on subsequent years are sometimes deliberately avoided to maintain nonconfounded results (Berberet and McNew 1986). This approach produces well replicated and clear results, but perhaps only a portion of the true impact of the target species is known. EILs or recommendations are still being developed based exclusively on within-year effects on yield (Bishop et al. 1982, Cuperus et al. 1983, Jensen et al. 1991).

Cumulative Effects

Not surprisingly, some infestations that do not appear to present a serious problem after a single year of infestation ultimately may cause economic losses over multiple seasons. Barnes and Moffitt (1978) examined the effects of European red mite and walnut aphid on walnut productivity but found no significant losses until the third year of the study. Barnes and Moffitt suggest that there is potential for very serious errors when examining chronic effects from arthropods. If the studies had been restricted to two successive years of infestation, they would have concluded no effect from mite and aphids, whereas their data ultimately demonstrated just the opposite. Walker et al. (1992) also showed that infestations of the citrus bud mite failed to show any negative effects on yield until after 21 months in two orchards, and up to 48 months in another orchard. Because the feeding injury of the citrus bud mite is to the embryonic fruit tissue, no change in percentage of distorted fruit occurred until 13 months after initiation of acaricide treatments.

However, long-term studies by Hare et al. (1990, 1992) show no cumulative effects of another indirect pest, citrus red mite, on navel orange tree yields over a four-year period. Infestations were allowed to develop without control or were treated at accepted threshold levels with an acaricide. The effects of citrus red mite feeding were inconsistent over the first two years and did not appear to accumulate. Although yield reductions of 10% were noted, increases in fruit size and the higher prices given to larger fruit helped to offset yield reductions. Continued studies over the next two years did not show any cumulative or interseasonal effects. Levels of citrus red mite used in the study also did not result in significant or consistent changes in leaf photosynthetic rates (Hare and Youngman 1987). Given these results, the value of controlling citrus red mite was found to be dubious. Unfortunately, the number of papers examining the effects of chronic low levels of damage over time is relatively low and thus our ability to make generalizations is severely limited.

Patterns of progressively greater losses were shown for alfalfa and various herbivores, including alfalfa weevil. Increasingly rapid stand reductions were associated with combinations of chronic infestations of alfalfa weevil and other stress agents; e.g., weeds (Norris et al. 1984, Latheef et al. 1992). These reductions were associated in one case with reduced root activity and size, an important component of reserve storage (Walton 1983). A similar reduction in stand longevity was associated with chronic damage (Godfrey and Yeargan 1987). Alfalfa growers have long recognized the need to sustain a field over time rather than just reap the short-term benefits of a single year's yield.

Recovery from Damage

Several authors have examined the effects of herbivore damage after injury has ceased (Barnes and Moffit 1978, McNab and Jerie 1993, Welter et al. 1991). The rate of recovery varied from a single year (Welter et al. 1991, McNab and Jerie 1993) to more than a year (Barnes and Moffitt 1978). The effects of rates and duration of recovery on EIL were discussed and graphically illustrated in Welter et al. (1991). Decreased EILs were determined if the duration required until either the rate of complete recovery was increased or rates of recovery were slowed.

An economic analysis of interseasonal effects that considered future damage prevention for rangeland insects was presented by Torell et al. (1989). Nontraditional variables such as future pest-population projections, future control efficacy and duration, and discounting future forage-saving benefits were included in their analyses. Because of the many biological variables (survival rates, site-specific species composition, forage-destruction rates), the model suggested there would be a high degree of uncertainty when developing interseasonal EILs.

The Future

EILs *and Pesticide Resistance*

One of the more serious problems facing agriculture today is the development of pesticide resistance by the more important pest species (Georghiou 1986). Conceptually, insecticide-resistance management has long embraced strict adherence to ETs as one of the key strategies; their importance is "universally recognized" (Densholm and Rowland 1992).

Despite the discussion of insecticide resistance in the original work by Stern et al. (1959), pesticide resistance has not traditionally been included in the determination of specific EILs. However, pesticide resistance as an interseasonal problem very clearly influences the economics of our control tactics and strategies (Miranowski and Carlson 1986). Knight and Norton (1989) reviewed the economics of agricultural pesticide resistance, and the long-term costs to U.S. agriculture have been estimated at $118 million annually (Pimentel et al. 1980). Archibald (1984) projected losses for California cotton due to resistance at $45 to $120 per hectare. The cost of resistance clearly depends on the assumptions of the models used, the means of defining the cost, and who defines the cost.

The number of papers actually dealing with estimating specific costs of this important matter is relatively small. Most papers deal with the larger issues of resistance management and general costs of resistance and not with specific commodities, risks, or alternatives. As Knight and Norton (1989) point out, decisions that may be cost effective for individual growers may not be the optimal economic solution if a larger community is defined or if resistance is evolving on a regional scale. However, EILs and ETs have usually been established at the individual level if not at the individual farm level. Various authors have argued for more global and long-term approaches—to include, for example, environmental costs (Higley and Wintersteen 1992) or interseasonal threshold costs associated with weed seedbanks (Cousens et al. 1986)—but these approaches seem not to have evolved beyond theory. It is therefore not surprising that EILs have not, to my knowledge, incorporated insecticide resistance and its associated costs.

Induced Plant Defenses and EILs

Development of most EILs has considered multifactor relationships between pest and various plant conditions (e.g., phenological state, water-stress level, or fruit load); however, the effect has generally been assumed to be linear and unidirectional. The possibility of these being a feedback mechanism, from herbivore-feeding damage to the crop plant's "defenses," has been largely ignored, conceptually, for EILs.

The existence of induced defenses that are intra- or interseasonal on perennial plants has been known for quite a while (Baltenseilter et al. 1977, Haukioja and Neuvonen 1985, Karban 1985, Bergelson et al. 1986, Hartley and Lawton 1987). However, Karban and Myers (1989) also stated that, whereas changes from insect damage can be readily found, the effects of these changes on plant fitness, herbivore performance under field conditions, or as a general explanation of herbivore population dynamics are unclear. In an agricultural context, the debate as to the evolutionary significance of herbivory on plant defenses, and for plant changes as selected defenses against herbivory, remains somewhat less important. For EILs, what is more important is the question of whether these changes impact significantly on pestiferous populations, and, if so, if they are relatively predictable in expression. To take advantage of intra- and interseasonal changes in plant defense, Karban et al. (1991) used a more innocuous species of spider mite to affect long-term control of a second, more deleterious, species in grapes: late releases in the fall resulted in lower populations of the more deleterious species the following growing season. There is a dearth of similar studies, and the long-term effects of herbivory on crop-plant defenses remain relatively unknown, as are the potential consequences on crop productivity.

Given that long-term effects of herbivory may include prevention of future damage, the models discussed by Torell et al. (1989) should be applicable to induced defenses. In essence, we may be able to tolerate or accept damage within a growing season that exceeds immediate control costs, if future losses can be prevented at a level that is sufficiently great to justify the risk or uncertainty. Clearly, many of the variables strongly influencing the results of Torell et al. (1989) will also affect the incorporation of induced defenses into EIL determination (e.g., variability in control from induced defenses, duration of the defenses, and economic benefits projected from crop-loss prevention).

A discussion of interseasonal thresholds for insect pests requires the reader to consider three disciplines: entomology, plant biology, and economics. The combination of any two disciplines is a difficult enough undertaking, and here we have three. Although our knowledge of the potential long-term negative effects of arthropod damage on plant productivity is relatively clear, that of the overall pattern as to how important the problems are is less clear. The long-term negative effects of factors such as resistance management and the disruptive effects of control practices on secondary pests have been demonstrated on numerous occasions; conversely, the long-term benefits of management strategies such as rotational programs, elimination of pest overwintering sites, and potential use of induced secondary plant compounds have also been shown. The question that remains is: Can we quantify, incorporate, and ultimately model EILs, in their inherent variability, to make meaningful decisions or recommendations beyond generalities?

Perhaps it is not realistic even to attempt to incorporate so many factors at this point in time at the farm level. But it is realistic for the public and agricultural sectors to decide what areas merit greatest concern and to incorporate this into EIL development. The interseasonal-cost areas—insecticide resistance, environmental costs, and human health—will most likely change only after legislative change; as to the interseasonal economic benefits—the results of optimizing plant production over time or minimizing control costs—growers will most likely incorporate such factors into their decision making if the returns are big enough. The answers lie at the interface of multiple disciplines—that area most scientists verbally embrace yet rarely tread. To me, the difficulties presented by plant-reserve effects, the considerable length of time required per experiment, and uncertainties about future conditions suggest two strategies: short-term studies to develop guidelines by means of traditional experiments, coupled with more rapid, physiological experiments (Welter 1993); and long-term development of accurate plant modeling with herbivore subprograms.

In the short run, recommendations for the EIL should be based on the results from traditional, empirical studies of damage-yield relationships in normal agricultural settings. *Normal* agricultural conditions will obviously be defined by local experts using the most commonly planted cultivars in the usual growing conditions and using the usual agronomic practices. The tailoring of the threshold for different conditions (e.g., higher levels of water stress, different cultivars, or different species of related pests) can perhaps be most easily approached through assessing more rapid physiological consequences or even more-immediate growth responses (see, for example, Welter et al. 1989b, 1991). Such data provide some evaluation of the potentially mitigating variables, although they fail to provide the more substantial and conclusive evidence of direct experimentation that uses yield as the dependent variable. Given that growers must make pragmatic decisions every day, regardless of the presence or absence of thresholds based on data, we can provide insights to help with the making of informed decisions. Unfortunately, in the end, the number of permutations for even three or four variables in EIL in perennial crops becomes overwhelming, in an experimental context. This only makes the need for modeling progressively greater.

An area in which entomologists can continue to provide significant insights is in the definition and understanding of plant responses to arthropod damage—knowledge that can later be incorporated into plant-arthropod modeling. Uncertainty in estimating parameters will continue to limit the development of interseasonal thresholds, but we can be sure of one thing: there is a need to consider long-term effects.

References

Alverson, D. R., W. C. Stringer, and A. J. Vybiral. 1991. Effect of fall cutting and alfalfa weevil on alfalfa yields and forage quality. J. Agric. Entomol. 8:137–146.

Ames, G. K., D. T. Johnson, and R. C. Rom. 1984. The effect of European red mite feeding on the fruit quality of Miller Sturdeespur apple. J. Am. Soc. Hort. Sci. 109:834–837.

Archibald, S. O. 1984. A dynamic analysis of production externalities: Pesticide resistance in California cotton. Ph.D. Diss., Univ. of California–Davis.

Asquith, D., and L. A. Hull. 1979. Integrated pest management systems in Pennsylvania apple orchards, p. 203–222. *In* D. J. Boethel and R. D. Eikenbary (ed.) Pest management programs for deciduous tree fruits and nuts. Plenum, New York.

Bailey, P. 1979. Effect of late season populations of twospotted mite on yield of peach trees. J. Econ. Entomol. 72:8–10.

Baker, R. T. 1983. Effect of European red mite *(Panonychus ulmi)* on quality and yield of apples. 10th Int. Congress Plant Prot., Brighton, UK, Nov. 1983.

Baltenseiler, W., G. Benz, P. Bovey, and P. Deluchhi. 1977. Dynamics of larch bud moth populations. Annu. Rev. Entomol. 22:79–100.

Barnes, M., and K. L. Andrews. 1978. Effects of spider mites on almond tree growth and productivity. J. Econ. Entomol. 71:555–558.

Barnes, M., and H. Moffitt. 1978. A five-year study of the effects of the walnut aphid and the European red mite on Persian walnut productivity in coastal orchards. J. Econ. Entomol. 71:71–74.

Baufeld, P., and B. Freier. 1991. Artificial injury experiments on the damaging effect of *Leucoptera malifoliella* on apple trees. Entomologia. exp. appl. 61:201–209.

Beers, E. H., and L. A. Hull. 1990. Timing of mite injury affects the bloom and fruit development of apple. J. Econ. Entomol. 83:547–551.

Beers, E. H., L. A. Hull, and G. M. Greene. 1990. Effect of a foliar urea application and mite injury on yield and fruit quality of apple. J. Econ. Entomol. 83:552–556.

Beers, E. H., L. A. Hull, and J. W. Grimm. 1987. Relationships between leaf:fruit ratio and varying levels of European red mite stress on fruit size and return bloom of apple. J. Am. Soc. Hort. Sci. 112:608–612.

Berberet, R. C., and R. W. McNew. 1986. Reduction in yield and quality of leaf and stem components of alfalfa forage due to damage by larvae of *Hypera postica* (Coleoptera: Curculionidae). J. Econ. Entomol. 79:212–218.

Berberet, R. C., R. D. Morrison, and K. M. Senst. 1981. Impact of the alfalfa weevil, *Hypera postica* (Gyllenhal) (Coleoptera: Curculionidae), on forage production in nonirrigated alfalfa in the southern plains. J. Kans. Entomol. Soc. 54:213–218.

Bergelson, J., S. Flower, and S. Hartley. 1986. The effect of foliage damage on casebearing moth larvae, *Coleophora serratella,* feeding on birch. Ecol. Entomol. 11:241–250.

Bishop, A. I., P. J. Walter, R. H. Holtkamp, and B. C. Dominiak. 1982. Relationships between *Acyrthosiphon kondoi* and damage in three varieties of alfalfa. J. Econ. Entomol. 75:118–122.

Boivin, G., and R. K. Stewart. 1982. Identification and evaluation of damage to McIntosh apples by phytophagous mirids (Hemiptera: Miridae) in southwestern Quebec. Can. Entomol. 114:1037–1045.

Bolkan, H. A., J. M. Ogawa, R. E. Rice, R. M. Bostock, and J. C. Crane. 1984. Leaffooted bug (Hemiptera: Coreidae) and epicarp lesion of pistachio fruits. J. Econ. Entomol. 77:1163–1165.

Boucher, T. J., and D. G. Pfeiffer. 1989. Influence of Japanese beetle (Coleoptera: Scarabeidae) foliar feeding on 'Seyval Blanc' grapevines in Virginia. J. Econ. Entomol. 82:1.

Buntin, G. D., and L. P. Pedigo. 1985a. Development of economic injury levels for last-stage variegated cutworm (Lepidoptera: Noctuidae) larvae in alfalfa stubble. J. Econ. Entomol. 78:1341–1346.

Buntin, G. D., and L. P. Pedigo. 1985b. Dry-matter accumulation, partitioning and development of alfalfa regrowth after stubble defoliation by the variegated cutworm (Lepidoptera: Noctuidae). J. Econ. Entomol. 78:371–378.

Chapin, F. S. I., E. D. Schulze, and H. A. Mooney. 1990. The ecology and economics of storage in plants. Annu. Rev. Ecol. Syst. 21:423–448.

Chapman, P. J., S. E. Lienk, and O. F. Curtis Jr. 1952. Responses of apple to mite infestations: I. J. Econ. Entomol. 45:815–821.

Cousens, R., C. J. Doyle, B. J. Wilson, and G. W. Cussaus. 1986. Modelling the economics of controlling *Avena fatua* in winter wheat. Pestic. Sci. 17:1–12.

Crawley, M. 1989. Insect herbivores and plant population dynamics. Annu. Rev. Entomol. 34:531–564.

Cuperus, G. W., E. B. Radcliffe, D. K. Barns, and G. C. Marten. 1982. Economic injury levels and economic thresholds for pea aphid, *Acyrthosiphon pisum* (Harris), on alfalfa. Crop Prot. 1:453–463.

Cuperus, G. W., E. B. Radcliffe, D. K. Barnes, and G. C. Marten. 1983. Economic injury levels and economic thresholds for potato leafhopper (Homoptera: Cicadellidae) on alfalfa in Minnesota. J. Econ. Entomol. 76:1341–1349.

Densholm, I., and M. W. Rowland. 1992. Tactics for managing pesticide resistance in arthropods: Theory and practice. Annu. Rev. Entomol. 37:91–112.

Dunstan, G. G., and A. B. Stevenson. 1961. Pear leaf scorch and its relationship to the European red mite. J. Econ. Entomol. 54:918–920.

Engle, C. E., and M. M. Barnes. 1983. Cultural control of navel orangeworm in almond orchards. Calif. Agric. 37:19.

Faris, M. A., H. Baenziger, and R. P. Terhune. 1981. Studies on potato leafhopper damage in alfalfa. Can. J. Plant Sci. 61:625–632.

Feder, G. 1979. Pesticides, information, and pest management under uncertainty. Amer. J. Agric. Econ. 61:97–103.

Flaherty, D. L., and C. B. Huffaker. 1970. Biological control of Pacific mites and Willamette mites in San Joaquin Valley vineyards. Hilgardia 40:267–330.

Forshey, C. G., and D. C. Elfving. 1977. Fruit numbers, fruit size, and yield relationships in 'McIntosh' apples. J. Am. Soc. Hort. Sci. 102:399–402.

Georghiou, G. P. 1986. The magnitude of the resistance problem, p. 14–46. *In* Pesticide resistance: Strategies and tactics for management. National Academy Press, Washington DC.

Godfrey, L. D., and K. V. Yeargan. 1987. Effects and interactions of early season pests on alfalfa yield in Kentucky. J. Econ. Entomol. 80:248–256.

Hamilton, G. C., F. C. Swift, and R. Marini. 1986. Effect of *Aphis pomi* (Homoptera: Aphidae) density on apples. J. Econ. Entomol. 79:472–478.

Hardman, J. M., H. J. Herbert, K. H. Sanford, and D. Hamilton. 1985. Effect of populations of the European red mite, *Panonychus ulmi,* on the apple variety Red Delicious in Nova Scotia. Can. Entomol. 117:1257–1265.

Hare, J. D., J. E. Pherson, T. Clemens, J. L. Menge, C. W. Coggins Jr., T. W. Embleton, and J. L. Meyers. 1990. Effects of managing citrus red mite (Acari: Tetranychidae) and cultural practices on total yield, fruit size, and crop value of 'Navel' orange. J. Econ. Entomol. 83:976–984.

Hare, J. D., J. E. Pherson, T. Clemens, J. A. Menge, C. W. Coggins Jr., T. W. Embleton, and J. L. Meyers. 1992. Effect of citrus red mite (Acari: Tetranychidae) and cultural practices on total yield, fruit size, and crop value of 'Navel' orange: Years 3 and 4. J. Econ. Entomol. 85:486–495.

Hare, J. D., and R. R. Youngman. 1987. Gas exchange of orange *(Citrus sinensis)* leaves in response to feeding injury by the citrus red mite (Acari: Tetranychidae). J. Econ. Entomol. 82:204–208.

Hartley, S. E., and J. H. Lawton. 1987. Effects of different types of damage on the chemistry of birch foliage, and the responses of birch feeding insects. Oecologia 74:432–437.

Haukioja, E., and S. Neuvonen. 1985. Induced long-term resistance of birch foliage against defoliators: Defensive or incidental? Ecol. 66:1303–1308.

Higley, L. G., and W. K. Wintersteen. 1992. A novel approach to environmental risk assessment of pesticides as a basis for incorporating environmental costs into economic injury levels. Am. Entomologist 38:34–39.

Hluchy, M., and Z. Pospisil. 1992. Damage and economic injury levels of eriophyid and tetranychid mites on grapes in Czechoslovakia. Expt. Appl. Acarol. 14:95–106.

Hower, A. A. 1989. Potato leafhopper as a plant stress factor on alfalfa. *In* History and perspective of potato leafhopper (Homoptera: Cicadellidae) research. Misc. Publ. Entomol. Soc. Am., Lanham MD.

Hower, A. A., and P. W. Flinn. 1986. Effects of feeding by potato leafhopper nymphs (Homoptera: Cicadellidae) on growth quality of established stand alfalfa. J. Econ. Entomol. 79:779–784.

Hoyt, S. C., L. K. Tanigoshi, and R. W. Browne. 1979. Economic injury level studies in relation to mites on apple. Recent Adv. Acarol. I:3–12.

Hull, L. A., and E. H. Beers. 1990. Validation of injury thresholds for European red mite (Acari: Tetranychidae) on 'Yorking' and 'Delicious' apple. J. Econ. Entomol. 83:2026–2031.

Jensen, B. M., J. I. Wedberg, and D. B. Hogg. 1991. Assessment of damage caused by tarnished plant bug and alfalfa bug (Hemiptera: Miridae) on alfalfa grown for forage in Wisconsin. J. Econ. Entomol. 84:1024–1027.

Johnson, H. W. 1936. Effect of leafhopper yellowing upon the carotene content of alfalfa. Phytopathol. 26:1061–1063.

Jubb, G. L., L. Danko, and C. W. Haeseler. 1983. Impact of Erythroneura comes Say (Homoptera: Cicadellidae) on caged 'Concord' grapevines. Environ. Entomol. 12:1576–1580.

Karban, R. 1985. Induced responses of cherry trees to periodical cicada oviposition. Oecologia 59:226–231.

Karban, R., and G. M. English-Loeb. 1990. A "vaccination" of Willamette spider mites (Acari: Tetranychidae) to prevent large populations of Pacific spider mites on grapevines. J. Econ. Entomol. 83:2252–2257.

Karban, R., G. English-Loeb, and P. Verdegaal. 1991. Vaccinating grapevines against spider mites. Calif. Agric. 45:19–21.

Karban, R., and J. H. Myers. 1989. Induced plant responses to herbivory. Annu. Rev. Ecol. Syst. 20:331–348.

Kindler, S. D., W. R. Kehr, R. L. Ogden, and J. M. Schalk. 1973. Effect of potato leafhopper injury on yield and quality of resistant and susceptible alfalfa clones. J. Econ. Entomol. 66:1298–1302.

Kinn, D. N., J. L. Joos, R. L. Doutt, J. T. Sorensen, and M. J. Foskett. 1974. Effects of *Tetranychus pacificus* and quality of grapes in North Coast vineyards of California. Environ. Entomol. 3:601–606.

Knight, A. L., and G. W. Norton. 1989. Economics of agricultural pesticide resistance in arthropods. Annu. Rev. Entomol. 34:293–313.

Kouskolekas, C., and G. C. Decker. 1968. A quantitative evaluation of factors affecting alfalfa yield reduction caused by the potato leafhopper attack. J. Econ. Entomol. 61:921–927.

Laing, J. E., D. L. Calvert, and C. B. Huffaker. 1972. Preliminary studies of effects of *Tetranychus pacificus* on yield and quality of grapes in the San Joaquin Valley, California. Environ. Entomol. 1:658–663.

Latheef, M. A., R. C. Berberet, J. F. Stritzke, J. L. Caddell, and R. W. McNew. 1992. Productivity and persistence of declining alfalfa stands as influenced by the alfalfa weevil, weeds, and early first harvest in Oklahoma. Can. Entomol. 124:135–144.

Lienk, S. E., P. J. Chapman, and O. F. Curtis Jr. 1956. Responses of apple trees to mite infestations: II. J. Econ. Entomol. 49:350–353.

Light, W. I. S. G., and F. A. B. Ludlam. 1972. The effects of fruit tree red spider mite *(Panonychus ulmi)* on yield of apple trees in Kent. Plant Pathol. 21:175–181.

McClernan, W. A., and R. P. Marini. 1986. European red mite on yield, fruit quality, and growth of peach trees. HortSci. 21:244–246.

McNab, S. C., and P. H. Jerie. 1993. Flowering, fruit set, and yield response of

'Bartlett' pear to leaf-scorch damage by twospotted spider mite (Acari: Tetranychidae). J. Econ. Entomol. 86:486–493.

McNally, P. S., and D. Farnham. 1985. Effects of Willamette mite (Acari: Tetranychidae) on Chenin Blanc and Zinfandel grape varieties. J. Econ. Entomol. 78:371–378.

Michaud, O. D., G. Boivin, and R. K. Stewart. 1989. Economic threshold for tarnished plant bug (Hemiptera: Miridae) in apple orchards. J. Econ. Entomol. 82:1722–1728.

Miranowski, J. A., and G. A. Carlson. 1986. Economic issues in public and private approaches to preserving pest susceptibility, p. 436–445. *In* Pesticide resistance: Strategies and tactics for management. National Academy Press, Washington DC.

Norgaard, R. B. 1976. The economics of improving pesticide use. Annu. Rev. Entomol. 21:45–60.

Norris, R. J., W. R. Cothran, and V. E. Burton. 1984. Interactions between winter annual weeds and Egyptian alfalfa weevil (Coleoptera: Curculionidae) in alfalfa. J. Econ. Entomol. 77:43–52.

Pedigo, L. P., S. H. Hutchins, and L. G. Higley. 1986. Economic injury levels in theory and practice. Annu. Rev. Entomol. 31:341–368.

Pfeiffer, D. G., M. W. Brown, and M. W. Varn. 1989. Incidence of spirea aphid (Homoptera: Aphididae) in apple orchards in Virginia, West Virginia, and Maryland. J. Entomol. Sci. 24:145–149.

Pimentel, D., D. Andow, R. Dyson-Hudson, R. Gallanan, S. Jacobson, M. Irish, S. Kroop, A. Moss, I. Schreiner, M. Shepard, T. Thompson, and B. Vinzant. 1980. Environmental and social costs of pesticides: A preliminary assessment. Oikos 34:126–140.

Plant, R. E. 1986. Uncertainty and the economic threshold. J. Econ. Entomol. 79:1–6.

Poos, F. W., and H. W. Johnson. 1936. Injury to alfalfa and red clover by the potato leafhopper. J. Econ. Entomol. 29:325–331.

Purcell, M., and S. C. Welter. 1991. Effect of *Calocoris norvegicus* (Hemiptera: Miridae) on pistachio yields. J. Econ. Entomol. 84:114–119.

Shaw, M. C., M. C. Wilson, and C. L. Rhykerd. 1986. Influence of phosphorus and potassium fertilization on damage to alfalfa, *Medicago sativa* L., by the alfalfa weevil, *Hypera postica* (Gllyenhall) and potato leafhopper, *Empoasca fabae* (Harris). Crop Prot. 5:245–249.

Stern, V. M., R. F. Smith, R. van den Bosch, and K. S. Hagen. 1959. The integrated control concept. Hilgardia. 28:81–101.

Summers, C. G., and R. L. Coviello. 1984. Impact of *Acyrthosiphon kondoi* (Homoptera: Aphididae) on alfalfa: Field and greenhouse studies. J. Econ. Entomol. 77:1052–1056.

Tedders, W. L., and B. W. Wood. 1985. Estimate of influence of feeding by *Monelliopsis pecanis* and *Monellia caryella* (Homoptera: Aphididae) on the fruit, foliage, carbohydrate reserves, and tree productivity of mature 'Stuart' pecans. J. Econ. Entomol. 78:642–646.

Torell, L. A., J. J. Davis, E. W. Huddleston, and D. C. Thompson. 1989. Economic injury levels for interseasonal control of rangeland insects. J. Econ. Entomol. 82:1289–1294.

Trumble, J., D. M. Kolodny-Hirsch, and I. P. Ting. 1993. Plant compensation for arthropod herbivory. Annu. Rev. Entomol. 38:93–119.

Uyemoto, J. K., J. M. Ogawa, R. E. Rice, H. R. Teranishi, R. M. Bostock, and W. M. Pemberton. 1986. Role of several true bugs (Hemiptera) on incidence and seasonal development of pistachio fruit epicarp lesion disorder. J. Econ. Entomol. 79:395–399.

Varn, M. W., and D. G. Pfeiffer. 1989. The effect of rosy apple aphid and spirea aphid (Homoptera: Aphididae) on dry matter accumulation and carbohydrate concentration in young apple trees. J. Econ. Entomol. 82:565–569.

Walker, G. P., A. L. Voulgaropoulos, and P. A. Phillips. 1992. Effect of citrus bud mite (Acari: Eriophyidae) on lemon yields. J. Econ. Entomol. 85:1318–1329.

Walton, P. D. 1983. Production and management of cultivated forages. Reston Publishing, Reston VA.

Weinbaum, S. A., I. Klein, F. E. Broadbent, W. C. Micke, and T. T. Muraoka. 1984. Effects of time of nitrogen application and soil texture on the availability of isotopically labelled fertilizer to reproductive and vegetative tissue of mature almond trees. J. Am. Soc. Hort. Sci. 89:117–122.

Welter, S. C. 1991. Traditional and eco-physiological approaches to determining the impact of spider mites on perennial crops. Agric. Development in the Am. Pac. Conf., Honolulu HI, HITAHR.

Welter, S. C. 1993. Responses of plants to insects: Eco-physiological insights, p. 773–778. *In* D. R. Buxton, R. Shibles, R. A. Forsberg, B. L. Blad, K. H. Asay, G. M. Paulson, and R. F. Wilson (ed.) International crop science I. Crop Sci. Soc. of Am., Madison WI.

Welter, S. C., M. M. Barnes, I. P. Ting, and J. T. Hayashi. 1984. Impact of various levels of late-season spider mite (Acari: Tetranychidae) impact on almond growth and yield. Environ. Entomol. 13:52–55.

Welter, S. C., R. Freeman, and D. S. Farnham. 1991. Recovery of 'Zinfandel' grapevines from feeding damage by Willamette spider mite (Acari: Tetranychidae): Implications for economic injury level studies in perennial crops. Environ. Entomol. 20:104–109.

Welter, S. C., P. S. McNally, and D. S. Farnham. 1989a. Willamette mite (Acari:

Tetranychidae) impact on grape productivity and quality: A re-appraisal. Environ. Entomol. 18:408–411.

Welter, S. C., P. S. McNally, D. Farnham, and R. Freeman. 1989b. Effect of Willamette mite and Pacific spider mite (Acari: Tetranychidae) on grape gas exchange. Environ. Entomol. 18:953–957.

Westigard, P. H., P. B. Lombard, and J. H. Grim. 1966. Preliminary investigations of the effect of feeding on various levels of two-spotted spider mite on its Anjou pear host. Proc. Am. Soc. Hort. Sci. 89:117–122.

Wieres, R. W., J. R. VanKirk, W. D. Gerling, and F. M. McNicholas. 1985. Economic losses from the tarnished plant bug on apple in eastern New York. J. Agric. Entomol. 2:256–263.

Wilson, H. K., and S. S. Quisenberry. 1986. Impact of feeding by alfalfa weevil larvae (Coleoptera: Curculionidae) and pea aphid (Homoptera: Aphididae) on yield and quality of first and second cuttings of alfalfa. J. Econ. Entomol. 79:785–789.

Wilson, M. C., R. L. Davis, and G. G. Williams. 1955. Multiple effects of leafhopper infestation on irrigated and non-irrigated alfalfa. J. Econ. Entomol. 48:323–326.

Wilson, M. C., M. C. Shaw, and M. A. Zajac. 1989. Interaction of the potato leafhopper and environmental stress factors associated with economic damage thresholds for alfalfa. *In* History and perspective of potato leafhopper (Homoptera: Cicadellidae) research. Misc. Publ. Entomol. Soc. Am., Lanham MD.

Wood, B. W., and W. L. Tedders. 1982. Effects of an infestation of blackmargined aphid on carbohydrates in mature 'Stuart' pecan. HortSci. 17:236–238.

Zwick, R. W., G. J. Fields, and W. M. Mellenthin. 1976. Effects of mite population density on 'Newtown' and 'Golden Delicious' apple tree performance. J. Am. Soc. Hort. Sci. 101:123–125.

Leon G. Higley and Wendy K. Wintersteen

14
Thresholds and Environmental Quality

One of the tenets of pest management is the principle of sustainability. Specifically, pest management should address the following:

economic sustainability through minimizing the economic impact of pests

ecological sustainability through employing management tactics so as to minimize selection pressure

environmental sustainability through minimizing the impact of management tactics on the environment.

The economic injury level (EIL) is an essential tool in pest management, providing, as it does, an objective basis for making management decisions. As a cost-benefit criterion, the EIL weighs economic risk against economic gain, and in this way helps ensure economic viability in the management of pest problems.

Indirectly, the EIL may also address issues of ecological and environmental sustainability, through reducing the use of management tactics. Ecological sustainability, as it relates to tactics, is a key issue, and it is receiving much attention—but usually under the umbrella of resistance management and not in the context of EILs. This lack of emphasis on EILs in resistance management is not surprising, given that EILs are used in a single-decision context and resistance management depends on questions of multiple use of tactics. The need for consideration to be given to EILs in combination with resistance management and in multiple-decision frameworks lies beyond the scope of this chapter; it will suffice to say that the need is growing and is, as yet, unmet. In contrast, EILs and environmental sustainability is receiving attention, and that will be the focus of this chapter.

Given that the EIL is most useful for curative management tactics (typically pesticides, because there are few other curative approaches available), use of EILs greatly influences the frequency of use of curative tactics. In the absence of EILs, the risk of excessive losses from pests is likely to be weighed more strongly than the cost of unnecessary pesticide use (which, in comparison, is less of an economic risk). Consequently, we expect the use of EILs to reduce pesticide use, and such has been the outcome observed in various systems when EILs have first become available.

However, this environmental benefit provided by the use of EILs is a consequence of the economically conservative bias of producers when they face risks posed by pest attack: the environmental benefits do not derive from an intrinsic consideration of risk in the EIL. Where the EIL does not significantly alter pesticide-use frequency, such as with very high-value crops or very severe pests, we would not expect an environmental benefit. Indeed, *increased* pesticide use with EILs is possible, if growers do not know the extent of economic losses from pests prior to the use of EILs. This can occur, for example, with forages, which often are not as intensively managed as are row-crops.

Overall, however, the premise that EILs reduce pesticide use is probably a fair generalization, because most pest-management situations do not fit these exceptions. The key here is to recognize the relationship of EILs to environmental sustainability: on a practical level, EILs can help maintain environmental quality through reduced frequency of use in most situations; on a theoretical level, the EIL does not address environmental questions.

This characterization leads to an important question: Is it possible, explicitly, to consider environmental risk in pest management decision making, and if so, how? We believe the answer to this question is yes, through environmental EILs (Higley and Wintersteen 1992), although our approach to this question is not without controversy. We will discuss environmental EILs at length later in the chapter. More generally, we may ask how we can be more responsive to issues of environmental quality in pest management. On this point, we can identify at least four approaches, three of which involve the EIL.

Historically, two approaches dominate the question of environmental quality and pest management: (1) frequency of use and (2) choice of management tactic. *Frequency of use* involves the issue of reducing pesticide use. As noted above, the environmental benefit derived from using EILs has been associated with this issue; indeed, the EIL provides the only formal criterion for making use/no-use decisions. In addition, we must note that considerations of pesticide resistance are playing an increasingly important part in decisions on pesticide use, and use of alternative (nonpesticidal) tactics is important for reducing aggregate pesticide use.

Choice of tactic usually is associated with identifying alternatives to pesticides. Considerable research effort is devoted to promoting and developing alternatives—a reflection of the importance given to this approach to reducing environmental risk. Arguably, the most successful alternatives are host-plant resistance, biological control, and cultural management practices. Unfortunately, most of these procedures are preventive, and there is a continuing problem in identifying alternative therapeutic tactics.

One area where choice of tactic has been successful in reducing environmental risk is in choice of pesticide. Later-generation pesticides provide a real improvement in environmental safety—a factor that is sometimes overlooked. The ongoing effort to develop benign pesticides and new, more selective modes of action suggests that pesticide choice will be an increasingly important issue. For the most part, formal evaluations for assessing environmental risk or for making choices between tactics (even between different pesticides) have not been developed. However, there is recognition of distinctions between pesticides—some having high, some low toxicities, some having long, some short half-lives, and some posing particular threats to an aspect of the environment, such as groundwater. These distinctions, however, are not made as part of the EIL. A consideration of pesticide properties could be used to develop formal pesticide-selection procedures and two such systems (Becker et al. 1989, Kovach et al. 1992) have been proposed.

Interestingly, the growing recognition of the importance of environmental issues in pest management has led not only to more direct consideration of pesticide risk but also to possible risks associated with other tactics, such as biological control (Lockwood 1993a,b, Carruthers and Onsager 1993).

Growing concerns regarding the environment, and philosophical questions regarding the role of pest management in relationship to environmental quality, have led to two additional approaches, both related to EILs. The first is a proposal by Pedigo and Higley (Higley and Pedigo 1993, Pedigo and Higley 1992) that the elements of the EIL should be used as a conceptual framework to reduce environmental risk. The second is the construction of an environmental EIL (Higley and Wintersteen 1992), in which environmental risk is estimated in monetary terms and combined with other economic considerations in the EIL. These approaches are complementary; indeed, Pedigo and Higley (1992) list the environmental EIL (EEIL) as an option in modifying variable C (management costs) to reflect environmental risk. We will consider each approach in more detail.

The EIL as a Framework for Reducing Environmental Risk

Because pest management is focused on reducing the impact of pest activities to tolerable levels, the EIL is an essential criterion for characterizing what is tolerable, albeit in economic terms. The argument Pedigo and Higley (1992) offer is that, by examining each component of the EIL with respect to environmental issues, it may be possible to make pest-management decisions more responsive to environmental concerns. For the most part, their approach is philosophical; the EIL and its variables are used as a mechanism for considering how environmental quality can be better maintained in the context of pest management. However, in supporting the development of environmental EILs (*sensu* Higley and Wintersteen 1992), which include an economic consideration of environmental risk, their approach also is operational: environmental risk is directly considered in the decision process.

The underlying goal with environmental considerations of the EIL is to find ways to increase EILs. It must be remembered that with a larger EIL more pest activity can be tolerated and the need for management is reduced. However, this goal must be achieved in a given economic context. For example, low crop values or high management costs will increase the EIL, but they also impact producer profitability. Although it may not be possible to avoid such impacts, potential effects on profitability are an important consideration in addressing environmental risk.

The specific variables to be considered are those of the EIL:

$$EIL = C/VDIK,$$

where C = management costs, V = crop value, D = yield loss per unit injury, I = injury per pest, and K = proportion of injury prevented by management. Increasing values in the numerator *(C)* or decreasing values in the denominator (V, D, I, and K) will have the desired effect of increasing the EIL, and thereby reducing the need for management action. Reducing the need for management is an important, but indirect, response to issues of environmental risk; however, some modifications of EIL variables (C and K) not only increase the EIL but also directly address environmental risk.

Pedigo and Higley (1992) and Higley and Pedigo (1993) discuss specific modifications of these variables; here we will only briefly review some possibilities. Increases in C could be associated with assigning costs to different levels of environmental risk. One possibility in this regard is differential taxation of pesticides based on environmental risk (Higley et al. 1993); another is estimating environmental costs associated with different risks (the Higley and Wintersteen 1992 approach). Both of these provide direct consideration of environmental risk, but they share the limitation that they could reduce producer profitability.

Changes in crop value *(V)* seem to offer relatively few opportunities for manipulation, because costs are outside the control of the producer, and reducing market value is an unwelcome outcome from a producer's standpoint. Rather than reducing costs, increasing crop values in association with choice of management tactic is another possibility. Premiums for pesticide-free produce or crops are available in some instances, but such premiums seem to be a limited option for addressing environmental risks in general.

Reduction in injury per pest *(I)* has been attempted with respect to insect vectors of disease, through altering their vector competency. Because reduction in injury involves the modification of entire pest populations, this method's practicability is suspect. However, this approach may have merit with weeds, because slight modifications in weed competitiveness early in the season may dramatically reduce later competition and potential yield losses. One approach proposed is to use extremely low rates of herbicides to impair weed competitiveness (D. Mortensen, personal communication).

Increasing plant tolerance to injury (reducing D) is another option that seems to offer considerable potential. Tolerance and compensation to injury in plants are well documented phenomena (Pedigo et al. 1986, Trumble et al. 1993), although the mechanisms are poorly understood. As an approach, tolerance has the tremendous theoretical advantage of not producing any selection pressure on pest populations; consequently, it offers sustainability far in excess of many other procedures. Specific approaches to tolerance include producing genetically based differences in plant responses to injury or competition and development of relevant cultural practices. Pedigo and Higley (1992) point out that our understanding of plant stress from pest attack is relatively rudimentary, therefore, as our understanding of stress improves, it seems likely that new opportunities for enhancing tolerance will be discovered.

A final option is to address K, the proportion of injury prevented by management. Pedigo and Higley (1992) argue that, in practice, most pesticide users strive for a K close to 1, resulting in some excessive use of pesticides. They propose that values of K less than 1 may be appropriate in some instances and that even where a K near 1 is desired, reduced rates of pesticides may achieve acceptable results. Reductions in pesticide application rates directly address environmental risk. Pedigo and Higley suggest that most rates are too conservative and that substantial rate reductions would still provide K values at acceptable levels. However, it is important not to overstate the potential impact of rate reductions on risk. Certainly, rate reductions reduce potential environmental contamination, but if a population (insect, bird, human, or other species) responds to pesticide concentrations over a 10-fold to 100-fold range (e.g., the typical range between an LD_5 and an LD_{95}), the rate reductions must be of a comparable range to reduce risk significantly. Because rate reductions of this magnitude are not practicable in most instances, the ultimate impact of rate reductions on risk is limited.

The Environmental EIL

The environmental EIL, as defined by Pedigo and Higley (1992) and Higley and Wintersteen (1992), involves the direct consideration, through the EIL, of environmental risk in the pest-management process. Specifically, Higley and Wintersteen considered objective pesticide-risk criteria and combined these with subjective criteria on relative rankings of risk and monetary values of avoiding different risk levels (through an economic survey procedure called contingent valuation) to estimate environmental costs of different pesticides. These costs were then combined with management costs in the EIL equation to produce an EEIL. Thus, the environmental EIL represents a combination of environmental and economic criteria. This is a significant departure from previous attitudes toward environmental risk in pest management, in that environmental risk is weighed equally with economic risk.

Another unique feature of the environmental EIL is that it provides a mechanism for considering environmental risk at a micro level: it allows individual producers to make choices based on environmental risk. In the main, environmental risks from pesticides are addressed at a national or regional level, through pesticide legislation and regulation; however, legislation and regulation cannot ameliorate all environmental hazards posed by pesticides. Information is needed by individual pesticide users to allow for direct considerations of environmental risk in their selecting pesticides, and in their choosing whether or not to use a pesticide. Additionally, information on environmental risks of pesticides is needed for use in macro level decisions about pesticides. This is particularly pertinent for decisions on pesticide registration, where economic information on pesticide use could be combined with information on environmental risks to provide a more balanced and insightful assessment. Thus, estimates of environmental costs of pesticides are potentially useful not only in establishing environmental EILs but also in addressing regulatory issues about pesticides.

Environmental Risk and Environmental Costs

Risks. Given the importance of the issue, surprisingly few studies have been made to rank environmental risks from pesticides; however, models for characterizing potential pesticide contamination of surface and ground water have been developed and tested in a number of studies. Particularly noteworthy in this regard is the GLEAMS (Groundwater Loading Effects of Agricultural Management Systems) model (Leonard et al. 1987), which simulates the runoff and leaching potential of pesticides. GLEAMS has been extensively validated and used to characterize pesticide risk to groundwater (e.g., Leonard and Knisel 1989, Leonard et al. 1988, Leonard et al.

1990). Becker et al. (1989) used the GLEAMS model to develop a ranking of potential risks (low, medium, and high) to surface and groundwater for common agricultural pesticides.

Although the GLEAMS model and the Becker et al. implementation of it are useful in characterizing risks of pesticides to water, neither address other environmental risks from pesticides, such as to nontarget organisms. More recently, Kovach et al. (1992) developed a rating system for pesticides that included considerations of toxicity, soil and plant residue half-lives, and potential for water contamination. In many respects, the Kovach et al. ranking system is similar to that of Higley and Wintersteen (1992), although Kovach et al. calculated an environmental-impact quotient, rather than environmental costs. Comprehensive discussions of issues involved in pesticides risks to the environment and issues associated with reducing those risks are provided in Pimentel and Lehman (1993).

Environmental Costs. Criteria for assessing environmental risk of pesticides in the Higley and Wintersteen (1992) model is based on GLEAMS information (Becker et al. 1989) for risks to water, and on toxicity and persistence data for risks to nontarget organisms and humans. A critical difference in the Higley and Wintersteen model, however, is that it addresses the issue of relating costs to environmental risk. Establishing environmental costs is essential on at least two points.

First, the utility of risk assessment (and our opportunities for using assessments to mitigate risk) is dramatically improved by expressing risk in economic terms. Like risk indices, environmental costs also can be used to describe relative risk, such as in selecting pesticides; however, unlike risk indices, expressing risk on an environmental cost basis allows for the direct application of risk information into economically based tools for evaluating pesticides. These include the direct incorporation of environmental-cost information into pest-management decision tools (EILs) and the calculation of regional or national environmental costs associated with use of a given pesticide (such as in determining the environmental impact of pesticide cancellations and replacements).

Secondly, the contingent-valuation procedure used in determining environmental costs allows for the inclusion of nonexpert opinion in the evaluation of risk, without biasing scientifically based risk estimates. Peterson and Higley (1993) discuss the substantial and growing literature on risk perception that points to a more encompassing view of risk by lay people than was previously appreciated. As Slovic (1987) states: "Lay people sometimes lack certain information about hazards. However, their basic conceptualization of risk is much richer than that of the experts and reflects legitimate concerns that typically are omitted from expert risk assessments." Because estimates of environmental costs of pesticides are necessarily subjective, it seems most appropriate that assessments on environmental costs should be made not solely from expert opinion but also from lay sources.

Contingent Valuation. Estimating environmental costs of pesticides is challenging. If it were possible to remove all the deleterious effects of pesticide use, the cost of this repair would be the environmental cost of pesticide use. However, such remediation costs cannot be used to assign environmental costs to pesticides, because we cannot quantify all impacts of pesticide use, nor can we remedy all pesticide effects (for example, how can we purify contaminated ground water or replace all nontarget organisms affected by pesticides?). Consequently, environmental costs for pesticides must be evaluated indirectly. Economic procedures have been developed to estimate the cost of nonmarket goods like environmental quality. Procedures named travel-cost, hedonic-price, and contingent valuation (Anderson and Bishop 1986) have been developed and used to estimate costs that cannot be measured directly. Of the available methods, only contingent valuation seems appropriate to the problem of establishing environmental costs for the single use of a pesticide.

Contingent valuation is a formal procedure for estimating, through opinion surveys, the value of nonmarket goods. The legitimacy of contingent valuation for estimating such costs has been debated at length; however, some recent studies indicate that it works as well as, or better than, alternative methods (Anderson and Bishop 1986, Cummings et al. 1986, Heberlein and Bishop 1986, Smith et al. 1986). Among the key issues in using contingent valuation are choosing the appropriate survey audience and avoiding strategic and hypothetical biases in responses (responses made to deliberately or inadvertently bias the survey results) (Anderson and Bishop 1986). More recently, Mitchell and Carson (1989) examined the theoretical foundations of the technique, and Arrow et al. (1993) reviewed the overall reliability of the method and proposed a set of guidelines and issues that might be addressed in preparing a contingent-valuation survey.

Despite its limitations, contingent valuation has been put to a number of entomological uses, including work by John et al. (1987) on mosquitoes, work by Larew et al. (1984), Raupp et al. (1988), Coffelt and Shultz (1990, 1993), and Sadof and Alexander (1993) on aesthetic pests, and work by Reiling et al. (1988) on black flies; however, none of these studies address environmental costs. In the context of EILs, contingent valuation is an appropriate foundation for aesthetic thresholds, given their intrinsic subjectivity, and similarly, we believe it provides one of the few methods for addressing the problem of environmental costs. Also, unlike most other uses of contingent valuation, questions on environmental costs associated with pesticide use are not completely hypothetical: environmental costs are used in calculating environmental EILs, which include real costs in additional yield loss or premiums for safer pesticides that a user would incur.

Calculating an Environmental EIL

Higley and Wintersteen (1992) detail the procedures for calculating environmental costs and determining EEILs. First, pesticides were ranked as posing high, moderate, low, or negligible risk to different areas of the environment, based upon physical and biological properties. In the case of insecticides, they considered eight aspects of the environment: surface water, groundwater, aquatic organisms, birds, mammals, beneficial insects, acute human toxicity, and chronic human toxicity. Risks to surface and groundwater were ranked based upon persistence and potential for leaching. Risks to organisms were based upon LC_{50} or LD_{50} values and some consideration of persistence.

The second step was to determine how to weigh these eight environmental categories. Are risks to each of these categories equally important? This question was answered by asking respondents to give relative rankings for the eight categories. A third step was to assign a dollar cost to each level of risk: high, moderate, or low. This was done through contingent-valuation survey, by asking respondents how much it would be worth to avoid a given level of risk for a single application of a pesticide.

In this study, Higley and Wintersteen ranked environmental categories and established monetary costs for different levels of risks by conducting a contingent-valuation survey of 8,000 field-crop producers in the north-central United States. They obtained a return rate of 21.8%, and virtually all respondents answered questions regarding relative importance of specific environmental categories. On a 1-to-10 point scale measuring increasing importance, the average ranking across all environmental categories was 8.6. The categories naming acute human toxicity, chronic human toxicity, surface water, and groundwater had the highest rankings, but there was not much discrimination between any of the categories. Regarding how much they would be willing to pay to avoid three increasing levels of risk, only 66% of the respondents indicated they would be willing to pay to avoid risks, although greater than 98% recognized the importance of environmental quality. Thus, virtually all the surveyed producers recognized the importance of environmental risks from pesticides, but only two-thirds of them accepted the premise that they should be willing to pay to avoid increasing risk. For a single application of a pesticide on a per-acre basis, producers indicated that they would be willing to pay $12.54 to avoid a high risk, $8.76 to avoid a moderate risk, and $5.79 to avoid a low risk.

These data were used to determine environmental costs for each pesticide in each environmental category. Specifically, the relative importance of a given category was multiplied by the dollar value associated with the risk of that category. This produced a cost for a given category. These costs in each of the eight categories were then summed across the pesticide to produce a total environmental cost of a single use of the pesticide.

The information thus produced can then be used in two ways:

A listing of pesticides and their environmental costs provides an index to the relative total environmental risk of a pesticide; consequently, if a producer was interested in choosing a pesticide that had the least environmental risk, he would look for the one that had the lowest environmental cost.

The information can be combined with data on cost and efficacy of a pesticide in the calculation of an EEIL.

A conventional EIL is

$$EIL = C/VDIK.$$

The environmental EIL is similar to this formula, however, it adds a term for the environmental costs *(EC)*. Specifically, the formula for the EEIL is

$$environmental\ \text{EIL} = (C+EC)/VDIK.$$

The environmental EIL combines conventional information on the economics of pesticide use with the additional consideration of environmental risks. In using such an EIL, a producer would consider both the direct economic costs and benefits of uses of a pesticide, as well as indirect environmental costs associated with pesticide use. In practical terms, the EEIL implies that, for pesticides posing greater environmental risks, we tolerate a greater number of pests before we take action. Also, just as a pesticide can be selected based on its having the lowest environmental cost, so it might be selected based on its producing the lowest EEIL. Choosing pesticides based on EEILs allows a consideration not only of environmental risk but also of cost and efficacy of the pesticide.

Higley and Wintersteen (1992) present environmental costs and EEILs based on risk data for insecticidal active ingredients. More recently, we have made calculations based on risk data for formulated product (Higley and Wintersteen *in press*). Table 14.1 presents environmental-cost values for common field-crop insecticides based on formulated product. From these environmental cost data, it is possible to calculate environmental EILs and ETs, as illustrated in Table 14.2. Additionally, cost data can be used in a worksheet or computer spreadsheet to allow calculations for specific (rather than generic) costs, expected market values, and expected yields (Table 14.3).

The development of environmental costs for formulated products is an important modification to improve the estimates of risk. We are working on other improvements in how risks are assigned; these will include more explicit characterizations of dose and persistence. It is also possible to provide more site-specific indications of risk by considering soil type, potential for surface-water contamination, application method, and related factors. Incorporating site specificity into risk estimates can improve the strength of those estimates (at the expense of additional complexity), although, as with any risk estimator, there will still be uncertainty.

Table 14.1. Environmental risks and costs of formulated field-crop insecticides, based on relative risk coefficients and risk costs from Higley and Wintersteen (1992). Individual costs are the product of the risk coefficient times the risk cost. NR = negligible risk, LR = low risk, MR = moderate risk, and HR = high risk.

| | | Risk Level and Environmental Cost by Environmental Category |||||||| |
Insecticide	Trade name	Surface Water	Ground-water	Aquatic Nontarget	Avian Nontarget	Mammalian Nontarget	Beneficial Insect	Acute Human	Chronic Human	Total Cost
acephate	Orthene 75S	LR 0.73	LR 0.77	LR 0.68	MR 1.02	LR 0.65	LR 0.69	LR 0.79	LR 0.79	6.14
	Orthene 80seed	LR 0.73	LR 0.77	LR 0.68	MR 1.02	LR 0.65	LR 0.69	LR 0.79	LR 0.79	6.14
	Orthene 90S	LR 0.73	LR 0.77	LR 0.68	MR 1.02	LR 0.65	LR 0.69	LR 0.79	LR 0.79	6.14
aldicarb	Temik 15G	LR 0.73	HR 1.68	HR 1.48	HR 1.46	HR 1.42	HR 1.50	HR 1.71	LR 0.79	10.76
azinphos-methyl	Guthion 2S	HR 1.59	LR 0.77	HR 1.48	HR 1.46	HR 1.42	HR 1.50	HR 1.71	MR 1.19	11.12
	Guthion 2L	HR 1.59	LR 0.77	HR 1.48	HR 1.46	HR 1.42	HR 1.50	HR 1.71	MR 1.19	11.12
	Guthion 35% WP	HR 1.59	LR 0.77	HR 1.48	HR 1.46	HR 1.42	HR 1.50	HR 1.71	MR 1.19	11.12
	Guthion 50% WP	HR 1.59	LR 0.77	HR 1.48	HR 1.46	HR 1.42	HR 1.50	HR 1.71	MR 1.19	11.12
B.t.-kurstaki	DiPel ES	NR 0	LR 0.77	LR 0.68	NR 0	NR 0	NR 0	NR 0	LR 0.79	2.25
	DiPel 2x	NR 0	LR 0.77	LR 0.68	NR 0	NR 0	NR 0	NR 0	LR 0.79	2.25
	DiPel 10G	NR 0	LR 0.77	LR 0.68	NR 0	NR 0	NR 0	NR 0	LR 0.79	2.25
	Javelin WG	NR 0	LR 0.77	LR 0.68	NR 0	NR 0	NR 0	NR 0	LR 0.79	2.25
carbaryl	Sevin XLR Plus	MR 1.11	LR 0.77	HR 1.48	MR 1.02	LR 0.65	MR 1.05	LR 0.79	LR 0.79	7.67

(continued)

Table 14.1 continued

Insecticide	Trade name	Surface Water	Ground-water	Aquatic Nontarget	Avian Nontarget	Mammalian Nontarget	Beneficial Insect	Acute Human	Chronic Human	Total Cost
	Sevin 4-oil	MR 1.11	LR 0.77	HR 1.48	MR 1.02	MR 0.99	MR 1.05	MR 1.19	LR 0.79	8.41
	Sevin 80s	MR 1.11	LR 0.77	HR 1.48	MR 1.02	MR 0.99	MR 1.05	MR 1.19	LR 0.79	8.41
	Sevin 50w	MR 1.11	LR 0.77	HR 1.48	MR 1.02	MR 0.99	MR 1.05	MR 1.19	LR 0.79	8.41
	Sevin 4F	MR 1.11	LR 0.77	HR 1.48	MR 1.02	MR 0.99	MR 1.05	MR 1.19	LR 0.79	8.41
carbofuran	Furadan 4F	LR 0.73	HR 1.68	HR 1.48	HR 1.46	HR 1.42	HR 1.50	HR 1.71	LR 0.79	10.76
	Furadan 15G	LR 0.73	HR 1.68	HR 1.48	HR 1.46	MR 0.99	HR 1.50	MR 1.19	LR 0.79	9.82
chlorpyrifos	Lorsban 4E	HR 1.59	LR 0.77	HR 1.48	HR 1.46	MR 0.99	HR 1.50	MR 1.19	MR 1.19	10.18
	Lorsban 15G	HR 1.59	LR 0.77	HR 1.48	HR 1.46	LR 0.65	HR 1.50	LR 0.79	MR 1.19	9.44
diazinon	D-z-n diazinon 4E	MR 1.11	MR 1.17	HR 1.48	HR 1.46	LR 0.65	HR 1.50	LR 0.79	LR 0.79	8.95
	D-z-n diazinon 14G	MR 1.11	MR 1.17	HR 1.48	HR 1.46	LR 0.65	HR 1.50	LR 0.79	LR 0.79	8.95
	D-z-n diazinon 50w	MR 1.11	MR 1.17	HR 1.48	HR 1.46	LR 0.65	HR 1.50	LR 0.79	LR 0.79	8.95
	D-z-n diazinon AG500	MR 1.11	MR 1.17	HR 1.48	HR 1.46	LR 0.65	HR 1.50	LR 0.79	LR 0.79	8.95
dimethoate	Cygon 400	LR 0.73	MR 1.17	HR 1.48	HR 1.46	LR 0.65	MR 1.05	LR 0.79	MR 1.19	8.53
disulfoton	Di-Syston 15%G	MR 1.11	LR 0.77	HR 1.48	HR 1.46	HR 1.42	MR 1.05	HR 1.71	MR 1.19	10.19
	Di-Syston 8	MR 1.11	LR 0.77	HR 1.48	HR 1.46	HR 1.42	MR 1.05	HR 1.71	MR 1.19	10.19
esfenvalerate	Asana	HR 1.59	LR 0.77	HR 1.48	MR 1.02	MR 0.99	HR 1.50	MR 1.19	LR 0.79	9.34

Table 14.1 continued

Risk Level and Environmental Cost by Environmental Category

Insecticide	Trade name	Surface Water	Ground-water	Aquatic Nontarget	Avian Nontarget	Mammalian Nontarget	Beneficial Insect	Acute Human	Chronic Human	Total Cost
	Asana XL	HR 1.59	LR 0.77	HR 1.48	MR 1.02	LR 0.65	HR 1.50	LR 0.79	LR 0.79	8.60
ethoprop	Mocap 10%G	MR 1.11	HR 1.68	HR 1.48	MR 1.02	MR 0.99	HR 1.50	MR 1.19	MR 1.19	10.16
	Mocap 20%G	MR 1.11	HR 1.68	HR 1.48	MR 1.02	MR 0.99	HR 1.50	MR 1.19	MR 1.19	10.16
	Mocap EC	MR 1.11	HR 1.68	HR 1.48	MR 1.02	HR 1.42	HR 1.50	HR 1.71	MR 1.19	11.11
fenvalerate	Pydrin 2.4	HR 1.59	LR 0.77	HR 1.48	LR 0.67	LR 0.65	HR 1.50	LR 0.79	LR 0.79	8.25
fonofos	Dyfonate 4-EC	HR 1.59	MR 1.17	HR 1.48	HR 1.46	HR 1.42	HR 1.50	HR 1.71	MR 1.19	11.52
	Dyfonate II 10G	HR 1.59	MR 1.17	HR 1.48	HR 1.46	MR 0.99	HR 1.50	MR 1.19	MR 1.19	10.58
	Dyfonate II 20G	HR 1.59	MR 1.17	HR 1.48	HR 1.46	HR 1.42	HR 1.50	HR 1.71	MR 1.19	11.52
malathion	Cythion 57%EC	LR 0.73	LR 0.77	MR 1.04	LR 0.67	LR 0.65	LR 0.65	LR 0.79	LR 0.79	6.14
methidathion	Supracide 2E	MR 1.11	MR 1.17	MR 1.04	MR 1.02	MR 0.99	HR 1.50	MR 1.19	MR 1.19	9.21
methomyl	Lannate	MR 1.11	HR 1.68	HR 1.48	HR 1.46	HR 1.42	HR 1.50	HR 1.71	LR 0.79	11.14
	Lannate L	MR 1.11	HR 1.68	HR 1.48	HR 1.46	HR 1.42	HR 1.50	HR 1.71	LR 0.79	11.14
	Lannate LV	MR 1.11	HR 1.68	HR 1.48	HR 1.46	MR 0.99	HR 1.50	MR 1.91	LR 0.79	10.20
methyl parathion	Penncap M	MR 1.11	LR 0.77	HR 1.48	HR 1.46	LR 0.65	MR 1.05	LR 0.79	MR 1.19	8.51

(continued)

Table 14.1 continued

Risk Level and Environmental Cost by Environmental Category

Insecticide	Trade name	Surface Water	Ground-water	Aquatic Nontarget	Avian Nontarget	Mammalian Nontarget	Beneficial Insect	Acute Human	Chronic Human	Total Cost
oxydemeton methyl	Metasystox-R	LR 0.73	HR 1.68	HR 1.48	HR 1.46	MR 0.99	MR 1.05	MR 1.19	MR 1.19	9.78
permethrin	Ambush	HR 1.59	LR 0.77	HR 1.48	LR 0.67	LR 0.65	HR 1.50	LR 0.79	LR 0.79	8.25
	Ambush 25w	HR 1.59	LR 0.77	HR 1.48	LR 0.67	LR 0.65	HR 1.50	LR 0.79	LR 0.79	8.25
	Ambush 25w-wsp	HR 1.59	LR 0.77	HR 1.48	LR 0.67	LR 0.65	HR 1.50	LR 0.79	LR 0.79	8.25
	Pounce 1.5G	HR 1.59	LR 0.77	HR 1.48	LR 0.67	LR 0.65	HR 1.50	LR 0.79	LR 0.79	8.25
	Pounce 3.2EC	HR 1.59	LR 0.77	HR 1.48	LR 0.67	LR 0.65	HR 1.50	LR 0.79	LR 0.79	8.25
	Pounce 25wp	HR 1.59	LR 0.77	HR 1.48	LR 0.67	LR 0.65	HR 1.50	LR 0.79	LR 0.79	8.25
phorate	Thimet 15-G	HR 1.59	LR 0.77	HR 1.48	HR 1.46	HR 1.42	HR 1.50	HR 1.71	LR 0.79	10.72
	Thimet 20-G	HR 1.59	LR 0.77	HR 1.48	HR 1.46	HR 1.42	HR 1.50	HR 1.71	LR 0.79	10.72
phorate + flucythrinate	AAStar	HR 1.59	LR 0.77	HR 1.48	HR 1.46	HR 1.42	HR 1.50	HR 1.71	LR 0.79	10.72
phosmet	Imidan 50-wp	MR 1.11	LR 0.77	HR 1.48	LR 0.67	MR 0.99	MR 1.05	MR 1.19	MR 1.19	8.46
propargite	Comite	HR 1.59	LR 0.77	HR 1.48	LR 0.67	LR 0.65	LR 0.69	LR 0.79	MR 1.19	7.85
tefluthrin	Force 1.5G	MR 1.11	LR 0.77	HR 1.48	LR 0.67	LR 0.65	HR 1.50	LR 0.79	LR 0.79	7.77
terbufos	Counter 15G	MR 1.11	LR 0.77	HR 1.48	HR 1.46	HR 1.42	MR 1.05	HR 1.71	LR 0.79	9.79
thiodicarb	Larvin 3.2	MR 1.11	LR 0.77	MR 1.04	LR 0.67	MR 0.99	MR 1.05	MR 1.19	MR 1.19	8.02
trichlorfon	Dylox 80%sp	LR 0.73	HR 1.68	HR 1.48	HR 1.46	MR 0.99	HR 1.50	MR 1.19	MR 1.19	10.23

Table 14.2. Example environmental ETs for second generation European corn borer on pretassel stage corn (all calculations based on 125 bu/a expected yield, $2.35/bu expected value/bu, yield loss per larva [D*I] of 6.6%, and an expected control of 50%).

Insecticide	Rate/a	Insecticide+ Application Costs	Environmental Costs	Total Cost	ET (larvae/plant)	Environmental ET (larvae/plant)
Ground Application ($3.80/a)						
Ambush 2EC	6.4 oz	$ 9.39	$ 8.25	$17.64	0.78	1.46
Asana XL .66EC	7.8 oz	$11.59	$ 8.60	$20.19	0.96	1.67
Dipel 10G	10.0 lb	$16.90	$ 2.25	$19.15	1.39	1.58
Dyfonate II 20G	5.0 lb	$14.65	$11.52	$26.17	1.21	2.16
Furadan 4F	2.0 pt	$18.75	$10.76	$29.51	1.55	2.43
Lorsban 4E	2.0 pt	$15.19	$10.18	$25.37	1.25	2.09
Lorsban 15G	5.0 lb	$12.70	$ 9.44	$22.14	1.05	1.83
Penncap-M	2.0 pt	$ 9.30	$ 8.51	$17.81	0.77	1.47
Pounce 1.5G	6.7 lb	$11.57	$ 8.25	$19.82	0.96	1.64
Pounce 3.2EC	4.0 oz	$ 9.39	$ 8.25	$17.64	0.77	1.46
Aerial Application ($5.00/a)						
Ambush 2EC	6.4 oz	$10.59	$ 8.25	$18.84	0.87	1.55
Asana XL .66EC	7.8 oz	$12.79	$ 8.60	$21.39	1.06	1.77
Dipel 10G	10.0 lb	$18.10	$ 2.25	$20.35	1.49	1.68
Dyfonate II 20G	5.0 lb	$15.85	$11.52	$27.37	1.31	2.26
Furadan 4F	2.0 pt	$19.95	$10.76	$30.71	1.65	2.53
Lorsban 4E	2.0 pt	$16.39	$10.18	$26.57	1.35	2.19
Lorsban 15G	5.0 lb	$13.90	$ 9.44	$23.34	1.15	1.93
Penncap-M	2.0 pt	$10.50	$ 8.51	$19.01	0.87	1.57
Pounce 1.5G	6.7 lb	$12.77	$ 8.25	$21.02	1.05	1.73
Pounce 3.2EC	4.0 oz	$10.59	$ 8.25	$18.84	0.87	1.55

Table 14.3. Sample environmental EIL calculation worksheet.

Insect:	*European corn borer: 2nd generation, pretassel stage corn*

Insecticide and application method: *Dipel 10G, ground application*

1. Insecticide cost/acre = rate * insecticide cost/unit:

    *10 lb/a * $1.31/lb = $13.10/a*

2. Total insecticide cost = application cost + insecticide cost/acre:

    $3.80/a + $13.10/a = $16.90/a

3. Environmental cost: *$2.25*[a]

4. Conventional EIL =

$$\text{EIL} = \frac{\text{Total insecticide cost}}{\text{crop value * yield loss per insect density * expected control}}$$

$$EIL = \frac{\$16.90/a}{\$2.35/bu * (125\ bu/a * 0.066\ loss/larva/plant) * 0.5\ control}$$

$$= 1.74\ larvae/plant$$

5. ET = Typically set as a % of the EIL (e.g. 80%) for time delays in management:

    *0.80 * 1.74 larvae/plant = 1.39 larvae/plant*

6. EEIL =

$$\frac{\text{Total insecticide cost + environmental cost}}{\text{crop value * yield loss per insect density * expected control}}$$

$$EEIL = \frac{\$16.90 + \$2.25}{\$2.35/bu * (125\ bu/a * 0.066\ loss/larva/plant) * 0.5\%\ control}$$

$$= 1.98\ larvae/plant$$

7. EET = Typically set as a % of the EEIL (e.g. 80%) for time delays in management:

    *0.80 * 1.98 larvae/plant = 1.58 larvae/plant*

[a] Taken from Table 14.1.

Critiques of Environmental EILs

Higley and Wintersteen (1992) is an early attempt to place environmental costs on pesticide use; in fact, a review of economic evaluations of IPM (Norton and Mullen 1994) did not identify any other studies on environmental costs of pesticides. Indeed, Norton and Mullen noted the lack of work in this area and emphasized the importance of such research. There is increasing interest in techniques for pesticide selection, which necessarily focus on environmental risks. Becker et al. (1989) includes procedures for pesticide selection based on water quality; Kovach et al. (1992) studies a broad array of environmental effects; however, neither of these approaches attempts to assign environmental costs. Until more work is available on assessing environmental costs and establishing environmental EILs, it is difficult to form a balanced assessment of the approach.

Some initial criticisms of the procedure have been made. Hutchins and Gering (1993) argue against the environmental EIL on philosophical grounds, based on how environmental costs are assigned and how environmental risks are assessed. They maintain that environmental costs should not be included in the EIL, because such costs are already incurred through the registration process and are therefore already reflected in the C variable. They further argue that it is inappropriate to incorporate opinion data from contingent valuation surveys because the EIL should remain "perception neutral."

We disagree with both arguments, but it must be pointed out that our disagreement is largely a matter of philosophy and opinion. One point, however, is less subjective: we do not accept that C already incorporates environmental costs because of pesticide registration. There are no substantial differences in registration costs based on degree of safety, nor do we see differences in pesticide costs based on differential registration costs: the price of a pesticide is largely set by production and marketing considerations. In the United States, the Environmental Protection Agency (EPA) has recently established procedures for fast-track registration of "biorational" pesticides, which are presumed to be safer; hence, some savings in registration costs may occur with these compounds. Nevertheless, faster registration procedures are unlikely to influence pesticide costs to an appreciable degree, compared with production and marketing considerations. Hutchins and Gering argue that any tactic meeting regulatory standards should be acceptable, and they imply that further discrimination on environmental grounds is unwarranted. We believe there is a need to consider environmental risk beyond pesticide regulation, because registered pesticides still pose some environmental risk.

A similar difference in philosophies occurs regarding the subjectivity of environmental costs determined by contingent valuation. Hutchins and Gering offer an example of incorrect grower perceptions on yield losses from insect injury. They argue by analogy that incorporating perceptions on environmental risk will render the EIL less accurate. We find their argument to be flawed: yield losses from insects are knowable; they can be objectively determined. In contrast, environmental risk assessment is intrinsically subjective. The importance of recognizing and incorporating public perspectives into risk assessments is discussed in the environmental-costs section of this chapter and elsewhere (Higley and Wintersteen 1992, Peterson and Higley 1993). Some aspects of environmental risk (dose, persistence, etc.) are objective, but weighting these factors or scaling risk is not. For instance, are acute risks more or less important than chronic risks? Are risks to all areas of the environment to be weighted equally? How do the risks from long persistence compare with risks from high toxicity? There are no objective answers to such questions; they are intrinsically subjective. But their subjectivity does not lessen their importance.

Experts may have opinions on such issues, but so may the public, and we believe a combination of these views must be the basis of any meaningful risk assessment. Consequently, we do not regard incorporating public perceptions on risk into environmental costs and their associated EEILs as weakening the EIL. Our position is that public input is a necessity in addressing environmental risk.

Another criticism was offered by Dushoff et al. (1994). Although most of their comments were focused on the Kovach et al. (1992) approach, they also question aspects of the Higley and Wintersteen (1992) study. Dushoff et al. argue that single-index methods for evaluating environmental risk are flawed conceptually and structurally, but we believe their arguments largely represent differences of opinion rather than real evidence of flaws. Regarding Higley and Wintersteen (1992), Dushoff et al. question the validity of relative-risk rankings made by a survey of growers (and yet they criticize Kovach et al. for not explicitly examining such weighting factors). They also question the monetary cost derived from the contingent valuation survey and how the results were used to calculate environmental EILs.

None of these criticisms provide any specific argument in support of their contention that the surveys and the analysis were conducted inappropriately. Dushoff et al. argue that, because respondents did not greatly discriminate between the relative importance of different environmental categories, and because high-risk costs were only twice those of low-risk costs (despite a more than tenfold difference in toxicity), the survey must be flawed. We find this a spurious argument. The fact that the survey results do not agree with Dushoff et al.'s a priori views—and, for that matter, with ours—highlights the importance of incorporating public attitudes in risk estimates. If anything, it points to the legitimacy of the survey procedure.

Dushoff et al. also question how pesticide-application rate is addressed in risk assessment. Kovach et al. assume a linear relationship between application rate and risk (half the rate represents half the risk). Higley and Wintersteen (1992) assign risk on a per-application basis, with little or no explicit consideration of application rate. We agree with Dushoff et al. that linear relationships between application rate and risk do not seem appropriate. Our view is that application rate should be accommodated on a qualitative basis. Given that a population may respond to pesticides over at least a 100-fold range in concentration (as we discussed in the section on K in this chapter), unless application-rate differences are in this range, we think modifications in risk estimates based on rate are unjustified.

Implementing Environmental EILs

The implementation of traditional EILs brought a fiscal discipline to pesticide applications. Producers understood the logic of weighing economic costs with economic benefits. Assuming that all labeled pesticides provided similar control, producers made applications, and chose between competing products, largely on the basis of the pesticide and application costs.

The addition of environmental costs into traditional EILs would require another layer of information and would result in two main outcomes: less pesticide use and improved or environmentally conscious pesticide selection. Less pesticide would be used because the addition of environmental costs would alter the EIL equation in favor of tolerating more pests before applying management measures. Pesticide selection would be improved as producers considered the application's environmental and direct costs, or total cost. Pesticides with the highest environmental costs would be at a disadvantage in the EEIL model, and their use would decline.

To use EEILs, producers would need ready access to accepted environmental-cost information from a credible source. Environmental-cost information could be included as part of label information, marketing information, or educational material provided by agricultural agencies. This might be similar to the presence of fuel-mileage ratings on automobiles, or the annual energy-cost estimate applied to major home appliances. In the latter instance particularly, previously unconsidered costs became a vital piece of information for consumers and a marketing device for manufacturers.

It has been difficult to determine the percentage of growers who use ETS to make pest-management decisions; however, the recent report, *Adoption of Integrated Pest Management in U.S. Agriculture,* summarizes information on the use of scouting and ETs for the U.S. production of fruits, vegetables, and field crops, including corn and soybeans. The report found that scouting and ETs were being used to determine insecticide applications on 29% of the corn acres; however, 15% of the overall corn acreage was treated with an insecticide without using ETs. The report was unable to determine ET use on soybeans.

Because use of ETs depends on scouting, the modest use of ETs may be associated with deficiencies in scouting skills and uncertain pest identification. In addition, some insect-monitoring programs are complex and require the use of computer programs to predict when fields should be scouted. The *Adoption of IPM* report stated that only 65% of corn acres were scouted for weeds, diseases, and insects; the report did not offer a breakdown for each pest.

Additional factors may explain the modest use of ETS. First, the use of ETS requires current information on pest populations, product performance and application cost, and potential crop value and yield. Some growers may not be able to assess these variables accurately. Second, information on ETs may not be available to the grower during the decision-making process, and even if the information is available, the grower may not feel competent in using the ET calculation worksheet or software program. Third, in some cases the insect population or damage may be so severe that a grower feels that calculating an ET is not necessary.

For those pesticide users who would use EEILs, one would expect the majority to choose products based on the lowest total costs, assuming equal control among products. However, some may respond to the EEIL information by choosing the product with the least environmental costs, although the total costs may be higher. Others may use the EEIL formula to monitor pest levels for a specific product, even though it has a higher total cost. As a practical matter, the final product selection may be influenced by factors such as weather, chemical class, ease of application, site-specific conditions, and regulations.

Regulatory agencies, such as the EPA or the USDA, may play an important role in implementing EEILs. The establishment of accurate environmental costs is fundamentally important for farm producers and chemical manufacturers. These agencies may be the only groups to have the regulatory and research abilities to set those costs and to approve labeling.

Once established, product environmental costs may lead to changes in regulating the pesticide industry. For instance, legislation may be introduced that would require pesticides to be less than a specific environmental cost. Taxes based on a product's environmental costs may be assessed, or products with high environmental costs may be assigned a special tax, much as gas-guzzler cars are today. If such laws were enacted, manufacturers would be forced to consider environmental costs more closely in their development of new products.

The use of EEILs may result in the creation of a new class of consumer food products. Currently, stores distinguish between organically grown products and all others; there is no middle ground. With the implementation of EEILs, marketers could claim a new range of environmental friendliness: "Products grown with low-environmental-costs crop protectants."

Many efforts to improve environmental quality in IPM will undoubtedly continue to focus on reductions in pesticide use, development of safer pesticides, and research on alternative tactics. Nevertheless, as a framework for addressing environmental quality, the EIL clearly offers some important opportunities. As a new approach, the environmental EIL clearly merits further examination and the authors of this chapter are working on refining objective criteria for risk ratings and for modifying ratings to reflect site-specific characteristics. Additional work is needed in developing contingent-valuation cost estimates, with input from agricultural and nonagricultural audiences, to prove more robust environmental-cost estimates. This is especially important in light of new work (Arrow et al. 1993) on approaches for avoiding bias in contingent-valuation surveys. Validation studies for EEILs are another obvious need.

The development of the environmental EILs is in its infancy, and its ultimate usefulness remains untested. But, in that it provides a direct consideration of environmental risk in pest-management decision making, the EEIL promises to improve the responsiveness of IPM to environmental sustainability.

References

Anderson, G. D., and R. C. Bishop. 1986. The valuation problem, p. 89–137. *In* D. W. Bromley (ed.) Natural resource economics: Policy problems and contemporary analysis. Kluwer Nijhoff, Boston.

Arrow, K., R. Solow, P. R. Portney, E. E. Leamer, R. Radner, and H. Schuman. 1993. Report of the NOAA panel on contingent valuation. Federal Register 58: 4601–4614.

Becker, R. L., D. Herzfeld, K. R. Ostlie, and E. J. Stamm-Katovich. 1989. Pesticides: Surface runoff, leaching, and exposure concerns. Minn. Extension Serv. Publ. AG-BU-3911.

Carruthers, R. I., and J. A. Onsager. 1993. Perspective on the use of exotic natural enemies for biological control of pest grasshoppers (Orthoptera: Acrididae). Environ. Entomol. 22:885–903.

Coffelt, M. A., and P. B. Schultz. 1990. Development of an aesthetic injury level to decrease pesticide use against orangestriped oakworm (Lepidoptera: Saturniidae) in an urban pest management project. J. Econ. Entomol. 83:2044–2049.

Coffelt, M. A., and P. B. Schultz. 1993. Quantification of an aesthetic injury level and threshold for an urban pest management program against orangestriped oakworm (Lepidoptera: Saturniidae). J. Econ. Entomol. 86:1512–1515.

Cummings, R. G., D. S. Brookshire, and W. D. Schulze. 1986. Valuing environmental goods: An assessment of the contingent valuation method. Rowman and Allanheld, Totowa NJ

Dushoff, J., B. Caldwell, and C. L. Mohler. 1994. Evaluating the environmental effect of pesticides: A critique of the environmental impact quotient. Am. Entomologist 40:180–184.

Heberlein, T. A., and R. C. Bishop. 1986. Assessing the validity of contingent valuation: Three field experiments. Science of the Total Environment 56:99–107.

Higley, L. G., and L. P. Pedigo. 1993. Economic injury level concepts and their use in sustaining environmental quality. Agric. Ecosystems Environ. 46:233–243.

Higley, L. G., and W. K. Wintersteen. 1992. A novel approach to environmental risk assessment of pesticides as a basis for incorporating environmental costs into economic injury levels. Am. Entomologist 38:34–39.

Higley, L. G., and W. K. Wintersteen. 1994. Modeling and managing environmental risks from pest management practices. Proc. World Soybean Res. Conf. V., Thailand. *in press.*

Higley, L. G., M. R. Zeiss, W. K. Wintersteen, and L. P. Pedigo. 1992. National pesticide policy: A call for action. Am. Entomologist 38:139–146.

Hutchins, S. H., and P. J. Gehring. 1993. Perspective on the value, regulation, and objective utilization of pest control technology. Am. Entomologist 39:12–15.

John, K. H., J. R. Stoll, and J. K. Olson. 1987. An economic assessment of the benefits of mosquito abatement in an organized mosquito control district. J. Am. Mosq. Control Assoc. 3:8–14.

Kovach, J., C. Petzold, J. Degni, and J. Tette. 1992. A method to measure the environmental impact of pesticides. New York's Food and Life Sci. Bull., No. 139. Cornell Univ., Ithaca NY.

Larew, H. G., J. J. Knodel-Montz, and S. L. Poe. 1984. Leaf miner damage. How much is too much? Greenhouse Manager 3(8):53–55.

Leonard, R. A., and W. G. Knisel. 1989. Groundwater loadings by controlled-release pesticides. A gleams simulation. Trans. Am. Soc. Agric. Eng. 32:1915–1922.

Leonard, R. A., W. G. Knisel, F. M. Davis, and A. W. Johnson. 1988. Modeling pesticide transport with GLEAMS. Proc. Planning Now for Irrigation and Drainage Conf., Am. Soc. Agric. Eng, Lincoln NE, 1–21 July 1988.

Leonard, R. A., W. G. Knisel, F. M. Davis, and A. W. Johnson. 1990. Validating GLEAMS with field data for fenamiphos and its metabolites. J. Irrigation and Drainage Eng. 116:24–35.

Leonard, R. A., W. G. Knisel, and D. A. Still. 1987. GLEAMS : Groundwater loading effects of agricultural management systems. Trans. Am. Soc. Agric. Eng. 30:1403–1418.

Lockwood, J. A. 1993a. Benefits and costs of controlling rangeland grasshoppers (Orthoptera: Acrididae) with exotic organisms: Search for a null hypothesis and regulatory compromise. Environ. Entomol. 22:904–914.

Lockwood, J. A. 1993b. Environmental issues involved in biological control of rangeland grasshoppers (Orthoptera: Acrididae) with exotic agents. Environ. Entomol. 22:503–518.

Mitchell, R. C., and R. T. Carson. 1989. Using surveys to value public goods: The contingent valuation method. Resources for the Future, Washington DC.

Mortensen, D. A. Personal communication. Dept. of Agronomy, Univ. of Nebraska.

Norton, G. W. and J. Mullen. 1994. Economic evaluation of integrated pest management programs: A literature review. Virginia Coop. Extn. Serv. Publ. 448-120, Virginia Polytechnic Inst. and State Univ.

Pedigo, L. P., and L. G. Higley. 1992. The economic injury level concept and environmental quality: A new perspective. Am. Entomologist 38:34–39.

Pedigo, L. P., S. H. Hutchins, and L. G. Higley. 1986. Economic injury levels in theory and practice. Annu. Rev. Entomol. 31:341–368.

Peterson, R. K. D., and L. G. Higley. 1993. Communicating pesticide risks. Am. Entomologist 39:206–211.

Pimentel, D., and H. Lehman. 1993. The pesticide question: Environment, economics, and ethics. Chapman and Hall, New York.

Raupp, M. J., J. A. Davidson, C. S. Koehler, C. S. Sadof, and K. Reichelderfer. 1988. Decision-making considerations for aesthetic damage caused by pests. Bull. Entomol. Soc. Am. 34:27–32.

Reiling, S. D., K. J. Boyle, M. L. Phillips, V. A. Trefts, and M. W. Anderson. 1988. The economic benefits of late season black fly control. Maine Agric. Exp. Stn. Tech. Bull.:822.

Sadof, C. S., and C. Alexander. 1993. Limitations of cost-benefit-based aesthetic injury levels for managing twospotted spider mites (Acari: Tetanychidae). J. Econ. Entomol. 86:1516–1521.

Smith, V. K., W. H. Desvousges, and A. Fisher. 1986. A comparison of direct and indirect methods for estimating environmental benefits. Am. J. Agric. Econ. 68:280–290.

Slovic, P. 1987. Perception of risk. Science 236:280–285.

Trumble, J. T., D. M. Kolodny-Hirsch, and I. P. Ting. 1993. Plant compensation for arthropod herbivory. Annu. Rev. Entomol. 38:93–120.

Scott H. Hutchins

15
Thresholds Involving Plant Quality and Phenological Disruption

Development and use of economic-injury levels (EILs) and economic thresholds (ETs) to manage pests traditionally has focused on the preservation of crop or host yield, which typically is characterized as quantity of biomass or grain production. Plant quality, however, frequently is of critical economic importance to the successful use and/or marketing of a commodity. Managing the impact of pest-induced injury to plant quality must, therefore, be considered within the scope of a value-based production system.

This chapter will characterize *quality* as a value-added consideration and present proposals on how best to incorporate this parameter within the EIL model. The ET will not be addressed specifically, but it represents the level at which a control strategy must be implemented to prevent continued losses (to quality) from exceeding the EIL (see Pedigo et al. 1986).

Although a uniform characterization of crop quality is not possible because of the situational nature and/or destination of the harvested commodity, specific qualitative attributes of plants/crops can be characterized and managed. To that end, factoring quality into the EIL model can be accomplished in at least two ways. The most straightforward approach for incorporating standards of crop quality from a pest-management standpoint is the *deminimus approach*. Here, the objective is to determine, a priori, the desired level of crop quality and to target pest-management strategies to manage pest intensity accordingly. By contrast, the *feed-value approach* is necessarily more complex, because the objective here is to maximize the production of a feed input for animal production. From a plant-protection perspective, therefore, assessing the simultaneous impact of pest-induced injury to *both* the biomass and qualitative factors (e.g., energy and protein) of feed and forage crops becomes the management objective. Both the deminimus and feed-value approaches are amenable to the use of EILs as the primary decision tool for pest management.

The deminimus approach to managing crop quality requires either a regulated level of acceptable commodity quality (e.g., USDA grade) or other similar anthropocentric valuation of quality (e.g., cosmetic damage) for a finished commodity. In either case, quality becomes an intrinsic consideration of the EIL model when managing pest influence on biomass yield.

The feed-value approach to managing crop quality requires knowledge regarding the physiology of plant biomass development, the development and relative value of qualitative attributes in plants, the relationship of qualitative attributes and feed utilization by animal consumers, and the temporal development of both biomass and qualitative attributes. Each of these considerations ultimately will affect the formulation of a pest-management program designed to optimize the overall feed value of a crop to the eventual animal consumer. Because of the complexity of the feed-value approach (versus the deminimus approach), prerequisite knowledge on nutrient development in plants and subsequent utilization in animals is necessary. The balance of the introduction to this chapter, therefore, will provide the baseline information required to develop quality-based EILs using the feed-value approach.

Pests should be managed (using the EIL-decision model) based on their impact on resultant animal growth. In the case of forages (e.g., alfalfa), certain qualitative attributes are widely recognized as indications of ultimate utility to the animal consumer. The character and nutritive value of forages are determined primarily by two factors: proportion of plant cell wall and its corresponding degree of lignification (Van Soest 1982). The feeding value to the animal is limited by the daily intake of digestible nutrients and the efficiency with which these digested nutrients can be used for metabolic processes (Barnes and Gordon 1972). Particular emphasis is given to the assessment of available energy, because forage rations frequently are limited by energy content rather than nutritional content per se (Gordon et al. 1961). In other words, plant characteristics that increase rate or digestibility of feed intake (i.e., nutritive value) are favorable.

Ruminants, even with their specialized digestive system and microbial symbiosis, are not capable of totally converting plant material into energy.

The percent digestibility represents the proportion of a feed that is available for absorption by the ruminant. Digestibility not only varies among plant species but also among plant parts and at different phenological stages (Buxton et al. 1985). The primary factor regulating digestibility of a feed relates to the level of lignification at the time of consumption. For example, alfalfa fed to a ruminant before bloom is much more digestible than the same growth after it reaches reproductive maturity. Lignin limits the extent of digestion but has comparatively little influence on the rate of digestion (Smith et al. 1971). The dense cell-wall and lignin fractions, which dominate mature stems, are not susceptible to the enzymatic action of gastric chemicals or symbiotic organisms in the rumen. Alfalfa leaves, however, which contain a lower concentration of cell walls, are relatively high in digestibility.

Feed intake is a critical aspect of forage quality. The species of animal, the animal's physiological status, the animal's energy demand, and the animal's individual preference (Van Soest 1982) all affect intake. It is often assumed that intake and digestibility of forages are directly related; in fact, however, intake is dependent upon the structural volume (cell-wall content), while digestibility is dependent upon both structural volume and its availability to digestion, as determined by lignification and other factors. Indeed, forages never provide for a metabolic-limiting situation (i.e., set point) in animals because limiting factors of feed quality impose a lower level of possible feed intake.

The total assessment of high-quality alfalfa goes beyond the digestive nutrient content and is compounded by a potential for being consumed at greater levels, a faster rate of digestibility, and perhaps a more efficient conversion of digested energy to productive energy (Barnes and Gordon 1972). For forages, the association with animal intake depends on plant structure. Cellulose, for example, is more closely associated with intake than digestibility, while lignin is more closely related to digestibility (Barnes and Gordon 1972). The total cell-wall concentration (versus cell-soluble concentration) is generally considered the most consistent cell fraction related to intake. This is not unexpected, because the cell wall contains structural components of the plant within which all other components are contained. Apparently, the expression of animal desire for a particular feed is greater at lower cell-wall concentrations. Moreover, if the principal effect of rumination is to collapse and release the highly digestible intercellular spaces within the forage cell walls, perhaps a high level of cell-wall content is counterproductive (i.e., the period of rumination is extended) (Van Soest 1982). Regardless of the mechanism, voluntary intake clearly is a key quality attribute, as it may account for two-thirds of the variability of animal growth-rate performance (Byers and Ormiston 1962).

For most situations, forage quality is assessed on a total-herbage basis, although digestibility varies among plant parts. Buxton et al. (1985) documented these differences and further determined that the lower stem component was less digestible than the apical portions. In a separate study, Buxton and Hornstein (1986) determined that cell-wall concentration was low in leaves and greatest in the basal stem segments. Hence, the qualitative value of the lower portions of an alfalfa canopy is reduced per unit of mass.

This characterization of quality for feed purposes is important as a precursor to recognizing the objective of plant protection and focusing a pest-management decision model. The variables of the EIL model as presented in Chapter 4 must be considered individually for managing quality through both the deminimus and feed-value approaches.

Integrating Plant Quality into EIL Variables

Values for the variables of the EIL model may require modification to account for expectations of certain qualitative yield components when using either the deminimus or the feed-value approaches to managing quality. The specific nature of the modification, however, is dependent upon the circumstances of the pest/crop management situation. In all cases, the biological variables must incorporate the complete nature of pest-induced injury to plants and the economic variables must capture and integrate the financial consequence in order to make accurate pest-management decisions.

Injury per Pest

Pest injure plants without regard to an anthropocentric or nutritional valuation of crop quality. To quantify the impact of pest populations on pest-induced injury, as well as the resultant impact of injury on host physiology, the biological pest-to-crop relationship must be characterized. Consistent with this need, categories of injury to plants have been developed (Pedigo et al. 1986) as a means of establishing a common plant-physiological basis for managing guilds of pests. The specific categories include leaf-mass consumers, fruit feeders, stand reducers, assimilate sappers, turgor reducers, and architecture modifiers (see Chapter 3). The extent to which quality attributes are considered within each of the plant-injury categories and, therefore, the injury per pest *(I)* EIL variable, is dependent upon how quality is affected by host physiology.

Standards for crop quality at harvest, whether they be cosmetic standards or incremental regulatory "grade" standards, represent a significant consideration when designing EILs for pest-induced injury using the deminimus approach. The primary need is to characterize how pest activity is injurious to plant physiology, in terms of crop quality becoming compromised and measurably affected. Leaf-mass consumers affecting soybean, for example, primarily will affect the biomass yield of the crop, with quality losses to seed grade or feed value being of distant secondary concern. By contrast, stink bug species within a soybean fruit-feeding guild have their primary influence on seed quality by affecting economic grade (Fig. 15.1). Leaf-mass consumers of ornamental plants may affect the aesthetic value of the crop as a primary quality (i.e., marketability) concern without reducing the actual number or size of flowering parts. Injury to the fruit (postflowering) component of ornamental plants, however, generally is of little consequence. If crop-quality standards (versus biomass loss) represent the factor limiting revenue production through the plants, then the rate of injury via quality loss becomes of primary concern. In essence, therefore, the deminimus approach to managing crop quality uses the single most-value-limiting physiological function of pest-induced injury (for each injury guild) for *I*.

As with the deminimus approach, the primary requirement for determining *I* with the feed-value approach is to understand and model the influence of each pest by injury category. The relationship of alfalfa weevil, *Hypera postica* (Gyllenhal), larval intensity to leaf-mass loss, for example, must be determined prior to assessing the impact of known levels of leaf-tissue loss on plant quality. Hence, values for *I* should be determined for both biomass (i.e., kg/ha) and quality (e.g., digestible energy [Mcal] per unit of herbage biomass) plant parameters. The feed-value approach is unique, however, in that pest influence to both yield and quality must be managed simultaneously.

Figure 15.1 Photograph of soybean damage resulting from feeding injury by green stink bug (*Acrosternum hilare* [Say]) nymphs. Damage of this nature results in reduced grade and price. Photograph courtesy of D. J. Boethel.

Damage per Unit of Injury

Damage is considered to be measurable reduction in plant growth, development, or reproduction resulting from categorical injury. In the case of quality, damage per unit of injury *(D)* may include a measurable loss in marketable grade–cosmetic utility or incremental reduction in feed value as a result of pest-induced physiological injury. In either situation, the focus of research is to develop the damage curve (see Chapter 3) with the appropriate dependent (i.e., value-loss component) and independent (i.e., physiological injury category) variables in order to assess pest influence on crop value. When using the deminimus approach, in which discrete grades of quality are established, those criteria (e.g., cosmetic grade) should be modeled and compared to yield-loss functions in order to determine the relative magnitude of damage types. The most value-limiting damage function should be the one used for managing pests.

Assessment of *D* when using the feed-value approach requires knowledge of how pests simultaneously affect both plant biomass and plant quality components under a continuum of injury intensities. As presented previously, structural components (e.g., stems) that contribute to plant establishment, competition for nutrients, and biomass must be considered as tradeoffs with nonstructural components (e.g., leaves) that contribute a disproportionately high level of digestible energy and protein per unit of biomass. To manage *both* biomass and quality, a nutrient-yield determination for the feed-value approach is recommended. Here, determination of digestible energy yield (Mcal/ha), for example, is calculated as the product of biomass yield (kg/ha; leaf plus stem components) and digestible energy (Mcal/kg; calculated as described by National Research Council 1978). For protein yield (kg/ha) determinations, biomass production (kg/ha) yield can be multiplied by percentage of crude protein.

When using the feed-value approach for managing quality, the integrated rate of nutrient yield loss per pest should be determined and used as the basis for the damage curve. In many instances, especially with assimilate-removing pests, the specific relationship is difficult to assess. In these cases, the *I* and *D* variables are intrinsic and may require a regression approach using a known population of pests to measure loss of nutrient yield for predictive purposes.

The two economic variables of the EIL model, *market value (V)* and *management costs (C)*, may require special consideration when managing quality as a primary or integral concern. Indeed, the "valuation" of various attributes of production represents the basis for making pest-management decisions.

Management Costs

Determination of *C* for pests within an injury guild generally is straightforward once the appropriate control tactic has been selected. In cases where the deminimus approach is used, management costs are determined as with any EIL-based decision; when using the feed-value approach to managing forage quality, special consideration of management costs may be necessary. The potential complexity is founded in the fact that early cutting of a forage crop may be a viable pest-control tactic. Here, the *C* is not only cost of management through mechanical cutting (materials plus labor), but also the opportunity cost of additional yield, had early cutting not occurred. Another potential complication of early cutting is the potential long-term effect on stand longevity by prematurely removing herbage before root carbohydrate levels are sufficient to support regrowth.

Market Value

The relationship of plant quality to market value is key in order to make accurate pest-management decisions using the EIL model. In situations where the deminimus approach is appropriate, quality standards generally are well known and accepted in the marketplace. The pest manager should utilize the commodity price consistent with the desired grade of production. For example, citrus producers planning to market top-grade or Fancy-grade produce should use the appropriate contract price accordingly. If the citrus is being produced for the relatively lower-value Choice-grade or Juice-grade markets, then that anticipated price should be used.

Assessing market value for crops when the feed-value approach is used can be complex, because animal-utilization is the actual "value" of the crop. Forages grown for on-farm animal consumption, for example, have no actual "commodity" exchange, but obviously they do contribute to the overall animal-production enterprise. One approach is to determine the cost of replacing digestible energy or protein with substitute feeds. This approach, although useful as a means to estimate V for on-farm feeds, may not fully capture the total value of forages (e.g., roughage) versus purchased feeds.

Quality-Sensitive EILs Using the Deminimus Approach

The incorporation of quality using the deminimus approach requires knowledge of injury and damage relationships with qualitative components of yield. If, for example, produce is to be sold by grade, then the effect of pest intensity on grade must be assessed via research. With cosmetic injury, the damage boundary (see Chapter 4) will be equal to the stage of first observation. Figure 15.2 presents a stylized damage curve for visible damage to a commodity that is market-sensitive to cosmetic damage. Because the actual physical yield response has a relatively long period of tolerance, the depreciation of marketable grade becomes the key factor affected by many pests. When using the deminimus approach, therefore, a pest manager should focus on the most sensitive relationship (i.e., the steepest slope) when formulating priority tactics for control of pests.

An example of where the deminimus approach may be used is with the influence of peanut-pod feeding by lesser cornstalk borer (LCB; *Elasmopalpus lignosellus* [Zeller]) on development of aflatoxin (product of *Asper-*

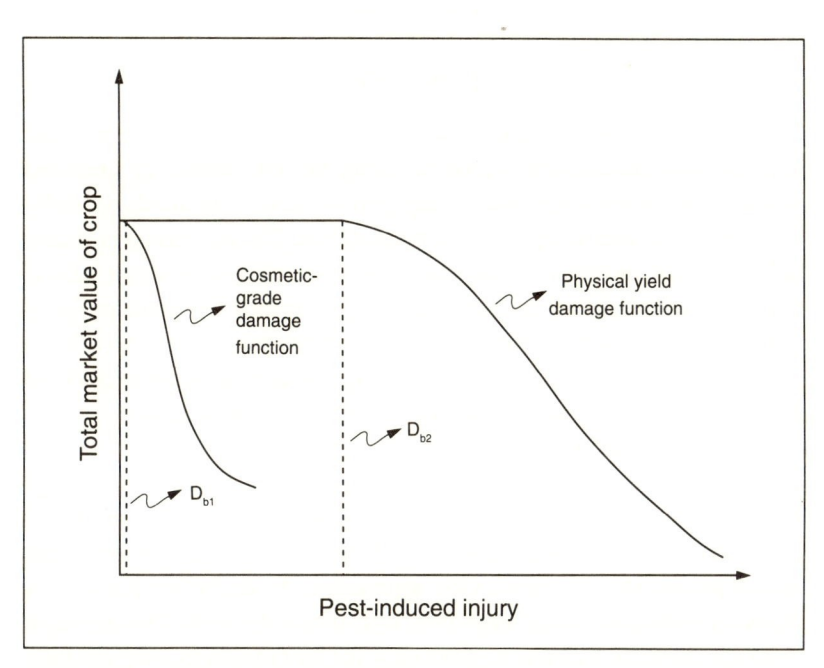

Figure 15.2. Curve showing relationship of pest-induced injury to harvest-time market value for a crop sensitive to cosmetic damage. Because the cosmetic-grade damage function is more sensitive to injury than the physical-yield damage function, pest-induced injury to the cosmetic value should be managed as a primary concern, using the deminimus approach. D_b represents the damage boundary for cosmetic (D_{b1}) and physical yield (D_{b2}).

gillus flavus fungus) in peanuts. In this situation, unacceptable levels of aflatoxin will severely limit the marketable quality and grade (i.e., price) of the commodity. The assessment of biological variables requires knowledge of *I* and *D* as determined through research. The relationship of LCB density on number of damaged pods represents *I*, and the relationship of number of damaged pods to production of aflatoxin-free yield represents *D*. Using the linear portion of the *I* and *D* functions, the relationship of LCB to aflatoxin-free yield (under specific environmental conditions) can be determined for incorporation within the EIL model.

Because varying grades of peanut quality are used, the market value for the desired grade should be determined a priori and used in the EIL model. If high-grade (i.e., little or no presence of aflatoxin) peanuts command a price *(V)* of $0.74/kg with expected management costs *(C)* of $148/ha, then the gain threshold would be 200 kg/ha. If low-grade peanuts are appropriate, *V* would change to approximately $0.14/kg, resulting in a gain threshold of 1057 kg/ha. The EIL now represents the population of LCB that compromises the selected acceptable level (i.e., deminimus level) of aflatoxin-free yield per regulated grading standards. For peanuts targeted as high-grade, any LCB population that does not put the peanut yield at risk of a reduced grade or otherwise reduce biomass yield up to 200 kg/ha (whichever comes first) should not be controlled.

Although many examples of the deminimus approach to managing crop-quality loss from pests exist, very few have established credible assessments for the biological variables of the EIL. This probably is a reflection of the general intolerance of cosmetic or regulatory-grade loss by the produce and ornamental marketing channels (on behalf of the retail customer), which results in a low or even trivial gain threshold. In order to assure that qualitative components are incorporated within the pest-management decision, quantitative cost-to-benefit determinations should be made. Therefore, moving from empirical knowledge of *I* and *D* when using the deminimus approach is a key research need.

Quality-Sensitive EILs Using the Feed-Value Approach

Unlike the deminimus approach, which maximizes yield at a predetermined expectation of quality, the feed-value approach focuses on optimizing the balance of biomass and quality yield. The use of EILs is amenable to the optimization requirement, but, as already noted, requires independent knowledge of the biological pest and crop interactions for both quality and biomass, as well as economic valuation of the yield.

The influence of potato leafhopper (PLH), *Empoasca fabae* (Harris), on alfalfa yield and quality underscores the complex nature of the feed-value approach. The PLH affects alfalfa by extracting assimilates, which results in reduced stem height and reduced conversion of simple carbohydrates to protein. The result of this physiological injury is reduced biomass (primarily as a result of the reduction in stem height), reduced leaf protein, and slightly increased digestibility (as a result of increased leaf:stem ratios from the reduced stem component [Hutchins et al. 1989]).

Using the nutrient-yield approach to combine biomass and quality con-

Table 15.1. Relationship of various potato leafhopper, *Empoasca fabae* (Harris), densities on digestible energy yields of alfalfa (Hutchins and Pedigo 1990).

PLH Density[a]	Digestible energy yield (Mcal/ha)[b]
0	13,673
50	10,375
100	9,962
200	9,567

[a] PLH densities consist of adult infestations to cages made 24 h after cutting.

[b] Calculated as product of physical biomass yield and digestible energy per unit of herbage (National Research Council 1978).

siderations, Table 15.1 presents data relating the impact of four levels of PLH on digestible energy (DE, calculated per National Research Council 1978) yield. In order to describe the relationship of PLH density to DE yield for predictive purposes, however, a regression model is required:

Digestible energy yield = 12,433 Mcal/ha − 18(PLH/m^2)

(correlation coefficient: $r = 0.80$).

Because DE yield inherently considers the PLH influence on both biomass and quality yield components, the feed-value approach objective of optimization is achieved.

The economic impact of nutrient-yield loss for a forage crop such as alfalfa requires valuation based on costs of substitute animal feeds that provide similar levels of nutritional value. If management costs *(C)* for PLH are approximately $17.30/ha with soybean meal prices equal to $0.193/kg, a gain threshold of 89.64 kg/ha of soybean meal is necessary to justify management. Because the equivalency price reflects a nutrient substitution, the 89.64 kg of soybean meal is equivalent to about 320 Mcal of DE/ha (i.e., soybean meal contains 3.56 Mcal of DE/kg; National Research Council

1978). In this example, the DE value is assumed to be the critical measure of animal utility and, through the feed-value substitution approach, should be the focus of feed-value management for alfalfa. The EIL calculation on a feed-value basis provides the critical PLH population resulting in an economically unacceptable loss to on-farm alfalfa. In the alfalfa example presented, an EIL of approximately 18 PLH adults per square meter infested just after cutting is the critical level (i.e., the EIL).

Even though the substitution-feed approach provides the basis for accurately assessing individual quality attributes of forages, the unique value of forages may not be completely incorporated. Although grain crops will, on average, provide a higher rate of weight gain per day, the fiber provided by forages is necessary to maintain the microbial flora in the rumen and hence the health of the animal. Forages should not be totally eliminated in animal rations but should be utilized as a source of protein, energy, and fiber within a least-cost rationing strategy. Therefore, the value of alfalfa from one application to another will vary in relation to its proportionate contribution to the final formulated feed.

As an input, specific properties of a forage may be emphasized for some enterprises and less emphasized for others. For example, dairy producers commonly utilize alfalfa as a source of energy and protein to capitalize on its nutritional value; beef producers, on the other hand, are attracted to the high biomass yields that alfalfa generates over several harvests when grown for on-farm use. Each producer requires a sufficiency of both quality and quantity, but they differ as to ratio.

As a business goal, the input costs of producing any commodity should be minimized to the degree that they do not constrain production. One method frequently employed to achieve this objective is least-cost rationing as determined by a linear programming analysis. This technique, which seeks to minimize the cost of a feed ration while maintaining optimal animal-growth rates, has proven helpful as a management tool to reduce input costs in animal-production systems. In addition to providing the optimal mix of available feeds, least-cost ration modeling provides a wealth of data on production efficiency. For example, income penalties for using a nonoptimal mix of feeds, and the value of the last unit of a nutrient (i.e., value marginal product), are calculated as part of a sensitivity analysis and can be utilized for planning. Indeed, for pest-management purposes, these data can be used as a relative measure of feed utility (i.e., V)

and utilized within the EIL model for customized management of forages based on specific animal-growth and consumption needs.

Management Schemes for Pests Affecting Phenological Development

Although not originally identified as a category of physiological injury, phenological disruption of plant growth is a critical form of pest-induced injury, and one that clearly affects crop quality. Delay in the rate of maturity, in particular, can have a significant impact on both the biomass and quality of harvested crops. Delayed morphological development of alfalfa, for example, recently has been proposed as a possible mechanism of injury induced by insect pests (Hutchins and Pedigo 1990). If this form of injury is indeed prominent, standard evaluations of yield and quality among different levels (i.e., treatments) of pest density may be confounded by having plants of distinct phenology. In these situations, the variable of time-to-maturation becomes critical because of concomitant development of physiological factors that influence both biomass and quality in forages (e.g., cell-wall development, lignification). The management objective as it relates to the EIL model becomes one of describing pest influence in terms of days delay, that, in turn, relate to a known impact on feed value.

In the case of PLH feeding on alfalfa, delayed phenological development has clearly been identified as a key form of injury. Table 15.2 presents the documented effect of different infestation levels of PLH on the rate of alfalfa development when based on the predicted date of first bloom (e.g., determined from regression of PLH density on morphological stage over time). Table 15.3 presents the regression of PLH density on anticipated days of delay for alfalfa compared with uninfested plots. This relationship (r = 0.87; significant at $P = 0.01$) serves as the basis for developing PLH EILs using phenological disruption as the primary mechanism of insect-induced injury. Using this relationship, Table 15.4 presents the EILs for DE with soybean meal prices at two levels and management costs at two levels.

To summarize, the incorporation of plant quality, as well as the effect of pest-induced injury on qualitative attributes, is important in accurately developing usable EILs. With the choice based on the end use of the marketable commodity, either the deminimus or feed-value approach may be utilized to incorporate quality within the EIL. The deminimus approach

Table 15.2. Relationship of various potato leafhopper, *Empoasca fabae* (Harris), densities on date of first bloom for alfalfa (Hutchins and Pedigo 1990).

PLH Density[a]	Date of first bloom[b] (days after cutting)
0	36.1
50	46.2
100	46.9
200	51.7

[a] PLH densities consist of adult infestations to cages made 24 h after cutting.

[b] Date of first bloom predicted based on morphological development rates within treatments.

Table 15.3. Regression equation from Hutchins and Pedigo (1990) describing the relationship of potato leafhopper, *Empoasca fabae* (Harris), density to days of phenological delay (i.e., days past first bloom).[a]

$$\text{Days delay} = 0.016 \times (\text{PLH}/m^2)$$

Correlation coefficient: $r = 0.87$

[a] Slope forced through the origin after determining that the y-intercept term was not significantly different from zero. The correlation coefficient is significant ($P = 0.01$).

develops an a priori focus to management based on a predetermined standard, or grade, of produce. The feed-value approach develops an optimization focus, maximizing nutrient yield of the crop for the animal consumer. Of the two approaches, the feed-value approach is more complex: quality may be affected by both the nature of the crop and the nature of the consumer. Least-cost rationing, a feed-mix optimization technique, offers promise as a means to further refine feed-value EILs relative to specific nutritional needs of an animal species.

Examples of incorporating quality into the EIL have been presented, but examples generally are scarce or poorly defined. The biological variables of the EIL (I and D), in particular, must be developed and reported in order to enable EILs—and, therefore, pest management—to be focused on plant

Table 15.4. Economic time delays (days) and EILs for potato leafhopper, *Empoasca fabae* (Harris), calculated at two soybean meal (SBM) substitute values *(V)* and two management cost estimates *(C)*. Based on Hutchins and Pedigo (1990).

Substitute feed price (*V*)	C = $17.30/ha Time delay	EIL	C = $24.70/ha Time delay	EIL
SBM at $0.193/kg	0.85	7.23	1.21	11.39
SBM at $0.248/kg	0.65	6.16	0.93	8.79

quality. Indeed, all categories of physiological injury must be characterized within injury guilds, including pests of different classes (e.g., collective effects of insects, diseases, and nematodes on physiological turgor reduction). In addition, all biomass and quality responses must be quantified for each physiological category of pest-induced injury (i.e., development of damage curves), including the need to further develop research methodology for incorporating phenological disruption within the context of the EIL. Once these most basic relationships are understood, more sophisticated approaches to researching and developing quality-sensitive EILs and ETs can be undertaken.

References

Barnes, R. F., and C. H. Gordon. 1972. Feeding value and on-farm feeding, p. 601–630. *In* C. H. Hanson (ed.) Alfalfa science and technology. Am. Soc. Agron., Madison WI.

Buxton, D. R., and J. S. Hornstein. 1986. Cell-wall concentration and components in stratified canopies of alfalfa, birdsfoot trefoil, and red clover. Crop Sci. 26:180–184.

Buxton, D. R., J. S. Hornstein, W. F. Wedin, and G. C. Marten. 1985. Forage quality in stratified canopies of alfalfa, birdsfoot trefoil, and red clover. Crop Sci. 25:273–279.

Byers, J. H., and L. E. Ormiston. 1962. Nutrient value of forages. II. The influence of two stages of development of an alfalfa-borne grass hay on consumption and milk production. J. Dairy Sci. 45:693.

Gordon, C. H., J. C. Derbyshire, H. G. Wiseman, E. A. Kane, and C. G. Melin. 1961. Preservation and feeding value of alfalfa stored as hay, haylage and direct-cut silage. J. Dairy Sci. 44:1299–1311.

Hutchins, S. H., D. R. Buxton, and L. P. Pedigo. 1989. Forage quality of alfalfa as affected by potato leafhopper feeding. Crop Sci. 29:1541–1545.

Hutchins, S. H., and L. P. Pedigo. 1990. Phenological disruption and economic consequence of injury to alfalfa induced by potato leafhopper (Homoptera: Cicadellidae). J. Econ. Entomol. 83:1587–1594.

National Research Council. 1978. Nutrient Requirements of Dairy Cattle, Fifth Revised Ed. Natl. Academy of Sciences, Washington DC.

Pedigo, L. P., S. H. Hutchins, and L. G. Higley. 1986. Economic injury levels in theory and practice. Annu. Rev. Entomol. 31:341–368.

Smith, O. E., H. K. Goering, D. R. Waldo, and C. H. Gordon. 1971. In vitro digestion rate of forage cell wall components. J. Dairy Sci. 54:71–76.

Van Soest, P. J. 1982. Nutritional ecology of the ruminant. O&B Books, Corvallis OR.

Leon G. Higley and Larry P. Pedigo

Afterword: Pest Science at a Crossroads

As the preceding chapters illustrate, the EIL concept remains strong and continues to have great theoretical and practical value. Although other approaches to decision making have been developed, EILs and ETs continue to be among the most useful decision tools and support much of our activity in pest management. Nevertheless, in certain areas their use is somewhat limited; for example, it is unlikely that conventional thresholds will become more used for plant pathogens, unless there are radical new developments in therapeutic tactics for managing plant disease. Increased use of thresholds seems likely with respect to weeds, and further development of thresholds seems a certainty with insects.

Expansions of the original EIL concept are encouraging. Work on aesthetic thresholds is particularly noteworthy, as are new efforts in threshold development for veterinary pests and with perennial plants. Similarly, the development of environmental EILs is promising, although it is too early to judge whether these will be of lasting benefit.

Although the EIL is useful for determining pest status irrespective of management tactic, the practical use of EILs clearly is focused on therapeutic tactics and single-event decision making. Consequently, when we exceed these bounds and consider decision making for preventive action or for multiple-event decision making, the usefulness of EILs is greatly constrained. With regard to preventive tactics, the issue is very much one of making predictions of the probability of pest occurrence. To an extent, this question of prediction also occurs with therapeutic uses of EILs, through the development and use of ETs.

Afterword

Making predictions about pest events, whether in the context of therapeutic or preventive practices, is a difficult problem. Predictions are tied not only to biotic events but also to abiotic events, which in turn influence pest and plant biology. In particular, the tremendous influence of weather on pests and our relative inability to make long-term weather predictions severely constrain decision making. The implications of mathematical chaos theory only now are beginning to be widely appreciated in ecology; they do, however, have serious implications for making predictive decisions. Chaos theory suggests it may be impossible to make long-term weather predictions or even predictions for population-growth patterns. If this is true, there may always be an absolute constraint in our ability to make predictive decisions. Obviously, this area merits greater attention, not only in the context of EILs but as a general area of research concern in pest management.

A related practical constraint to the use of EILs and thresholds pertains to questions of sampling pests. Because decision making rests on assessments of pest populations, difficulties in making those assessments will inevitably impact the implementation of EILs and ETs. The many technical and practical issues associated with sampling in agroecosystems are discussed in Pedigo and Buntin (1994). As E. J. Bechinski points out in that volume, biological, economic, and social issues combine to influence the implementation of sampling programs (and therefore decision making). Our best EILs and ETs are of no use if the data they require cannot be, or will not be, obtained. Thus, the call in Bechinski (1994) for programs that "provide the requisite fine-grained data while still remaining uncomplicated enough to be adopted on the farm" is as much a requirement for successful implementation of thresholds as for sampling.

The issue of single- versus multiple-event decision making is also of growing importance. One important aspect of pest management that EILs do not explicitly address is the sustainability of management tactics. With the remarkable increase in pest resistance to pesticides, using decision making to help reduce the potential for resistance is essential. Chapter 13 discusses some aspects of this important question. It is difficult to know whether or not conventional EILs and ETs can be adapted to fit this goal, but striving to improve decision making for multiple-event management is an important challenge for pest management and should be addressed.

Other priorities within decision-making theory are not as pressing but are important nevertheless. These include a better representation of natural enemies in thresholds and greater development of procedures for calculating thresholds from EILs. Although calculation of the EIL may involve some experimental and practical difficulties, it does represent a solid value, easily determined once the variables are defined. In contrast, the ET, which is based on an EIL, is not easily calculated. Research efforts to provide more formal means for calculating ETs are extremely valuable, and we hope such efforts continue.

A further point relative to the usefulness of EILs and ETs relates to the available management tactics: the long association of thresholds with pesticides is a disturbing reflection of our reliance on pesticides as a therapeutic management tactic. Although new-generation pesticides avoid some of the environmental risks associated with earlier classes of pesticides serious concerns remain: we rely almost exclusively on pesticides for therapeutic management. Clearly this is an issue that merits greater attention, and the development of alternative therapeutic tactics should be a high priority.

It is disturbing to recognize the current state of the art in EIL and ET development (Chapter 10), particularly in entomology where the concept originated. It is true that there may be some practical restraints in developing EILs and ETs; it is also true that our current poor understanding of insect/plant relationships hinders threshold development. But these reasons do not explain why calculated thresholds are lacking for many important insect pests.

It is difficult for us to look at current threshold development and not conclude—to state our opinion—that the poor development of thresholds reflects philosophical differences among researchers and practitioners in pest management. In Chapter 1, we discussed the distinction we draw between control and management. It is our suspicion that the relatively poor development of EILs among insect pests is a reflection of control and management philosophies. Many workers ostensibly operate in pest management, but the continued heavy research emphasis on new tactics, rather than on how to use existing tactics, implies a greater emphasis on control than it does on true pest management. This conclusion seems well supported, when we consider the continued primacy of what we have termed the *silver-bullet fetish*. The current expression of the silver-bullet fetish—the emphasis on biotechnology and bioengineered plants—is in many ways analogous to the unwarranted promises that were given regarding pesticides, pheromones, and other new management tactics. Moreover, recent strategies proposed for the employment of tactics, particularly so-called "area-wide pest management," seem to have much more in common with a control philosophy than they do with management.

The conclusion we draw from this survey of threshold development and from our knowledge of the current research emphases among workers in pest science is that there is a compelling need for reexamination and recommitment to the basic tenets of pest management. We are left to ask if pest management is the operating paradigm of pest science. And further, if it is, what precisely is a statement of that paradigm? We believe pest science is at a crossroads. Unless we reexamine the tenets of pest management and move away from control, we will continue to follow the path that has misled us in the past.

One part of this reexamination must be a consideration of pest/host relationships, not merely of pest biology. We and our colleagues have argued in many places about the need for greater emphasis on pest/plant relationships, or, more generally, on biotic stress of plants. Many authors in this volume echo those arguments. Although work is slowly progressing in this area, there is no broad, concerted effort at understanding biotic stress—even though there is for issues such as population biology and abiotic stress. Unquestionably, the lack of emphasis given to understanding biotic stress contributes to the slow development of thresholds.

The contributions in this book reinforce this point. If we are to move forward, not merely in decision making but also in pest management, it is essential we have better understandings of pest/plant relationships. Ideally, such understandings must move toward the building of general models to account for stress responses of plants. Ultimately, if such an emphasis is successful, we see pest management moving toward a broader area of stress management, which would incorporate issues both of pest management and of managing abiotic stress.

The EIL concept and the ET have not yet reached their full potential—as we believe is shown in the contributions to this book—and there are many indications that this conventional cost-benefit analysis still has much to offer. With the pressing challenges in pest science, including resistance to management tactics, environmental concerns, new governmental policies, new pests, and growing demands on agriculture from the increasing global population, the need for work on thresholds has never been greater. Our hope is that these pages move us toward better decision making and a renewed emphasis on the cornerstones of our science.

References

Bechinski, E. J. 1994. Designing and delivering in-the-field scouting, p. 683–706. *In* L. P. Pedigo and G. D. Buntin, (ed.). Handbook of sampling methods for arthropods in agriculture. CRC, Boca Raton FL.

Pedigo, L. P., and G. D. Buntin. 1994. Handbook of sampling methods for arthropods in agriculture. CRC, Boca Raton FL.

Compiled by Robert K. D. Peterson

Selected Bibliography: Economic-Decision-Level Literature, 1959–1993

1. Adams, R. G., and L. M. Los. 1989. Use of sticky traps and limb jarring to aid in pest management decisions for summer populations of the pear psylla (Homoptera: Psyllidae) in Connecticut. J. Econ. Entomol. 82:1448–1454.
2. Andow, D. A., and K. Kiritani. 1983. The economic injury level and the control threshold. Jpn. Pestic. Inf. 43:3–9.
3. Anonymous. 1968. Economic principles of pest management. *In* Insect pest management and control, Chapter 18. National Academy of Sciences, Washington DC.
4. Appel, L. L., R. J. Wright, and J. B. Campbell. 1993. Economic injury levels for western bean cutworm, *Loxagrotis albicosta* (Smith) (Lepidoptera: Noctuidae) eggs and larvae in field corn. J. Kans. Entomol. Soc. 66:434–438.
5. Archer, T. L., and E. D. Bynum Jr. 1990. Economic injury level for the Banks grass mite (Acari: Tetranychidae) on corn. J. Econ. Entomol. 83:1069–1073.
6. Archer, T. L., and E. D. Bynum Jr. 1992. Economic injury level for the Russian wheat aphid (Homoptera: Aphididae) on dryland winter wheat. J. Econ. Entomol. 85:987–992.
7. Auld, B. A., and C. A. Tisdell. 1987. Economic thresholds and response to uncertainty in weed control. Agric. Syst. 25:219–227.
8. Ba-Angood, S. A., and R. K. Stewart. 1980. Economic thresholds and economic injury levels of cereal aphids on barley in southwestern Quebec. Can. Entomol. 112:759–764.
9. Bacheler, J. S., and J. R. Bradley Jr. 1989. Evaluation of bollworm action thresholds in the absence of the boll weevil in North Carolina: The egg concept, p. 308–311. *In* Proc. Beltwide Cotton Conf., Book 1.
10. Barnard, D. R. 1985. Injury thresholds and production loss functions for the

lone star tick, *Amblyomma americanum* (Acari: Ixodidae) on pastured preweaner beef cattle, *Bos taurus.* J. Econ. Entomol. 78:852–855.

11. Barnard, D. R., R. T. Ervin, and F. M. Epplin. 1986. Production system-based model for defining economic thresholds in preweaner beef cattle, *Bos taurus,* infested with lone star tick, *Amblyomma americanum* (Acari: Ixodidae). J. Econ. Entomol. 79:141–143.

12. Bauer, T. A., and D. A. Mortensen. 1992. A comparison of economic and economic optimum thresholds for two annual weeds in soybeans. Weed Technol. 6:228–235.

13. Bauer, T. A., D. A. Mortensen, G. A. Wicks, T. A. Hayden, and A. R. Martin. 1991. Environmental variability associated with economic thresholds for soybeans. Weed Sci. 39:564–569.

14. Bautista, R. C., E. A. Heinrichs, and R. S. Rejesus. 1984. Economic injury levels for the rice leaffolder, *Cnaphalocrocis medinalis* (Lepidoptera: Pyralidae): Insect infestation and artificial leaf removal. Environ. Entomol. 13:439–443.

15. Bechinski, E. J., and R. Hescock. 1990. Bioeconomics of the alfalfa snout beetle (Coleoptera: Curculionidae). J. Econ. Entomol. 83:1612–1620.

16. Bechinski, E. J., C. D. McNeal, and J. J. Gallan. 1989. Development of action thresholds for the sugarbeet root maggot (Diptera: Otitidae). J. Econ. Entomol. 82:608–615.

17. Bechinski, E. J., and R. L. Stoltz. 1985. Presence-absence sequential decision plans for *Tetranychus urticae* (Acari: Tetranychidae) in garden-seed beans, *Phaseolus vulgaris.* J. Econ. Entomol. 78:1475–1480.

18. Beers, E. H., L. A. Hull, and G. M. Greene. 1990. Effect of a foliar urea application and mite injury on yield and fruit quality of apple. J. Econ. Entomol. 83:552–556.

19. Bellows, T. S. Jr., J. C. Owens, and E. W. Huddleston. 1983. Model for simulating consumption and economic injury level for the range caterpillar (Lepidoptera: Saturniidae). J. Econ. Entomol. 76:1231–1238.

20. Ben-Huai Lye, R. N. Story, and V. L. Wright. 1988. Damage threshold of the southern green stink bug, *Nezara viridula,* (Hemiptera: Pentatomidae) on fresh market tomatoes. J. Entomol. Sci. 23:366–373.

21. Benedict, J. H., K. M. El-Zik, L. R. Oliver, P. A. Roberts, and L. T. Wilson. 1989. Economic injury levels and thresholds for pests of cotton. *In* R. E. Frisbie, K. M. El-Zik, and L. T. Wilson (ed.) Integrated pest management systems and cotton production. Wiley, New York.

22. Berry, R. E., and E. J. Shields. 1980. Variegated cutworm: Leaf consumption and economic loss in peppermint. J. Econ. Entomol. 73:607–608.

23. Blackshaw, R. P. 1986. Resolving economic decisions for the simultaneous control of two pests, diseases or weeds. Crop Prot. 5:93–99.

Selected Bibliography

67. Greene, G. L. 1972. Economic damage threshold and spray interval for cabbage looper control on cabbage. J. Econ. Entomol. 65:205–208.
68. Gupta, M. P., S. Ram, and B. D. Patil. 1982. Economic injury level of lucerne weevil, *Hypera variablilis* Herbst (Curculionidae: Coleoptera) on lucerne crop (*Medicago sativa* L.). J. Entomol. Res. 6:99–101.
69. Gutierrez, A. P., and R. Daxl. 1984. Economic thresholds for cotton pests in Nicaragua: Ecological and evolutionary perspectives. *In* G. R. Conway (ed.) Pest and pathogen control: Strategic, tactical and policy models. Wiley, New York.
70. Gutierrez, A. P., R. Daxl, G. L. Quant, and L. A. Falcon. 1981. Estimating economic thresholds for bollworm, *Heliothis zea* Boddie, and boll weevil, *Anthonomus grandis* Boh., damage in Nicaraguan cotton, *Gossypium hirsutum* L. Environ. Entomol. 10:872–879.
71. Gutierrez, A. P., and Y. Wang. 1979. An optimization model for *Lygus hesperus* (Heteroptera: Miridae) damage in cotton: The economic threshold revisited. Can. Entomol. 111:41–54.
72. Hall, D. C., and L. J. Moffitt. 1985. Application of the economic threshold for interseasonal pest control. West. J. Agric. Econ. 10:223–229.
73. Hall, D. C., and R. B. Norgaard. 1973. On the timing and application of pesticides. Am. J. Agric. Econ. 198–201.
74. Hall, D. G. IV, and G. L. Teetes. 1982. Yield loss-density relationships of four species of panicle-feeding bugs in sorghum. Environ. Entomol. 11:738–741.
75. Hallman, G. J., G. L. Teetes, and J. W. Johnson. 1984. Relationship of sorghum midge (Diptera: Cecidomyiidae) density to damage to resistant and susceptible sorghum hybrids. J. Econ. Entomol. 77:83–87.
76. Hare, J. D., and P. A. Phillips. 1992. Economic effect of the citrus red mite (Acari: Tetranychidae) on southern California coastal lemons. J. Econ. Entomol. 85:1926–1932.
77. Harrison, F. P. 1984. The development of an economic injury level for low populations of fall armyworm (Lepidoptera: Noctuidae) in grain corn. Florida Entomol. 67:335–339.
78. Haufe, W. O. 1982. Growth of range cattle protected from horn flies (*Haematobia irritans*) by ear tags impregnated with fenvalerate. Can. J. Anim. Sci. 62:567–573.
79. Headley, J. C. 1972. Defining the economic threshold, p. 100–108. *In* Pest control strategies for the future, Nat. Acad. Sci., Washington DC.
80. Higgins, R. A., L. P. Pedigo, and D. W. Staniforth. 1984. Effect of velvetleaf competition and defoliation simulating a green cloverworm (Lepidoptera: Noctuidae) outbreak in Iowa on indeterminate soybean yield, yield components, and economic decision levels. Environ. Entomol. 13:917–925.

24. Bode, W. M., and D. D. Calvin. 1990. Yield-loss relationships and economic injury levels for European corn borer (Lepidoptera: Pyralidae) populations infesting Pennsylvania field corn. J. Econ. Entomol. 83:1595–1603.
25. Bracken, G. K. 1987. Relation between pod damage caused by larvae of Bertha armyworm, *Mamestra configurata* Walker (Lepidoptera: Noctuidae), and yield loss, shelling, and seed quality. Can. Entomol. 119:365–369.
26. Breen, J. P., and G. L. Teetes. 1990. Economic injury levels for yellow sugarcane aphid (Homoptera: Aphididae) on seedling sorghum. J. Econ. Entomol. 83:1008–1014.
27. Buntin, G. D., and L. P. Pedigo. 1982. Foliage consumption and damage potential of *Odontota horni* and *Baliosus nervosus* (Coleoptera: Chrysomelidae) on soybean. J. Econ. Entomol. 75:1034–1037.
28. Buntin, G. D., and L. P. Pedigo. 1985. Development of economic injury levels for last-stage variegated cutworm (Lepidoptera: Noctuidae) larvae in alfalfa stubble. J. Econ. Entomol. 78:1341–1346.
29. Burton, R. L., D. D. Simon, K. J. Starks, and R. D. Morrison. 1985. Seasonal damage by greenbugs (Homoptera: Aphididae) to a resistant and a susceptible variety of wheat. J. Econ. Entomol. 78:395–401.
30. Burts, E. C. 1988. Damage threshold for pear psylla nymphs (Homoptera: Psyllidae). J. Econ. Entomol. 81:599–601.
31. Campbell, J. B., I. L. Berry, D. J. Boxler, R. M. Davis, D. C. Clanton, and G. H. Deutscher 1987. Effects of stable flies (Diptera: Muscidae) on weight gain and feed efficiency of feedlot cattle. J. Econ. Entomol. 80:117–119.
32. Cancelado, R. E., and E. B. Radcliffe. 1979a. Action thresholds for green peach aphid on potatoes in Minnesota. J. Econ. Entomol. 72:606–609.
33. Cancelado, R. E., and E. B. Radcliffe. 1979b. Action thresholds for potato leafhopper on potatoes in Minnesota. J. Econ. Entomol. 72:566–569.
34. Carlson, G. A., and J. C. Headley. 1987. Economic aspects of integrated pest management threshold determination. Plant Dis. 71:459–462.
35. Cartwright, B., J. V. Edelson, and C. Chambers. 1987. Composite action thresholds for the control of lepidopterous pests on fresh-market cabbage in the lower Rio Grande Valley of Texas. J. Econ. Entomol. 80:175–181.
36. Cartwright, B., T. G. Teaue, L. C. Chandler, J. V. Edelson, and G. Bentsen. 1990. An action threshold for management of the pepper weevil (Coleoptera: Curculionidae) on bell peppers. J. Econ. Entomol. 83:2003–2007.
37. Chalfant, R. B., W. H. Denton, D. J. Schuster, and R. B. Workman. 1979. Management of cabbage caterpillars in Florida and Georgia by using visual damage thresholds. J. Econ. Entomol. 72:411–413.
38. Cheshire, J. M. Jr., J. E. Funderburk, D. J. Zimet, T. P. Mack, and M. E. Gilreath. 1989. Economic injury levels and binomial sampling program for

Selected Bibliography

lesser cornstalk borer (Lepidoptera: Pyralidae) in seedling grain sorghum. J. Econ. Entomol. 82:270–274.

39. Chiang, H. C. 1982. Factors to be considered in refining a general model of economic threshold. Entomophaga 27:99–103.

40. Chowdhury, M. A., R. B. Chalfant, and J. R. Young. 1987. Ear damage in sweet corn in relation to adult corn earworm (Lepidoptera: Noctuidae) populations. J. Econ. Entomol. 80:867–869.

41. Coble, H. D. 1985. The development and implementation of economic thresholds for soybeans, p. 295–307. *In* R. E. Frisbie and P. L. Adkisson (ed.) Integrated pest management on major agricultural systems. Texas A&M University, College Station TX.

42. Coble, H. D., and D. A. Mortensen. 1992. The threshold concept and its application to weed science. Weed Technol. 6:191–195.

43. Coffelt, M. A., and P. B. Schultz. 1990. Development of an aesthetic injury level to decrease pesticide use against orangestriped oakworm (Lepidoptera: Saturniidae) in an urban pest management project. J. Econ. Entomol. 83: 2044–2049.

44. Coffelt, M. A., and P. B. Schultz. 1993. Quantification of an aesthetic injury level and threshold for an urban pest management program against orangestriped oakworm (Lepidoptera: Saturniidae). J. Econ. Entomol. 86:1512–1516.

45. Coop, L. B., and R. E. Berry. 1986. Reduction in variegated cutworm (Lepidoptera: Noctuidae) injury to peppermint by larval parasitoids. J. Econ. Entomol. 79:1244–1248.

46. Cousens, R. 1987. Theory and reality of weed control thresholds. Plant Prot. Q. 2:13–20.

47. Cousens, R., C. J. Doyle, B. J. Wilson, and G. W. Cussans. 1986. Modelling the economics of controlling *Avena fatua* in winter wheat. Pestic. Sci. 17:1–12.

48. Cuperus, G. W., E. B. Radcliffe, D. K. Barnes, and G. C. Marten. 1982. Economic injury levels and economic thresholds for pea aphid, *Acyrthosiphon pisum* (Harris), on alfalfa. Crop Prot. 1:453–463.

49. Cuperus, G. W., E. B. Radcliffe, D. K. Barnes, and G. C. Marten. 1983. Economic injury levels and economic thresholds for potato leafhopper (Homoptera: Cicadellidae) on alfalfa in Minnesota. J. Econ. Entomol. 76:1341–1349.

50. Davidson, A., and R. B. Norgaard. 1973. Economic aspects of pest control. Eur. Plant Prot. Org. Bull. 3:63–75.

51. Davis, P. M., and L. P. Pedigo. 1991. Economic injury levels for management of stalk borer (Lepidoptera: Noctuidae) in corn. J. Econ. Entomol. 84:290–293.

52. Davis, R. M., M. D. Skold, J. S. Berry, and W. P. Kemp. 1992. The economic

threshold for grasshopper control on public rangelands.
17:56-65.
53. Dillard, H. R., and R. C. Seem. 1990. Use of an action tl maize rust to reduce crop loss in sweet corn. Phytopath
54. Doyle, C. J., R. Cousens, and S. R. Moss. 1986. A model controlling *Alopecurus myosuroides* Huds. in winter whe 150.
55. Ervin, R. T., F. M. Eppling, R. L. Byford, and J. A. Hair. economic implications of lone star tick (Acari: Ixodidae) gain of cattle, *Bos taurus* and *Bos taurus* x *Bos indicu* 80:443-445.
56. Evans, D. C., and P. A. Stansly. 1990. Weekly economic armyworm (Lepidoptera: Noctuidae) infestation of corn J. Econ. Entomol. 83:2452-2454.
57. Eversmeyer, M. G., and C. L. Kramer. 1987. Component integrated pest management threshold determinations f Plant Dis. 71:456-459.
58. Ezulike, T. O., and R. I. Egwuatu. 1990. Determination of c green spider mite, *Moononychellus tanajoa* (Bondar) on Applic. 11:43-45.
59. Farrington, J. 1977. Economic thresholds of insect pest inf agriculture: A question of applicability. PANS 23:143-148.
60. Ferris, H. 1978. Nematode economic thresholds: Derivation theoretical consideration. J. Nematol. 10:341-350.
61. Ferris, H. 1987. Components and techniques of integrated threshold determinations for soilborne pathogens. Plant D
62. Flanders, K. L., E. B. Radcliffe, and D. W. Ragsdale. 1991. P spread in relation to densities of green peach aphid (Homo Implications for management thresholds for Minnesota seed Entomol. 84:1028-1036.
63. Fuller, B. W., T. E. Reagan, and J. L. Flynn. 1988. Economic sugarcane borer (Lepidoptera: Pyralidae) on sweet sorghum (L.) Moench. J. Econ. Entomol. 81:349-353.
64. Gerowitt, B., and R. Heitefuss. 1990. Weed economic thresh the Federal Republic of Germany. Crop Prot. 9:323-331.
65. Gholson, L. E. 1987. Adaptation of current threshold techni farm techniques. Plant Dis. 71:462-465.
66. Girma, M., G. E. Wilde, and T. L. Harvey. 1993. Russian wh ptera: Aphididae) affects yield and quality of wheat. J. Eco 594-601.

81. Higley, L. G., and L. P. Pedigo. 1993. Economic injury level concepts and their use in sustaining environmental quality. Agric. Ecosyst. Environ. 46:233–243.
82. Higley, L. G., and W. K. Wintersteen. 1992. A novel approach to environmental risk assessment of pesticides as a basis for incorporating environmental costs into economic injury levels. Am. Entomologist 38:34–39.
83. Hluchý, M., and Z. Pospísil. 1992. Damage and economic injury levels of eriophyid and tetranychid mites on grapes in Czechoslovakia. Exp. Appl. Acarol. 14:95–106.
84. Hoffmann, M. P., L. T. Wilson, F. G. Zalmon, and R. J. Hilton. 1990. Parasitism of *Heliothis zea* (Lepidoptera: Noctuidae) eggs: Effect on pest management decision rules for processing tomatoes in the Sacramento Valley of California. Environ. Entomol. 19:753–763.
85. Hopkins, A. R., R. F. Moore, and W. James. 1982. Economic injury level for *Heliothis* spp. larvae on cotton plants in the four-true-leaf to pinhead-square stage. J. Econ. Entomol. 75:328–332.
86. Hosny, M. M., C. P. Topper, G. M. Moawad, and G. B. El-Saadany. 1986. Economic damage thresholds of *Spodoptera littoralis* (Boisd.) (Lepidoptera: Noctuidae) on cotton in Egypt. Crop Prot. 5:100–104.
87. Hoy, C. W., C. E. McCullocoh, A. J. Sawyer, A. M. Shelton, and C. A. Shoemaker. 1990. Effect of intraplant insect movement on economic thresholds. Environ. Entomol. 19:1578–1596.
88. Hoyt, S. C., L. K. Tanigoshi, and R. W. Browne. 1979. Economic injury level studies in relation to mites on apple. Rec. Adv. Acarol. Vol. 1:3–12.
89. Hruska, A. J., and S. M. Gladstone. 1988. Effects of period and level of infestation of the fall armyworm, *Spodoptera frugiperda*, on irrigated maize yield. Florida Entomol. 71:249–254.
90. Hutchins, S. H., and J. E. Funderburk. 1991. Injury guilds: A practical approach for managing pest losses to soybean. Agric. Zool. Rev. 4:1–21.
91. Hutchins, S. H., and P. J. Gehring. 1993. Perspective on the value, regulation, and objective utilization of pest control technology. Am. Entomol. 39:12–15.
92. Hutchins, S. H., L. G. Higley, and L. P. Pedigo. 1988. Injury equivalency as a basis for developing multiple-species economic injury levels. J. Econ. Entomol. 81:1–8.
93. Hutchins, S. H., L. G. Higley, L. P. Pedigo, and P. H. Calkins. 1986. Linear programming model to optimize management decisions with multiple pests: An integrated soybean pest management example. Bull. Entomol. Soc. Am. 32:96–102.
94. Hutchins, S. H., and L. P. Pedigo. 1990. Phenological disruption and economic consequence of injury to alfalfa induced by potato leafhopper (Homoptera: Cicadellidae). J. Econ. Entomol. 83:1587–1594.

Selected Bibliography

95. Jackai, L. E. N., P. K. Atropo, and J. A. Odebiyi. 1989. Use of the response of two growth stages of cowpea to different population densities of the coreid bug, *Clavigralla tomentosicollis* (Stål.), to determine action threshold levels. Crop Prot. 8:422–428.
96. Jamjanya, T., and S. S. Quisenberry. 1988. Impact of fall armyworm (Lepidoptera: Noctuidae) feeding on the quality and yield of coastal bermudagrass. J. Econ. Entomol. 81:922–926.
97. Johnson, K. B., P. S. Teng, and E. B. Radcliffe. 1987. Coupling feeding effects of potato leafhopper, *Empoasca fabae* (Homoptera: Cicadellidae), nymphs to a model of potato growth. Environ. Entomol. 16:250–258.
98. Johnston, R. L., and G. W. Bishop. 1987. Economic injury levels and economic thresholds for cereal aphids (Homoptera: Aphididae) on spring-planted wheat. J. Econ. Entomol. 80:478–482.
99. Jordan, N. 1992. Weed demography and population dynamics: Implications for threshold management. Weed Technol. 6:184–190.
100. Judenko, E. 1972. The assessment of economic losses in yield of annual crops caused by pests, and the problem of the economic threshold. PANS 18:186–191.
101. Keerthisinghe, E. I. 1982. Economic thresholds for cotton pest management in Sri Lanka. Bull. Entomol. Res. 72:239–246.
102. Keerthisinghe, E. I. 1984. Fiducial inference in economic thresholds. Prot. Ecol. 6:85–90.
103. Kirby, R. D., and J. E. Slosser. 1984. Composite economic threshold for three lepidopterous pests of cabbage. J. Econ. Entomol. 77:725–733.
104. Knight, A. L., and L. A. Hull. 1989. Predicting seasonal apple injury by tufted apple bud moth (Lepidoptera: Tortricidae) with early-season sex pheromone trap catches and brood I fruit injury. Environ. Entomol. 18:939–944.
105. Koehler, C. S., and S. S. Rosenthal. 1975. Economic injury levels of the Egyptian alfalfa weevil or the alfalfa weevil. J. Econ. Entomol. 68:71–75.
106. Koehler, P. G., and D. Pimentel. 1973. Economic injury levels of the alfalfa weevil (Coleoptera: Curculionidae). Can. Entomol. 105:61–74.
107. Kogan, M. 1976. Evaluation of economic injury levels for soybean insect pests. *In* L. D. Hill (ed.) Proc. World Soybean Conf., Champaign, IL.
108. KOJIMA, A., AND K. EMURA. 1979. THRESHOLD DENSITY OF THE RICE LEAF BEETLE, *Oulema oryzae* Kuwayama, for insecticidal control. II. Estimation of the threshold density. Jap. J. Appl. Entomol. Zool. 23:1–10.
109. Kolodny-Hirsch, D. M., and F. P. Harrison. 1980. Foliar loss assessments and economic decision-making for the tobacco budworm on Maryland tobacco. J. Econ. Entomol. 73:465–468.
110. Kolodny-Hirsch, D. M., and F. P. Harrison. 1986. Yield loss relationships of

tobacco and tomato hornworms (Lepidoptera: Sphingidae) at several growth stages of Maryland tobacco. J. Econ. Entomol. 79:731–735.

111. Koyama, J. 1975. Studies on the diminution of insecticide application to the rice stem borer, *Chilo suppressalis* Walker. II. The economic injury level of the rice stem borer and its predictive estimation. Jap. J. Appl. Entomol. Zool. 19:63–69.

112. Koyama, J. 1978. Control threshold for the rice leaf beetle, *Oulema oryzae* Kuwayama (Coleoptera: Chrysomelidae). Appl. Entomol. Zool. 13:203–208.

113. Mack, T. P., C. B. Backman, and J. W. Drane. 1988. Effects of lesser cornstalk borer (Lepidoptera: Pyralidae) feeding at selected plant growth stages on peanut growth and yield. J. Econ. Entomol. 81:1478–1484.

114. Mailloux, G., and N. J. Bostanian. 1988. Economic injury level model for tarnished plant bug, *Lygus lineolaris* (Palisot de Beauvois) (Hemiptera: Miridae), in strawberry fields. Environ. Entomol. 17:581–586.

115. Maiteke, G. A., and R. J. Lamb. 1985. Spray timing and economic threshold for the pea aphid, *Acyrthosiphon pisum* (Homoptera: Aphididae), on field peas in Manitoba. J. Econ. Entomol. 78:1449–1454.

116. Marra, M. C., and G. A. Carlson. 1983. An economic threshold model for weeds in soybeans *(Glycine max)*. Weed Sci. 31:604–609.

117. Martin, P. B., B. R. Wiseman, and R. E. Lynch. 1980. Action thresholds for fall armyworm on grain sorghum and coastal bermudagrass. Florida Entomol. 63:375–405.

118. Michaud, O., R. K. Stewart, and G. Boivin. 1989. Economic injury levels and economic thresholds for the green apple bug, *Lygocoris communis* (Knight) (Hemiptera: Miridae), in Quebec apple orchards. Can. Entomologist 121:803–808.

119. Michaud, O. D., G. Boivin, R. K. Stewart. 1989. Economic threshold for tarnished plant bug (Hemiptera: Miridae) in apple orchards. J. Econ. Entomol. 82:1722–1728.

120. Michels, G. J. Jr., and C. C. Burkhardt. 1981. Economic threshold levels of the Mexican bean beetle on pinto beans in Wyoming. J. Econ. Entomol. 74:5–6.

121. Miyashita, T. 1985. Estimation of the economic injury level in the rice leaf roller, *Cnaphalocrocis medinalis* Guenee (Lepidoptera: Pyralidae). Jap. J. Appl. Entomol. Zool. 29:73–76.

122. Morgan, D. R., N. P. Tugwell, and J. L. Bernhardt. 1989. Early rice field drainage for control of rice water weevil (Coleoptera: Curculionidae) and evaluation of an action threshold based upon leaf-feeding scars of adults. J. Econ. Entomol. 82:1757–1759.

123. Morisak, D. J., D. E. Simonet, and R. K. Lindquist. 1984. Use of action thresholds for management of lepidopterous larval pests of fresh-market cabbage. J. Econ. Entomol. 77:476–482.

124. Mortensen, D. A., and H. D. Coble. 1991. Two approaches to weed control decision-aid software. Weed Technol. 5:445–452.
125. Mount, G. A., and J. E. Dunn. 1983. Economic thresholds for lone star ticks (Acari: Ixodidae) in recreational areas based on a relationship between CO_2 and human subject sampling. J. Econ. Entomol. 76:327–329.
126. Mumford, J. D., and G. A. Norton. 1984. Economics of decision making in pest management. Annu. Rev. Entomol. 29:157–174.
127. Nakasuji, F., and T. Matsuzaki. 1977. The control threshold density of the tobacco cutworm *Spodoptera litura* (Lepidoptera: Noctuidae) on eggplants and sweet peppers in vinyl-houses. Appl. Entomol. Zool. 12:184–189.
128. Noe, J. P. 1993. Damage functions and population changes of *Hoplolaimus columbus* on cotton and soybean. J. Nematol. 25:440–445.
129. Nordh, M. B., L. R. Zavaleta, and W. G. Ruesink. 1988. Estimating multidimensional economic injury levels with simulation models. Agric. Syst. 26:19–33.
130. Norgaard, R. B. 1976. Integrating economics and pest management. *In* Integrated pest management, Plenum, New York.
131. Norris, R. F. 1992. Case history for weed competition/population ecology: Barnyardgrass *(Echinochloa crus-galli)* in sugarbeets *(Beta vulgaris)*. Weed Technol. 6:220–227.
132. Norton, G. A. 1976. Analysis of decision making in crop protection. Agro-Ecosystems 3:27–44.
133. Norton, G. A., and D. E. Evans. 1974. The economics of controlling froghopper *(Aeneolamia varia saccharina* (Dist.) (Hom., Cercopidae)) on sugar-cane in Trinidad. Bull. Entomol. Res. 63:619–627.
134. O'Donovan, J. T. 1991. Quackgrass *(Elytrigia repens)* interference in canola *(Brassica campesstris)*. Weed Sci. 39:397–401.
135. Ogunlana, M. O., and L. P. Pedigo. 1974. Economic injury levels of the potato leafhopper on soybeans in Iowa. J. Econ. Entomol. 67:29–32.
136. Oliver, L. R. 1988. Principles of weed threshold research. Weed Technol. 2:398–403.
137. Onsager, J. A. 1984. A method for estimating economic injury levels for control of rangeland grasshoppers with malathion and carbaryl. J. Range Management 37:200–203.
138. Onstad, D. W. 1987. Calculation of economic-injury levels and economic thresholds for pest management. J. Econ. Entomol. 80:297–303.
139. Onstad, D. W., and R. Rabbinge. 1985. Dynamic programming and the computation of economic injury levels for crop disease control. Agric. Syst. 18:207–226.
140. Ordish, G., and D. Dufour. 1969. Economic bases for protection against plant diseases. Annu. Rev. Phytopathol. 7:31–50.

141. Oseto, C. Y., and G. A. Braness. 1980. Chemical control and bioeconomics of *Smicronyx fulvus* on cultivated sunflower in North Dakota. J. Econ. Entomol. 73:218–220.
142. Ostlie, K. R., and L. P. Pedigo. 1985. Soybean response to simulated green cloverworm (Lepidoptera: Noctuidae) defoliation: Progress toward determining comprehensive economic injury levels. J. Econ. Entomol. 78:437–444.
143. Ostlie, K. R., and L. P. Pedigo. 1987. Incorporating pest survivorship into economic thresholds. Bull. Entomol. Soc. Am. 33:98–102.
144. Palti, J., and R. Ausher. 1986. Crop value, economic damage thresholds, and treatment thresholds. *In* Crop protection monographs: Advisory work in crop pest and disease management, Springer-Verlag, New York.
145. Pedigo, L. P., R. B. Hammond, and F. L. Poston. 1977. Effects of green cloverworm larval intensity on consumption of soybean leaf tissue. J. Econ. Entomol. 70:159–162.
146. Pedigo, L. P., and L. G. Higley. 1992. The economic injury level and environmental quality: A new perspective. Am. Entomologist 38:12–21.
147. Pedigo, L. P., L. G. Higley, and P. M. Davis. 1989. Concepts and advances in economic thresholds for soybean entomology. Proc. World Soybean Res. Conf. IV, Vol. III:1487–1493.
148. Pedigo, L. P., S. H. Hutchins, and L. G. Higley. 1986. Economic injury levels in theory and practice. Annu. Rev. Entomol. 31:341–368.
149. Pedigo, L. P., and J. W. van Schaik. 1984. Time-sequential sampling: A new use of the sequential probability ratio test for pest management decisions. Bull. Entomol. Soc. Am. 30:32–36.
150. Peña, J. E. 1990. Relationships of broad mite (Acari: Tarsonemidae) density to lime damage. J. Econ. Entomol. 83:2008–2015.
151. Peña, J. E., K. Pohoronezny, V. H. Waddill, and J. Stimac. 1986. Tomato pinworm (Lepidoptera: Gelechiidae) artificial infestation: Effect on foliar and fruit injury of ground tomatoes. J. Econ. Entomol. 79:957–960.
152. Peterson, R. K. D., S. D. Danielson, and L. G. Higley. 1993. Yield responses of alfalfa to simulated alfalfa weevil injury and development of economic injury levels. Agron. J. 85:595–601.
153. Plant, R. E. 1986. Uncertainty and the economic threshold. J. Econ. Entomol. 79:1–6.
154. Poston, F. L., L. P. Pedigo, and S. M. Welch. 1983. Economic injury levels: Reality and practicality. Bull. Entomol. Soc. Am. 29:49–53.
155. Ram, S., and B. D. Patil. 1986. Economic injury level of major defoliator insect pests of fodder cowpea, *Vigna unguiculata* (L.) Walp. J. Entomol. Res. 10:99–202.
156. Raupp, M. J., J. A. Davidson, C. S. Koehler, C. S. Sadof, and K. Reichelderfer.

1987. Decision-making considerations for aesthetic damage caused by pests. Bull. Entomol. Soc. Am. 34:27–32.
157. Raupp, M. J., J. A. Davidson, C. S. Koehler, C. S. Sadof, and K. Reichelderfer. 1989. Economic and aesthetic injury levels and thresholds for pests of ornamental plants. Fla. Entomologist 72:403–407.
158. Raupp, M. J., C. S. Koehler, and J. A. Davidson. 1992. Advances in implementing integrated pest management for woody landscape plants. Annu. Rev. Entomol. 37:561–585.
159. Raworth, D. A. 1986. An economic threshold function for the twospotted spider mite, *Tetranychus urticae* (Acari: Tetranychidae), on strawberries. Can. Entomol. 118:9–16.
160. Riley, D. G., D. J. Schuster, and C. S. Barfield. 1992. Refined action thresholds for pepper weevil adults (Coleoptera: Curculionidae) in bell peppers. J. Econ. Entomol. 85:1919–1925.
161. Ring, D. R., and J. H. Benedict. 1991. Yield change functions and economic injury levels: *Helicoverpa zea* vs. *Heliothis virescens* injury to cotton. Proc. Beltwide Cotton Conf. 2:672–674.
162. Ring, D. R., J. H. Benedict, J. A. Landivar, and B. R. Eddleman. 1993. Economic injury levels and development and application of response surfaces relating insect injury, normalized yield, and plant physiological age. Environ. Entomol. 22:273–282.
163. Rinker, D. L., and R. J. Snetsinger. 1984. Damage threshold to a commercial mushroom by a mushroom-infesting phorid (Diptera: Phoridae). J. Econ. Entomol. 77:449–453.
164. Roberts, R. K., and R. M. Hayes. 1989. Decision criterion for profitable johnsongrass *(Sorghum halepense)* management in soybeans *(Glycine max)*. Weed Technol. 3:44–47.
165. Rogers, C. E. 1976. Economic injury level for *Contarinia texana* on guar. J. Econ. Entomol. 69:693–696.
166. Sadof, C. S., and C. M. Alexander. 1993. Limitations of cost-benefit-based aesthetic injury levels for managing twospotted spider mites (Acari: Tetranychidae). J. Econ. Entomol. 86:1516–1521.
167. Saito, T. 1983. Effect of artificial defoliation on growth and yield of soybean: Development of dynamic economic injury level and control threshold. Jap. J. Appl. Entomol. Zool. 27:203–210.
168. Sarup, P., V. K. Sharma, V. P. S. Panwar, K. H. Siddiqui, K. K. Marwaha, and K. N. Agarwal. 1977. Economic threshold of *Chilo partellus* (Swinhoe) infesting maize crop. J. Entomol. Res. 1:92–99.
169. Schaefers, G. A. 1980. Yield effects of tarnished plant bug feeding on June-bearing strawberry varieties in New York state. J. Econ. Entomol. 73:721–725.

170. Schaub, L., W. A. Stahel, J. Baumgartner, V. Delucchi. 1988. Elements for assessing mirid (Heteroptera: Miridae) damage threshold on apple fruits. Crop Prot. 7:118–124.
171. Schreiber, E. T., J. B. Campbell, S. E. Kunz, D. C. Clanton, and D. B. Hudson. 1987. Effects of horn fly (Diptera: Muscidae) control on cows and gastrointestinal worm (Nematode: Trichostrongylidae) treatment for calves on cow and calf weight gains. J. Econ. Entomol. 80:451–454.
172. Sears, M. K., R. P. Jaques, and J. E. Laing. 1983. Utilization of action thresholds for microbial and chemical control of lepidopterous pests (Lepidoptera: Noctuidae, Pieridae) on cabbage. J. Econ. Entomol. 76:368–374.
173. Sears, M. K., A. M. Shelton, T. C. Quick, J. A. Wyman, and S. E. Webb. 1985. Evaluation of partial plant sampling procedures and corresponding action thresholds for management of Lepidoptera on cabbage. J. Econ. Entomol. 78:913–916.
174. Senanayake, D. G., and N. J. Holliday. 1990. Economic injury levels for Colorado potato beetle (Coleoptera: Chrysomelidae) on 'Norland' potatoes in Manitoba. J. Econ. Entomol. 83:2058–2064.
175. Seshu Reddy, K. V., and K. O. Sum. 1991. Determination of economic injury level of the stem borer, *Chilo partellus* (Swinhoe) in maize, *Zea mays* L. Insect Sci. Applic. 12:269–274.
176. Sharma, H. C., and V. F. Lopez. 1989. Assessment of avoidable losses and economic injury levels for the sorghum head bug, *Calocoris angustatus* Leth. (Hemiptera: Miridae) in India. Crop Prot. 8:429–435.
177. Shelton, A. M., J. T. Andaloro, and J. Barnard. 1982. Effects of cabbage looper, imported cabbageworm, and diamondback moth on fresh market and processing cabbage. J. Econ. Entomol. 75:742–745.
178. Shelton, A. M., C. W. Hoy, and P. B. Baker. 1990. Response of cabbage head weight to simulated lepidoptera defoliation. Entomol. Exp. Appl. 54:181–187.
179. Shelton, A. M., M. K. Sears, J. A. Wyman, and T. C. Quick. 1983. Comparison of action thresholds for lepidopterous larvae on fresh-market cabbage. J. Econ. Entomol. 76:196–199.
180. Shields, E. J., D. I. Rouse, and J. A. Wyman. 1985. Variegated cutworm (Lepidoptera: Noctuidae): Leaf-area consumption, feeding site preference, and economic injury level calculation for potatoes. J. Econ. Entomol. 78:1095–1099.
181. Simonet, D. E., and D. J. Morisak. 1982. Utilizing action thresholds in smallplot insecticide evaluations against cabbage-feeding, lepidopterous larvae. J. Econ. Entomol. 75:43–46.
182. Smith, R. F. 1969. The importance of economic injury levels in the development of integrated pest control programs. Qualitas Plant. Matero. Veg. 17:81–92.

183. Sparks, A. N. Jr., and L. D. Newsom. 1984. Evaluation of the pest status of the threecornered alfalfa hopper (Homoptera: Membracidae) on soybean in Louisiana. J. Econ. Entomol. 77:1553–1558.
184. Steelman, C. D. 1979. Economic thresholds for mosquitoes. Mosq. News 39:724–729.
185. Sterling, W. 1984. Action and inaction levels in pest management. Tex. Agric. Exp. Stn. Bull. 1480:20pp.
186. Stern, V. M. 1965. Significance of the economic threshold in integrated pest control. Proc. FAO Symp. Integ. Pest Cont. 2:41–56.
187. Stern, V. M. 1967. Control of aphids attacking barley and analysis of yield increases in the Imperial Valley, California. J. Econ. Entomol. 60:485–490.
188. Stern, V. M. 1973. Economic thresholds. Annu. Rev. Entomol. 18:259–280.
189. Stern, V. M., R. Sharma, and C. Summers. 1980. Alfalfa damage from *Acyrthosiphon kondoi* and economic threshold studies in southern California. J. Econ. Entomol. 73:145–148.
190. Stern, V. M., R. F. Smith, R. van den Bosch, and K. S. Hagen. 1959. The integrated control concept. Hilgardia 29:81–101.
191. Stevenson., A. B. 1985. Early warning system for the carrot weevil (Coleoptera: Curculionidae) and its evaluation in commercial carrots in Ontario. J. Econ. Entomol. 78:704–708.
192. Stewart, J. G. G., and M. K. Sears. 1988. Economic threshold for three species of lepidopterous larvae attacking cauliflower grown in southern Ontario. J. Econ. Entomol. 81:1726–1731.
193. Stewart, R. K., and A. R. Khattat. 1980. Pest status and economic thresholds of the tarnished plant bug, *Lygus lineolaris* (Hemiptera [Heteroptera]: Miridae), on green beans in Quebec. Can. Entomol. 112:301–305.
194. Stone, J. D., and L. P. Pedigo. 1972. Development and economic-injury level of the green cloverworm on soybean in Iowa. J. Econ. Entomol. 65:197–201.
195. Story, R. N., A. J. Keaster, W. B. Showers, J. T. Shaw, and V. L. Wright. 1983. Economic threshold dynamics of black and claybacked cutworms (Lepidoptera: Noctuidae) in field corn. Environ. Entomol. 12:1718–1723.
196. Sylven, E. 1968. Threshold values in the economics of insect pest control in agriculture. Nat. Swedish Inst. Plant Prot. Contrib. 14:65–79.
197. Szmedra, P. I., M. E. Wetzstein, and R. W. McClendon. 1990. Economic threshold under risk: A case study of soybean production. J. Econ. Entomol. 83:641–646.
198. Talpaz, H., and R. E. Frisbie. 1975. An advanced method for economic threshold determination: A positive approach. Southern J. Agric. Econ. 7:19–25.
199. Thistlewood, H. M. A., R. D. McMullen, and J. H. Borden. 1989. Damage and

economic injury levels of the mullein bug, *Campylomma verbasci* (Meyer) (Heteroptera: Miridae), on apple in the Okanagan Valley. Can. Entomol. 121:1-9.

200. Thomas, G. D., D. B. Smith, and C. M. Ignoffo. 1979. Economic thresholds for insect control, p. 419-429. *In* W. B. Ennis, Jr. (ed.) Introduction to crop protection. Am. Soc. Agron., Crop Sci. Soc. Am., Madison WI.

201. Thornton, P. K., R. H. Fawcett, J. B. Dent, and T. J. Perkins. 1990. Spatial weed distribution and economic thresholds for weed control. Crop Prot. 9: 337-342.

202. Torell, L. A., J. H. Davis, E. W. Huddleston, and D. C. Thompson. 1989. Economic injury levels for interseasonal control of rangeland insects. J. Econ. Entomol. 82:1289-1294.

203. Tsuzuki, H., T. Asayama, M. Takimoto, J. Kayumi, and S. Kobayashi. 1983. Assessment of yield loss due to the rice water weevil, *Lissorhoptrus oryzophilus* Kuschel (Coleoptera: Curculionidae). II. Damage caused by adult and larval infestation and estimation of the tolerable injury level. Jap. J. Appl. Entomol. Zool. 27:252-260.

204. Villacorta, A., and P. L. Sanchez-Rodriquesm. 1984. Action threshold in the insecticide utilization on the coffee leaf miner management (*Perileucoptera coffeella* Guerin-Meneville 1842) in Parana State (Lepidoptera: Lyonetiidae). Ann. Soc. Entomol. Brasil 13:157-165.

205. Walgenbach, J. F., and J. A. Wyman. 1984. Dynamic action threshold levels for the potato leafhopper (Homoptera: Cicadellidae) on potatoes in Wisconsin. J. Econ. Entomol. 77:1335-1340.

206. Walker, G. P., A. L. Voulgaropoulos, and P. A. Phillips. 1992. Effect of citrus bud mite (Acari: Eriophyidae) on lemon yields. J. Econ. Entomol. 85:1318-1329.

207. Way, M. O., A. A. Grigarick, J. A. Litsinger, F. Palis, and P. Pingali. 1991. Economic thresholds and injury for insect pests of rice, p. 67-105. *In* E. A. Heinrichs and T. A. Miller (ed.) Rice insects: Management strategies. Springer Verlag, New York.

208. Weaver, S. E. 1986. Factors affecting threshold levels and seed production of jimsonweed (*Datura stramonium* L.) in soyabeans (*Glycine max* [L.] Merr.). Weed Res. 26:215-223

209. Weaver, S. E. 1991. Size-dependent economic thresholds for three broadleaf weed species in soybeans. Weed Technol. 5:674-679.

210. Weinzierl, R. A., R. E. Berry, and G. C. Fisher. 1987. Sweep-Net sampling for western spotted cucumber beetle (Coleoptera: Chrysomelidae) in snap beans: Spatial distribution, economic injury level, and sequential sampling plans. J. Econ. Entomol. 80:1278-1283.

Selected Bibliography

211. Welter, S. C., R. Freeman, and D. S. Farnham. 1991. Recovery of 'Zinfandel' grapevines from feeding damage by Willamette spider mite (Acari: Tetranychidae): Implications for economic injury level studies in perennial crops. Environ. Entomol. 20:104–109.
212. Welter, S. C., P. S. McNally, and D. S. Farnham. 1989. Effect of Willamette mite (Acari: Tetranychidae) on grape productivity and quality: A reappraisal. Environ. Entomol. 18:408–411.
213. Wilde, G., and J. Morgan. 1978. Chinch bug on sorghum: Chemical control, economic injury levels, plant resistance. J. Econ. Entomol. 71:908–910.
214. Wilson, L. J. 1993. Spider mites (Acari: Tetranychidae) affect yield and fiber quality of cotton. J. Econ. Entomol. 86:566–585.
215. Workman, R. B., R. B. Chalfant, and D. J. Schuster. 1980. Management of the cabbage looper and diamondback moth on cabbage by using two damage thresholds and five insecticide treatments. J. Econ. Entomol. 73:757–758.
216. Wright, L. W., and W. W. Cone. 1980. Economic damage and vine response from simulated cutworm damage to Concord grape buds. J. Econ. Entomol. 73:787–790.
217. Yencho, G. C., L. W. Getzin, and G. E. Long. 1986. Economic injury level, action threshold, and a yield-loss model for the pea aphid, *Acyrthosiphon pisum* (Homoptera: Aphididae) on green peas, *Pisum sativum*. J. Econ. Entomol. 79:1681–1687.
218. Yu, Y., H. J. Gold, R. E. Stinner, and G. G. Wilkerson. 1993. A model for a time-dependent economic threshold for chemical control of corn earworm on soybean. Agric. Syst. 43:439–458.
219. Zadoks, J. C. 1985. On the conceptual basis of crop loss assessment: The threshold theory. Annu. Rev. Phytopathol. 23:455–473.
220. Zungoli, P. A., and W. H. Robinson. 1984. Feasibility of establishing an aesthetic injury level for German cockroach pest management programs. Environ. Entomol. 13:1453–1458.

The Contributors

PAUL A. BACKMAN is a professor in the Department of Plant Pathology and director of the Biological Control Institute at Auburn University, Alabama, where he has conducted research on the epidemiology and control of peanut, soybean, and cotton diseases since 1971. His research has focused on the development of integrated management models for peanuts and soybeans, linking decision making to easy-to-monitor weather variables and weather forecasts. Current research relates to the development and integration of biological pest control practices into sustainable production systems.

G. DAVID BUNTIN is a professor of entomology at the Georgia Station of the University of Georgia. He has published more than 65 articles and book chapters on biology, sampling, plant injury, and management of insects in agricultural crops.

JOHN B. CAMPBELL is a professor of entomology at the University of Nebraska West Central Research and Extension Center in North Platte. He has published extensively on the economics and management of livestock insects.

HAROLD D. COBLE is a professor in the Crop Science Department at North Carolina State University. He is widely recognized for his pioneering work on weed thresholds and integrated pest management in row crops. His basic and applied research has resulted in numerous publications and software releases.

MICHAEL D. DUFFY is a professor in the Department of Economics and associate director for the Leopold Center for Sustainable Agriculture at Iowa State University. He is an extension economist in farm management, and his research interests include soil conservation, integrated pest management, and sustainable agriculture.

RONALD B. HAMMOND is an associate professor of entomology at the Ohio Agricultural Research and Development Center of the Ohio State University. He has conducted research on insect/plant relationships, host plant resistance, and impacts of conservation tillage on foliage and soil invertebrates.

LEON G. HIGLEY is an associate professor in the Department of Entomology at the University of Nebraska. He conducts research on plant stress physiology, insect ecology, and many areas of integrated pest management, including decision making. He is the author or coauthor of more than 50 scientific papers and is a contributing editor for *American Entomologist* and associate editor for *Agronomy Journal*.

SCOTT H. HUTCHINS is global development manager for insect management products with DowElanco in Indianapolis, Indiana. He has published or presented more than 65 articles on pest-management theory and implementation with specific emphasis on bioeconomics and EIL development.

JAMES C. JACOBI is a field development specialist for ISK Biosciences. Formerly, he was a senior research associate at Auburn University, Alabama, where he was involved with the development of the AU-Pnuts disease-management model for peanut, as well as having developed loss models for soilborne diseases of peanut. While engaged on these projects, he also developed data bases for cost-benefit analysis of numerous pesticides used in peanut.

DAVID A. MORTENSEN is an associate professor of agronomy at the University of Nebraska. He has published widely on weed thresholds and spatial distribution. He has also implemented the results of this research through WeedSOFT$_{sm}$, a bioeconomic weed-management decision aid for farmers.

LARRY P. PEDIGO is professor of entomology in the College of Agriculture of Iowa State University in Ames. He conducts research with insect-population dynamics and integrated pest management and teaches both undergraduate and graduate courses in entomology. He is the author or coauthor of more than 150 research publications and, before this one, has written or edited six books on aspects of entomology and pest management. His popular textbook, *Entomology and Pest Management*, has been adopted worldwide.

ROBERT K. D. PETERSON is a research biologist with DowElanco, where he serves as field development scientist for North American insect, nema-

The Contributors

tode, and disease management research and development. His research interests include physiological responses of plants to biotic stress, insect thermal development, economic decision levels, pest management theory, and pesticide-related issues.

MICHAEL J. RAUPP is a professor and chair of the Department of Entomology at the University of Maryland. He has published more than 80 scientific papers on the ecology and management of insect and mite pests.

CLIFFORD S. SADOF is an associate professor of entomology at Purdue University, Indiana. He has published numerous scientific articles and extension bulletins on the management of pests on ornamental plants.

GUSTAVE D. THOMAS is the research leader of the Midwest Livestock Insects Research Laboratory, Agricultural Research Service, U.S. Department of Agriculture, at Lincoln, Nebraska. He is also an adjunct professor of entomology at the University of Nebraska. He has authored publications on face flies, horn flies, stable flies, house flies, mosquitoes, and soybean insect pests.

STEPHEN C. WELTER is an associate professor in the Department of Environmental Science, Policy, and Management at the University of California–Berkeley. His research focuses on many aspects of integrated pest management, including plant-insect interactions and decision making, particularly for perennial crops. He also has worked and published extensively on plant ecophysiology, emphasizing stress induced by arthropod injury.

WENDY K. WINTERSTEEN is a professor and extension entomologist in the Department of Entomology at Iowa State University. She coordinates the integrated pest management, pesticide education, and pesticide impact assessment programs and conducts research on IPM and pesticide issues.

Index

abiotic stress, 24, 30, 35, 36, 294
Abutilon theophrasti, 102, 161, 163, 169
Acari, 161
acaricide, 236
acephate, 261
Aceria sheldoni, 161
Acrididae, 158
Acrostrenum hilare, 171, 280
action threshold, 42, 128
Acyrthosiphon kondoi, 158
Acyrthosiphon pisum, 158
Aeneolamia varia saccharina, 158
aerial densities, 118
aesthetic injury, 203, 204, 209, 214
aesthetic injury level (AIL), 68, 164, 166, 167, 204, 205, 206, 210, 212, 215, 219, 223
aesthetic pests, 19, 258
aesthetic quality, 213
aesthetics, 58, 68, 69, 175, 211, 219, 223
aesthetic thresholds, 162, 203, 258, 291
aesthetic value, 13, 166, 187, 204, 214, 224, 279
aflatoxin, 282, 283
African white stem borer, 32
age distribution, 174
age structure, 165
agroecosystems, 8
Agropyron repens, 100
Agrotis gladiaria, 159
Agrotis ipsilon, 159
AIL. *See* aesthetic injury level
aldicarb, 261
alfalfa, 3, 62, 131, 135, 139, 158, 159, 160, 162, 166, 170, 228, 235, 237, 276, 277, 284, 286, 287, 288
alfalfa snout beetle, 139
alfalfa weevil, 46, 135, 166, 170, 237, 279
allelopathy, 95
almonds, 232, 233
Alopercurus myosuroides, 103, 161

Alphitobuis diaperinus, 192
Amaranthus hybridus, 99
Amaranthus tuberculatus, 161, 163
Amathes c-nigrum, 159
Amblyomma americanum, 161, 162, 181
Amblyomma maculatum, 182
American arborvitae, 160
Anisota senatoria, 161, 208
annual systems, 228
Antherigona socata, 52
Anthonomus eugenii, 158
Anthonomus grandis, 158
Anthracnose, 121
Anticarsia gemmatalis, 159, 171
aphids, 26, 43, 122, 130, 158, 234, 235, 236. *See also* rosy apple aphid; tulip tree aphid; walnut aphid
apple, 49, 81, 120, 157, 159, 161, 223
architecture modification, 27
area-wide pest management, 294
Argyresthis cuperssella, 207, 209
artificial infestations, 130, 131
artificial populations, 130
Asana XL.66EC, 265
assimilate removal, 27
assimilate sappers, 27, 155
AU-PNUTS Advisory, 121
AUSIMM, 66, 123
Avena fatua, 102, 161, 163
azinphosmethyl, 261

Bacillus thuringiensis-kurstaki, 261
Baliosus nervosus, 158
barley, 96, 158, 169. *See also* winter barley
bean, common, 49, 157, 158, 159, 161, 162. *See also* soybean; snap bean
bean leaf beetle, 29, 172
bedbug, 191
beetles. *See* alfalfa snout beetle; bean leaf beetle; cereal leaf beetle; Colorado potato beetle; elm leaf beetle; green

317

beetles (cont.)
  June beetle; Japanese beetle; litter beetle; Mexican bean beetle; mountain pine beetle; southern pine beetle; western spotted cucumber beetle
beet mild yellows virus, 117
beet yellows virus, 117
bell pepper, 158, 160
bermuda grass, 130
bioeconomic models, 101, 103, 106
biological control, 3, 4, 7, 124, 140, 171, 223, 224, 251
biomass, 101, 279, 281, 284, 285, 286, 287, 289
biotechnology, 293
biotic stress, 23, 25, 36, 67, 168, 169, 173, 294
biotypes, 227
biting midges, 193
black blow fly, 193
black flies, 180, 189, 258
black grass, 103
Blatella germanica, 158, 162, 207
Blissus leucopterus leucopterus, 159
BLITECAST model, 120
blow fly, 193
blue gamma grass, 161
bluetongue virus, 189, 193
Bos indicus, 182
Bos taurus, 182
Botrytis leaf blight, 122
Bovicola bovis, 184
bovine herpesvirus, 186
bovine keratoconjunctivitis, 185
brown spot, 119, 121, 169
Brucella abortus, 186
burning bush, 215
bush euonymus, 213

C (variable; management costs in EIL), 12, 58, 61, 139, 140, 167, 168, 213, 215, 216, 228, 240, 252, 253, 267, 281, 284, 289
cabbage, 32, 160, 162
cabbage butterfly, 157
cabbage looper, 135, 157
cabbage looper equivalents, 133
calculated threshold, 152
Calepitrimerus vitis, 161

CALEX/Cotton model, 66
Calocoris angustatus, 159
Camplyomma verbasci, 159
canola, 159, 161, 163
carbaryl, 261
carbofuran, 262
carrot, 159
cassava, 161
caterpillar, 18, 43, 45, 212, 213
cattle, 158, 161, 181, 183, 184, 185, 186, 188, 194
cattle grubs, 180, 183
cattle lice, 183, 184
cauliflower, 133, 134, 160
Cecidomyia spp., 207
Cecidomyiidae, 158
Ceratoma trifurcata, 158
Ceratopogonids, 188
Cercopidae, 158
Cercospora arachidicola, 120
Cercosporidium personaturm, 120
cereal grains, 104
cereal leaf beetle, 132, 133
Cerecospora apii, 118
Chamecyparis spp., 207
chaos theory, 292
chemical control, 5, 6, 7, 77, 115, 116
chicken body louse, 191
chicken mite, 191
chickens, 190
Chilo partellus, 160
Chilo suppressalis, 160
Chlorodchroa ligata, 160
chlorpyrifos, 262
Choristoneura occidentalis, 207
Christmas trees, 69, 211
Chrysanthemum, 207
Chrysomelidae, 158
Cicadellidae, 158
Cimex lectularius, 191
citrus bud mite, 231, 236
citrus red mite, 236, 237
Clavigralla tomentosicollis, 33, 158
Cnaphalocrocis medinalis, 160
coastal bermudagrass, 160
Coccinellidae, 158
cockles, 192

Index

cockroaches, 209, 210, 221. *See also* German cockroaches
coffee, 159
Colmerus vitis, 161
Colorado potato beetle, 32
commodity grade, 138
common cocklebur, 98, 99, 100
common grub, 183
common sunflower, 102
compensation, 31, 32, 34, 137, 254
competition, 95, 96, 97; index of, 65, 96, 104; interspecific, 99; intraspecific; 99; load, 104; rankings, 65. *See also* species-relative competitiveness; weed competitiveness
conifer, 167, 214
consumption rates, 29, 135, 136
containment practices, 90
Contarinia sorghicola, 158
Contarinia texana, 134, 158
contingent valuation, 140, 167, 168, 204, 211, 255, 256, 257, 258, 259, 267, 268, 271
control, 5, 6, 75, 179, 281, 293, 294
corn, 62, 76, 99, 100, 102, 106, 117, 139, 141, 159, 160, 161, 162, 170, 206, 265, 266, 270
corn earworm, 157, 171
corn rootworm, 81, 131
cosmetic damage, 283. *See also* damage
cost-benefit, 220, 223, 249, 294
cost functions, 9
costs. *See* economic costs; environmental costs; management, costs of; risk, costs of; social costs
Cotinus nitida, 207, 209
cotton, 32, 50, 66, 76, 117, 132, 157, 158, 159, 160, 161, 162, 170, 228, 238
cowpea, 33, 158, 160
critical duration, 93
crop equivalents, 96
crop-growth-stage-specific EILs, 170, 173
crop phenology, 46
cropping cycles, 227
crop production practices, 64
crop value, 12
Culicidae, 158
Culicoides, 180, 193

Culicoides variipennis, 193
cultivation, 154
cultural management, 77, 251
curative action. *See* therapeutic tactics
curative herbicides, 154
Curculionidae, 158
CV ratio, 220, 221, 222
Cylindrocladium black rot, 116
Cylindrocladium crotalariae, 116
cypress tip miner, 209

D (variable; damage or yield loss per unit injury in EIL equation), 12, 30, 58, 134, 156, 167, 168, 170, 174, 213, 214, 216, 217, 251, 253, 280, 281, 283, 284, 288
damage, 12, 16, 25, 30, 128, 129, 156, 162, 169, 173, 210, 214, 219, 221, 228, 229, 230, 234, 237, 239, 240, 270, 280, 282; boundary, 14, 42, 47, 92, 282, 283; cosmetic, 283; curve, 14, 30, 31, 32, 33, 34, 280, 281, 289; economic, 4, 10, 11, 22, 211; functions, 93, 137, 141, 169, 240; potential, 41; simulation, 134, 135; threshold, 92, 97, 128
damaging stage, 52
Datura stramonium, 99, 161, 163
DDT, 6
decision delays, 46
decision-making theory, 80
deep plowing, 115
defoliation, 60, 135, 136, 155, 156
delayed effects, 230, 232
deminimus approach, 275, 276, 278, 279, 280, 281, 282, 284, 287
Dendroctonus frontalis, 207
Dendroctonus ponderosae, 207
density, 97, 98, 101
Dermanyssus gallinae, 191
descriptive ET, 48, 49, 55, 166
desensitization, 31, 32
Diabrotica spp., 131
Diabrotica undecimpunctata, 158
Diabrotica undecimpunctata undecimpunctata, 129
diamondback moth, 157
Diatraea saccharalis, 160
diazinon, 262
dichotomous ET, 48, 51, 55, 166

differentiating injury, 16
digestibility, 276, 278, 284
digestible energy, 281, 282, 285
dimethoate, 262
dipel, 265, 266
Diptera, 180, 194
direct injury. *See under* injury
Diruaphis noxia, 158
discrimination analysis, 10
disease, 26, 48, 59, 62, 66, 91, 114, 180, 187, 195, 253, 270; forecasting, 120, 122; loci, 164; management, 10, 24, 114; severity, 116, 117, 121, 122, 164; warnings, 119
disulfoton, 262
dyfonate, 265
dynamic action threshold, 174
dynamic EIL, 173
Dysaphis plantaginea, 49

Earias vittella, 159
early leaf spot, 120
economic benefits, 269
economic costs, 269
economic damage. *See under* damage
economic injury, 17, 64
economic injury level (EIL), xi, 4, 7, 9, 10, 11; aesthetic pests and, 68, 166–167, 203–226, biological basis of, 22–40; definition of, 11–15, 93–94; environmental quality and, 69, 70, 167, 168, 249–274; future, 55, 56, 141, 142, 171–175, 291–294; history of, 3–5, 152, 153, 154; insects and, 128–149, 155–162; interseasonal, 227–248; limitations of, 58–73; multiple pests and, 27, 28, 65–67; multiple-species, 16, 28, 164, 168, 169; multiple-stress, 58; past injury and, 17–19 (*see also* injury, past); plant pathogens and, 114–127, 161, 163–164; replanting and, 17; veterinary pests and, 179–202; weeds and, 89–113, 161, 163. *See also* crop-growth-stage-specific EILs; dynamic EIL; environmental EIL; hybrid EIL; stand-loss EIL; static EIL
economic models, 80
economic optimum threshold (EOT), 93, 102
economic parameters, 22

economic risk, 64, 249, 250. *See also* risk
economic threshold (ET), 4, 10, 11, 41, 42, 80, 91, 93, 102, 103, 117, 164; calculation of, 49–54, 61–62; definition of, 4, 9–11, 41, 42; descriptive, 49–54; deterministic derivation of, 50–51; dichotomous, 51–54; fixed, 48, 49, 55, 166; history of, 3–5, 152–154; limitations of, 55–56, 61–65; multidimensional, 173–174; objective, 165, 166; objective determination of, 49–54; risk aversion and, 64–65; sampling and, 62–63, 292; stochastic derivation of, 50–51; subjective, 48, 165; subjective determination of, 49–54. *See also* thresholds
economic types, 53
economic value, 166
ectoparasites, 180, 181. *See also* parasites
eggplant, 160
EIL. *See* economic injury level
Elasmopalpus lignosellus, 160, 282
elm leaf beetle, 209
Elytrigia repens, 161
Empoasca fabae, 157, 158, 169, 284, 285, 288, 289
endoparasites, 194. *See also* parasites
environmental conservation, 5, 8
environmental costs, 5, 8, 13, 19, 69, 70, 168, 238, 240, 253, 256, 257, 258, 259, 260, 267, 268, 269, 270, 271, 272
environmental EIL, 164, 167, 204, 252, 254, 255, 258, 260, 266, 267, 268, 269, 270, 271, 272, 291
environmental-impact quotients, 70, 256
environmental impacts, 84
environmental problems, 7
Environmental Protection Agency, 267
environmental quality, 3, 8, 175, 249, 250, 252, 257, 271
environmental risk, 69, 153, 175, 250, 251, 252, 253, 254, 255, 256, 259, 260, 261, 267, 268, 272, 293, 294. *See also* risk
environmental sustainability, 3, 5, 6, 7, 249, 250, 272
Eotetranychus willamettei, 161
Epilachna varivestis, 158, 172
EPIPRE system, 122
equivalency. *See under* injury

Index

eradication, 90
Eriophyidae, 161
Erwinia stewartii, 117
Erythroneura comes, 235
esfenvalerate, 262
ET. *See* economic threshold
ethoprop, 263
Euonymus, 221
Euonymus alatus, 161, 208, 219
Euonymus alatus 'compacta', 215
European corn borer, 140, 153, 265, 266
European red mite, 236
Euxoa ochrogaster, 159
expert systems, 66, 83
externalities, 79, 80, 140
eye worm, 186

face fly, 180, 185, 186, 195
fall armyworm, 32, 130, 157
Fallopia convolvulus, 105
fallowing, 115
feed-value approach, 275, 276, 278, 279, 280, 281, 284, 285, 286, 287, 288
fenvalerate, 263
filth flies, 189, 190, 191
fitness, 94
fixed threshold, 105. *See also* economic threshold
flooding, 115
flies. *See* black flies; blow fly; face fly; filth flies; green bottle fly; Hessian fly; horn fly; house fly; sheep bot fly; sorghum shoot fly; stable fly
fonofos, 263
food safety, 79, 84
forage, 139
forecasts, 121
forest, 207; injury to, 211; pests in, 69
frogeye leaf spot, 121
fruit, 34, 70, 231, 270; destruction of, 27; load, 233, 234; scarring of, 27; tree, 229, 232, 233
fungicide, 117, 119, 120, 121, 122

gain threshold, 12, 52, 284
Galium aparaine, 105
Gelechiidae, 159
genetic shifts, 130

geographic information systems (GIS), 101
German cockroaches, 162
giant foxtail, 99, 100
Gillus flavus, 283
GLEAMS model, 255, 256
Glycine max. See soybean
GOSSYM/COMAX model, 66
gotch ear, 182
grape, 49, 157, 159, 161, 231, 232, 234, 235, 239
grape root borer, 129
grasshoppers, 32
green bottle fly, 193
greenbug, 32
green cloverworm, 44, 45, 49, 50, 51, 52, 53, 54, 157, 162, 165, 171
greenhouses, 213
green June beetle, 209
green stink bug, 171, 280
growth rates, 45
guar, 158
Gulf Coast ticks, 182
gypsy moth, 69

Haematobia irritans, 159, 162, 180, 182
Haematopinus suis, 189
hay, 139
heat stress, 181
Helianthus annuus, 102, 161, 163
Helicoverpa zea, 157, 159, 171
Heliothis armigera, 159
Heliothis injury, 32
Heliothis virescens, 159
Hemileuca oliviae, 139, 161
herbicides, xi, 253; curative, 154; postemergence, 10, 106, 163; preemergence, 106, 154
HERB model, 66, 104, 105
Hessian fly, 32
Heterodera glycines, 161
hog louse, 189, 190
Hoplolaimus columbus, 161
horn fly, 162, 182, 195
host phenology, 45, 46, 170
host-plant resistance, 251
house fly, 187, 188, 190, 195
human health, 158, 161, 162, 240

hybrid EIL, 206, 210, 212, 213, 215, 218, 219, 220, 221, 222, 223, 224
Hypera brunneipennis, 159
Hypera postica, 46, 135, 159, 166, 279
Hypera variabilis, 159
hypersusceptive responses, 137, 138
Hypoderma bovis, 183
Hypoderma lineatum, 183

I (variable; injury in EIL equation), 12, 28, 29, 44, 58, 134, 156, 165, 167, 170, 174, 213, 214, 216, 253, 279, 281, 283, 284, 288
implementation delays, 45, 46
inaction threshold, 165
indirect assessment, 119
indirect injury. *See under* injury
infective propagules, 116
inherent impunity, 31, 32
injury, 11, 12, 14, 15, 16, 18, 22, 25, 26, 28, 32, 33, 41, 45, 60, 62, 65, 67, 128, 130, 134, 137, 138, 142, 153, 155, 162, 167, 168, 169, 170, 173, 174, 210, 211, 212, 214, 219, 221, 223, 224, 228, 233, 234, 235, 237, 253, 254, 268, 275, 276, 278, 280, 282, 284, 287; acute, 26, 29, 34; chronic, 26, 28, 29; direct, 34, 35, 232; equivalency, 16, 28, 44, 51, 61, 65, 165, 169; guilds, 27, 28, 35, 65, 169, 172, 279, 281, 289; indirect, 34, 35, 232; past, 13, 17, 19, 47; potential, 165; rates of, 28, 30, 46, 50, 55; types of, 27, 34, 156, 169, 173, 174. *See also* aesthetic injury; differentiating injury; economic injury; preventable injury; quantal injury; surrogate injury; total injury
inoculum, 114, 115; secondary, 117
insecticide, 117, 132, 133, 140, 156, 232, 270
insecticide resistance, 7, 195, 227, 232, 238, 240. *See also* pesticide: resistance
insects, 9, 10, 15, 65, 66, 91, 92, 123, 154, 169, 270, 291. *See also* aphids; beetles; flies; lice; mites; mosquitos; ticks; weevils
integrated farm management (IFM), 74, 83
integrated pest management (IPM), xi, 3–8, 74, 82–85, 128, 151, 179, 291–295
interference, 95

internal parasites, 183. *See also* parasites
interseasonal dynamics, 5
interseasonal effects, 237
interseasonal management, 227, 231
interseasonal models, 228
interseasonal thresholds, 227, 238
interspecific competition, 99
intervally-scaled-preference surveys, 210
intraspecific competition, 99
investment models, 9
ivermectin, 180
Ixodidae, 161

Japanese beetle, 235
jimsonweed, 99
Juniperus spp., 207

K (variable; proportion of management in EIL equation), 12, 59, 61, 140, 141, 167, 168, 170, 213, 215, 217, 220, 221, 223, 253, 254, 269
Keiferia lycopersicella, 159

landscape plants, 68
late blight of potato, 121
late leaf spot, 120
leaf area, 98
leaf-area index (LAI), 63
leafhoppers, 235
leaf-mass consumers, 27
leaf-mass reduction, 27
leaf mining, 27
leaf photosynthetic-rate reduction, 27
leaf rust, 123
leaf skeletonizing, 27
leaf spot, 122
least-cost rationing, 286, 288
legumes, 235
lemon, 161, 231
Lepidoptera, 156
lepidopteran larvae, 130, 131, 134
Leptinotarsa decemlineata, 136, 158
Leptoglossus phyllopus, 158
lesser cornstalk borer, 282
lesser mealworm, 192
Leucocytozoon smithi, 189
lice. *See* cattle lice; chicken body louse; hog louse

Index

lime, 161
Linognathus vituli, 184
Liriodendron tulipifera, 208
Liromyza trifolii, 207
Lissorhoptrus oryzophilius, 159
litter beetle, 191, 192
livestock, 169, 179, 187; confined, 179
lone star tick, 162, 181
Lorsban, 265
Loxagrotis albicosta, 159
Lygaeidae, 159
Lygocoris communis, 159
Lygus hesperus, 159
Lygus lineolaris, 159
Lyonetiidae, 159

Macrosiphum avenae, 158
Macrosiphum liriodendri, 208
maize. *See* corn
malathion, 263
Maliarpha separatella, 32
Mamestra configurata, 159
management, 5, 9, 75, 293, 294; costs of, 22; models, 124; tactics, 22, 41, 43, 45, 46, 47, 140, 223, 291, 292, 293, 294. *See also* area-wide pest management; cultural management; disease management; integrated farm management; integrated pest management; interseasonal management; preventive management; prophylactic weed-management; protective population management
Manduca quinquemaculata, 161
Manduca sexta, 135, 161
mange, 189
mange mite, 189
Maracanthus stramineus, 191
marginal analysis, 80
market surveys, 211
market value, 22, 138
mean-risk analysis, 171
mechanical techniques, 77
Megaselia halterata, 160
Melophagus ovinus, 192
Membracidae, 159
mesophyll feeders, 155
Metapolophium dirhodum, 158
methidathion, 263

methomyl, 263
methyl parathion, 263
Mexican bean beetle, 171, 172
microclimate, 131
microsclerotia, 116, 117
Miridae, 159, 231, 235
mites, 26, 154, 156, 192, 234, 235, 236. *See also* chicken mite; citrus bud mite; citrus red mite; European red mite; mange mite; northern fowl mite; Pacific mite; spiter mite; twospotted spider mite; Willamette mite
models. *See* AUSIMM; bioeconomic models; BLITECAST model; CALEX/Cotton model; economic models; GLEAMS model, GOSSYM/COMAX model; HERB model; interseasonal models; investment models; management models; NEBHERB model; plant modeling; postemergence model; remedial weed-control models; SIRATAC model
monocyclic pathogens, 115
Mononychellus tanajoa, 161
Moraxella bovis, 185, 186, 195
mosquitoes, 180, 188, 258
mountain pine beetle, 211
multidimensional economic threshold, 173–174
multiple-event decision, 291, 292
multiple-pest management system, 123
multiple pests. *See* pests: multiple
multiple-species EIL. *See* economic injury level
multiple species interaction, 92
multiple-stress EIL, 58. *See also* economic injury level
multispecies, 101
Musca autumnalis, 162, 180, 185
Musca domestica, 187
Muscidae, 159
mushroom, 160
Myzus persicae, 158

natural enemies, 4, 10, 29, 44, 74, 132, 154, 165, 172, 293; conservation, 4
navel orangeworm, 232
NEBHERB model, 104
nematodes, 12, 32, 59, 62, 65, 66, 114, 123, 154, 184; soybean cyst, 81

Nezara viridula, 160, 171
Noctuidae, 156, 159, 235
nominal thresholds, 48, 152, 171
nontarget organisms, 256, 257
northern black blow fly, 193
northern fowl mite, 190
northern grub, 183
nuisance threshold, 162
nurseries, 204, 205, 212, 213, 215, 218
nutrient-yield approach, 284, 285
nut tree, 231, 233

oak, 161, 167, 214
oats, 132, 133. *See also* wild oats
objective ET. *See* economic threshold
Odontota horni, 158
Oebalus pugnax, 160
Oetrus ovis, 192
Oligonychus pratensis, 161
onions, 122
orangestriped oakworms, 212, 222
Ornithonyssus sylviarum, 190
Ostiorhynchus ligustici, 159
Ostrinia nubilalis, 140, 153, 160
Otitidae, 160
Oulema melanopus, 132, 133
Oulema oryzae, 158
overcompensation, 31, 32
oxydemeton methyl, 264

Pacific mite, 234
Pagria signata, 158
Panonychus citri, 161
Panonychus ulmi, 161
Papaipema nebris, 159
Parafilaria bovicola, 186
parasites, 29, 50, 76, 77. *See also* ectoparasites; endoparasites; internal parasites
past injury. *See* injury: past
pasture, 139
pathogens, 9, 23, 28, 30, 65, 66, 117, 118, 154
payoff matrices, 9
pea, 158
peanut, 116, 117, 120, 122, 160, 282, 283, 284
pear, 160

pecan, 234
Pectinophora gossypiella, 159
Penncap-M, 265
Pentatomidae, 160
perennial crops, 19
perennial legumes, 231, 232
perennial plants, 291
perennial systems, 228, 229
Peridroma saucia, 136, 160
Perileucoptera coffeella, 159
period threshold, 92, 93
permethrin, 264
pest-density-tolerance surveys, 210
pesticide, xi, 6, 10, 47, 69, 70, 75, 76, 77, 80, 123, 124, 139, 141, 174, 179, 204, 218, 223, 224, 230, 250, 251, 253, 254, 255, 256, 257, 259, 260, 267, 269, 271, 293; registration, 267; regulation, 268; resistance, 5, 76, 132, 154, 168, 171, 182, 183, 238, 240, 249, 251
pest-population dynamics, 47, 166, 170
pests, 15, 74, 75, 162; density of, 5, 14, 25, 132, 141, 165, 170, 209, 210, 214, 219, 228, 287; relationships with hosts, 10, 22; management of (*see* integrated pest management); mortality potentials of, 46; multiple, 65, 67, 122, 174, 175; sampling of, 6, 171; status of, 9, 10, 20, 22, 46; survival of, 61; types of, 59, 66, 67; veterinary, 179–202, 291
Phaenicia sericata, 193
phenological disruption, 27, 287, 288, 289
Philaenus spumarius, 131
Phormia regina, 193
phoroate, 264
phosmet, 264
PHYTOPROG system, 119
Phytothphora infestans, 122
Pieridae, 160
Pieris rapae, 157, 160
pigs, 189, 190
pine resin midge, 211
pinkeye, 185, 186, 195
Pinus radiata, 207
plant architecture modification, 27
plant canopy, 35
plant defenses, induced, 239
plant modeling, 240

plant pathogens, 9, 24, 26, 34, 43, 114–127, 291
plant phenology, 34, 35
Plathypena scabra, 45, 50, 53, 135, 136, 157, 160, 162, 165, 171
Platynota idaeusalis, 161
Plusia nigrisigna, 160
Plutella xylostella, 157, 160
Plutellidae, 160
polycyclic pathogens, 115, 117, 120
Polyphagotarsonemus latus, 161
pome fruit, 233
population assessment, 114
population density, 41, 62, 89, 128, 129, 130, 137, 204
postemergence herbicides, 10, 106, 163
postemergence model, 103
potato, 32, 119, 120, 136, 158, 160
potato cyst eelworm potato, 12
potato late blight, 119
potato leafhopper, 157, 169, 284, 285, 287, 288, 289
poultry, 189, 190, 191, 192
pounce, 265
predation, 50, 77
preemergence herbicides, 106, 154
preventable injury, 13, 17, 22, 47
preventive management, 9, 10, 24, 41, 48, 154, 163, 251, 291, 292
primary inoculum, 115, 117
profitability, 78, 79, 89, 107, 252
propagule, 117
propargite, 264
prophylactic treatments, 90, 204, 228, 231
prophylactic weed-management, 90, 105, 107
protective population management, 3
protein yield, 281, 282
Protophormia terraenovae, 193
Pseudomonas syringae pv. syringae, 119
Pseudoplusia includens, 160, 171
Psychidae, 160
psychological disutility, 211
Psylla pyricola, 160
Psyllidae, 160
Puccinia sorghi, 161
Pyralidae, 160
pyrethroid insecticides, 186
Pyricularia oryzae, 119

quackgrass, 100
quality, 275, 276, 277, 278, 279, 280, 281, 282, 283, 284, 286, 287, 288, 289. *See also* aesthetic quality; environmental quality; water quality
quantal injury, 34
Quecus spp., 208

range caterpillar, 139
rangeland, 32, 139, 158, 231, 237
RBWHIMS expert system, 66
relative-risk rankings, 268
remedial weed-control models, 90, 91
replacement-feed cost analysis, 139
replanting decision, 16
resistance. *See under* pesticide
resource depletion, 79, 84
Rhopalosiphum maidis, 158
Rhopalosiphum padi, 158
rice, 32, 49, 98, 158, 159, 160
rice blast, 119
risk, 65, 107, 230, 239, 257, 258, 268, 269; assessment of, 256, 268, 269; aversion to, 107; coefficients, 261; costs of, 261; indices of, 256. *See also* economic risk; environmental risk; mean-risk analysis; relative-risk rankings
root maggot, 32
rosy apple aphid, 49
rotation, 115
rye, winter, 105

sage, 16
sampling, 55, 62, 94, 129, 140, 166, 292. *See also* economic threshold, sampling and; pest sampling; sequential sampling; spore sampling; time-sequential sampling; weed population sampling
sanitation, 179
Sarcoptes scabiei, 189
Saturniidae, 161
Schizaphis graminum, 158
sclerotia, 116
Sclerotinia blight, 117
Sclerotinia minor, 117
Sclerotium rolfsii, 115
scouting, 75, 76, 104, 140, 270
secondary inoculum, 117

Index

seedbank, 94, 102. *See also* weed: seedbank
seed destruction, 27
seed longevity, 102
seed production, 102. *See also* weed: seed production
Septoria glycines, 169
Septoria leaf blotch, 118, 123
Septoria tritici, 118
sequential sampling, 52, 54, 62, 101, 166
Setaria faberi, 99
sheep, 181, 192, 193
sheep bot fly, 192, 193
sheep ked, 192
sheep tick, 192
Silent Spring, 3
silver-bullet fetish, 6, 7, 293
Simuliids, 188
Sipha flava, 158
SIRATAC model, 66
Sitobion avenae, 158
Smicronyx fulvus, 159
smooth pigweed, 99
snap bean, 119, 141
social attitudes, 107
social costs, 69, 140
societal impacts, 84
sod farms, 213
Solenopotes capillatus, 184
sorghum, 32, 49, 52, 99, 158, 159, 160, 162
Sorghum halepense, 161
sorghum shoot fly, 52
southern green stink bug, 171
southern pine beetle, 211, 222
southern stem rot, 116
soybean, 16, 28, 29, 32, 43, 44, 49, 51, 52, 54, 59, 60, 66, 76, 100, 102, 103, 104, 105, 121, 123, 135, 136, 139, 157, 158, 159, 160, 161, 163, 169, 171, 172, 206, 270, 279, 285
soybean cyst nematode, 81
soybean looper, 171
species-relative competitiveness, 99
Sphingidae, 161
spider mite, 43, 50, 132, 167, 216, 221, 233, 239
Spissistilus festinus, 159
Spodoptera exigua, 160

Spodoptera frugiperda, 130, 157, 160
Spodoptera littoralis, 160
Spodoptera litura, 43, 160
spore sampling, 118, 119, 164
spruce budworm, 211
stable fly, 180, 181, 187, 188, 190, 195
standard weed units, 96
stand-loss EIL, 16
stand reduction, 27
static EIL, 173
stem boring, 27
Stewart's wilt, 117
stochastic dominance, 81
Stomoxys calcitrans, 159, 187
stone fruit, 233
strawberry, 157, 159, 161
street trees, 212
stress, 15, 25, 26, 30, 35, 36, 60, 163, 254, 294; interactions, 67. *See also* abiotic stress; biotic stress; heat stress; water stress
stripe rust, 120
subjective ET. *See* economic threshold
substitution-feed approach, 286
sugar beet, 32, 98, 116, 160, 163
sugarcane, 158
sunflower, 159
surrogate injury, 134, 135, 136
surveys. *See* intervally-scaled-preference surveys; market surveys; pest-density-tolerance surveys; tolerance surveys
susceptive responses, 137
sustainability, 3, 6, 7, 168, 249, 254, 292
sustainable agriculture, 8
swine, 189, 190
systems approach, 82, 83

Tarsonemidae, 161
tefluthrin, 264
terbufos, 264
Tetanops myopaeformis, 160
Tetranychidae, 161
Tetranychus spp., 50
Tetranychus urticae, 157, 161, 172, 208, 216, 217, 219, 220
Thelazia, 186
therapeutic tactics, 9, 10, 20, 41, 61, 62, 154, 156, 165, 250, 251, 291, 293

Index

Therioaphis maculata, 3
thiodicarb, 264
thresholds, 4, 17, 36, 58, 90, 114, 116, 117, 118, 119, 120, 121, 133, 152, 153, 163, 165, 168, 172, 173, 183, 241, 293. *See also individual kinds of thresholds:* action; aesthetic; calculated; damage; dynamic action; economic; economic-optimum; fixed; gain; inaction; interseasonal; multidimensional economic; nominal; nuisance; period; visual
Thuja occidentalis, 208
Thuja spp., 207
Thyridopteryx ephemeraeformis, 160, 208
ticks, 156, 180, 181, 194. *See also* Gulf Coast ticks; lone star ticks; sheep tick
time-sequential sampling, 52, 54, 62, 166
tobacco, 159, 161
tolerance, 4, 17, 18, 30, 32, 137, 167, 168, 254, 282; surveys, 210
tolerant plant varieties, 8
tomato, 135, 159
Tortricidae, 161
total injury, 13, 14
transgenic plants, 7
transplants, 115
trichlorfon, 264
Trichoplusia ni, 135, 157, 160
Triticum aestivum, 103
tulip tree aphid, 212
turfgrass, 68, 207, 209
turgor reducers, 27, 155
turkeys, 189
twospotted spider mite, 157, 172, 215

Ulmus procera, 208

V (variable; crop value in EIL formula), 12, 58, 61, 138, 167, 213, 215, 216, 253, 281, 282, 284, 289
vectors, 63, 117, 162, 164, 179, 186, 193, 253
vegetables, 34, 70, 103, 270
velvetbean caterpillar, 171
velvetleaf, 102, 169

vertical resistance, 115
Verticillium dahliae, 117
Verticillium sp., 102
veterinary pests, 179–202, 291
visual thresholds, 107
Vitace polistiformis, 129

walnut, 236
walnut aphid, 236
water-balance disruption, 27
water quality, 84
water status, 172
water stress, 35
weather, 55, 76, 174
weed, 9, 15, 17, 24, 28, 30, 35, 48, 59, 62, 65, 66, 89, 101, 154, 161, 163, 169, 227, 237, 253, 270, 291; competitiveness, 15, 99 (*see also* competition); density, 95, 96, 97, 98, 101; distribution, 101, 106; duration, 95; emergence, 97; interference studies, 89; population, 91, 102; population sampling, 101; science, xi; seed production, 103; seedbank, 103, 106, 227, 238; size, 97; spatial patterns, 92; suppressive ability, 94
weevils, 235. *See also* alfalfa weevil
western spotted cucumber beetle, 129, 141
wheat, 32, 118, 122, 158, 161, 163. *See also* winter wheat
wild oats, 96, 102, 103
Willamette mite, 234
winter barley, 97, 105
winter rye, 105
winter wheat, 97, 102, 103, 105, 120. *See also* wheat
woody landscape plants, 209
woody ornamentals, 223

Xanthium strumarium, 98, 161, 163
Xanthogaleruca luteola, 208, 209

yellow rust, 123
yield loss, 6, 12, 14, 15, 19, 22, 25, 32, 33, 59, 64, 65, 94, 98, 100, 114, 152, 172

zero-threshold concept, 103

In the *Our Sustainable Future* series

Volume 1
Ogallala: Water for a Dry Land
John Opie

Volume 2
Building Soils for Better Crops: Organic Matter Management
Fred Magdoff

Volume 3
Agricultural Research Alternatives
William Lockeretz and Molly D. Anderson

Volume 4
Crop Improvement for Sustainable Agriculture
Edited by M. Brett Callaway and Charles A. Francis

Volume 5
Future Harvest: Pesticide-free Farming
Jim Bender

Volume 6
A Conspiracy of Optimism: Management of the National Forests since World War Two
Paul W. Hirt

Volume 7
Green Plans: Greenprint for Sustainability
Huey D. Johnson

Volume 8
Making Nature, Shaping Culture: Plant Biodiversity in Global Context
Lawrence Busch,
William B. Lacy, Jeffrey Burkhardt,
Douglas Hemken, Jubel Moraga-Rojel,
Timothy Koponen,
and José de Souza Silva

Volume 9
Economic Thresholds for Integrated Pest Management
Edited by Leon G. Higley and Larry P. Pedigo